ADVANCES IN COLLOID SCIENCE

Edited by **Mohammed Muzibur Rahman**
and **Abdullah Mohamed Asiri**

Advances in Colloid Science
http://dx.doi.org/10.5772/62716
Edited by Mohammed Muzibur Rahman and Abdullah Mohamed Asiri

Contributors

Tifeng Jiao, Utkirjon Sharopov, Duyang Zang, Chun-Feng Lai, Mehmet Cabuk, Gang Yang, Alexander Wittemann, Claudia Simone Plüisch, Alessandro Ponti, Anna M. Ferretti, Sara Mondini, Tiberius O. Cheche, Agustin E. Gonzalez, Sylwester Rzoska, Jane M. Morrell

Published by InTech

Janeza Trdine 9, 51000 Rijeka, Croatia

Notice

Publishing Process Manager Maja Bozicevic

Technical Editor SPi Global

Cover InTech Design team

First published November, 2016

Printed in Croatia

Additional hard copies can be obtained from orders@intechopen.com

Advances in Colloid Science, Edited by Mohammed Muzibur Rahman and Abdullah Mohamed Asiri

p. cm.

Print ISBN 978-953-51-2773-4
Online ISBN 978-953-51-2774-1

For EU product safety concerns:
IN TECH d.o.o., Prolaz Marije Krucifikse Kozulić 3, 51000 Rijeka, Croatia,
info@intechopen.com or visit our website at intechopen.com.

Meet the editors

Mohammed Muzibur Rahman received his MSc and PhD from Shahjalal University of Science and Technology (Bangladesh) and Chonbuk National University (South Korea) in 2001 and 2007, respectively. He worked as postdoctoral fellowship and assistant professor in different universities in South Korea, Japan, and Saudi Arabia (2007 to 2011). Presently, he is working as an associate professor in Chemistry Department at King Abdulaziz University, Saudi Arabia, since 2011. He published over 170 research articles in ISI journals, several books, patents, and book chapters. He is the editorial board member in different international journals. His research work was largely in the area of nanotechnology, sensors, surface chemistry, electrochemistry, instrumental science, nanomaterials, self-assembled monolayers, carbon nanotubes, colloids, photochemistry, μ-chips, devices, etc.

Abdullah Mohamed Asiri received his PhD from the University of Wales, College of Cardiff, UK, in 1995. He is the Head of the Chemistry Department at King Abdulaziz University since October 2009, and he is the founder and the Director of the Center of Excellence for Advanced Materials Research (CEAMR). He is a Professor of Organic Photochemistry. His research interest covers color chemistry, photochromic, thermochromic systems, coloring matters and dyeing of textiles, polymers, and plastics. He is the Editor in Chief of King Abdulaziz University Journal of Science. He is also a member of the Editorial Board of Pigment and Resin Technology (UK), Organic Chemistry in Sight (New Zealand), and Recent Patents on Materials Science (USA).

Contents

Preface

It gives us immense pleasure to introduce a book named *Advances in Colloid Science* based on state of the art of physical, materials, composites, and nanoparticles with their various potential and significant colloidal applications. These deal with physical aspects of the synthesis, growth, development and prospective applications of colloidal materials, self-assembly, liquid crystals, photonic crystals, quantum dots, composite particles, and hybrid polymers. The discussion of these aspects develops through the fundamentals and applied experimental routes in dispersible solution methods via the interaction of various solvents to finally include the interfacing of the colloidal science and technology world. Usually, colloidal material has certainly attained accomplishment of its classical essence and has taken new directions of growth, development, continuous changes, and various drug design applications in research and development sectors. The new route and emerging frontiers are branching out from time to time around this innovative colloidal technology phase on dispersed colloids. Advances in colloid materials with instrumentation for evaluating the theoretical models now enable us to implement quite comprehensively almost all the events that take place at the colloids at least at lower disperse into various solution phases. This work aims to bridge the gap between undergraduate, graduate, and researches in applied physical and optical behaviors as well as biomedical and drug design sciences, in order to initiate researchers into colloidal study in as straightforward way as possible and to introduce the scientists to the opportunities offered in the applied technological fields. We worked consistently to complete this work on *Advances in Colloid Science* under the InTech Open-Access Publisher. The colloid book offers a comprehensive but easy to follow theoretical context with practical investigation in colloidal science. It frameworks significant research methods and contemplates technological implementations. A straightforward acquaintance of the ideologies and principles of chemical, physical, nanomaterial, and drug-designed chemistry is anticipated. Hopefully, it will serve to a broad audience such as undergraduate, graduate, and postgraduate students at colleges and universities as well as scientists in the industry who require a comprehensive study in the colloidal science. We hope that this contribution would further enhance the applied colloid materials in physical and biomedical sciences, especially in bringing new entrants into the applied colloidal science and technology fields and help scientists to forward and develop their own field of specialization.

Prof. Mohammed Muzibur Rahman and Prof. Abdullah Mohamed Asiri
Center of Excellence for Advanced Material Research (CEAMR) and Chemistry Department
King Abdulaziz University
Jeddah, Saudi Arabia

Section 1

State-of-the-Art Physical

Chapter 1

Wetting and Drying of Colloidal Droplets: Physics and Pattern Formation

Ruoyang Chen, Liyuan Zhang, Duyang Zang and Wei Shen

Additional information is available at the end of the chapter

http://dx.doi.org/10.5772/65301

Abstract

When a colloidal droplet is deposited on a solid substrate at ambient condition, it will experience the processes of wetting and drying spontaneously. These ostensibly simple and ubiquitous processes involve numerous physics: droplet spreading and wetting, three-phase contact line motion, flow fields inside droplets, and mass transportation within droplets during drying. Meanwhile, the continuous evaporation of liquid produces inter- and/or intra-molecular interactions among suspended materials and builds up the internal stress within droplets. After drying, interesting and complex desiccation patterns form in the dried droplets. These desiccation patterns are believed to have wide applications, e.g., medical diagnosis. However, many potential applications are limited by the current understanding of wetting and drying of colloidal droplets. This chapter focuses on the complex physics associated with these processes and the pattern formation in the dried colloidal droplets. Moreover, potential applications of these desiccation patterns and prospective works of wetting and drying of the colloidal droplets are outlined in this chapter.

Keywords: colloidal droplets, wetting dynamics and statics, three-phase contact line motion, flow field, mass transport, desiccation patterns

1. Introduction

The formation of complex patterns during wetting and drying of colloidal droplets is commonly observed phenomenon in our daily life, which is also involved in numerous research and industrial fields, e.g., inkjet printing, spray coating, and medical practice. There are various interesting patterns formed by the dried colloidal droplets, e.g., cracking patterns, crystal patterns, ring-stain patterns, and uniform-deposition patterns. The morphological difference of these desiccation patterns is mainly attributed to the different physical processes experienced by colloidal droplets during wetting and drying; moreover, it is significantly influenced by the complex systems of colloidal fluids and the complicated interactions between the systems and the substrate. These desiccation patterns have potentials to be convenient and low-cost tools for health diagnosis. Therefore, it is intriguing and important to study the physics associated with wetting and drying of colloid droplets and the formation of desiccation patterns in these droplets. Applications of these patterns in medical diagnosis for human beings and animals and prospective work are also covered in this chapter.

2. Wetting and drying of colloidal droplets: physics

2.1. Wetting dynamics and statics

When a colloidal droplet is placed on an ideal solid substrate (flat, smooth and homogeneous), it is far from the state of equilibrium. The droplet will spread on the substrate to respond to the interaction between the droplet and the substrate, i.e., the cohesion of the droplet and the adhesion of the droplet to the substrate. The spreading behaviour of the droplet is mainly driven by three types of forces: surface tension, gravitational force, and viscous force. These acting forces can be quantitatively characterised by relevant dimensionless numbers. The capillary number (C_a) reflects the relationship between viscosity and surface tension:

$$C_a = \frac{\eta u}{\gamma} \tag{1}$$

where u is the characteristic velocity of the three-phase contact line of the colloidal droplet, η and γ are the dynamic viscosity and surface tension, respectively. When the capillary number is low ($C_a < 10^{-5}$), the drop spreading is dominated by surface tension; while the capillary number is high, the liquid surface tension is negligible compared to the viscosity force. The Bond number (B_o) represents the relationship between surface tension and gravity force as follows:

$$B_o = \frac{\rho g h^2}{\gamma} \tag{2}$$

where ρ and h are the density and height of the droplet, and g is the gravitational constant. A low Bond number ($B_o < 1$) indicates that the surface tension dominates the drop spreading.

After reaching equilibrium, the colloidal droplet stops spreading on the surface and forms a contact angle (θ_{eq}) at the three-phase contact line of the droplet (the drop/substrate/air interface), which can be expressed by the Young equation:

$$\cos \theta_{eq} = \frac{(\gamma_{sv} - \gamma_{sl})}{\gamma_{lv}} \tag{3}$$

where γ_{sv}, γ_{sl}, and γ_{lv} are the surface energies of solid-vapour (liquid vapour), solid-liquid, and liquid-vapour surfaces, respectively. However, from a microscopic view, there is a very thin precursor film with micro/nanometre length in front of the contact line. The thickness of the precursor film is nearly at a monomolecular level, which has been experimentally observed and qualitatively demonstrated in sessile drops [1]. This could be related to the interactions between spreading forces and surface energies [2]. The molecular precursor film could influence the contact angle of the sessile drop. Additionally, real surfaces are seldom automatically smooth, and the real contact angle has never been experimentally observed. Therefore, the apparent contact angle (θ^*) is used to describe the contact angle of the sessile drop on a real surface. If the apparent contact angle of a water droplet on the surface, e.g., polytetrafluoroethylene (Teflon), alkyl ketene dimmer (AKD) and polyethylene (PE), is higher than 90°, the surface is defined as a hydrophobic surface (low wettability). By contrast, the surface is regarded as the hydrophilic surface (high wettability) such as glass, when $\theta^* < 90°$.

After reaching the equilibrium, the geometric shape of the sessile drop can be estimated by the capillary length (λ_c):

$$\lambda_c = \sqrt{\frac{\gamma}{\rho g}} \tag{4}$$

If the contact radius (R) of the sessile drop is less than the capillary length, the gravitational force is negligible; the droplet will form a spherical cap shape. Therefore, the height (h), volume (V) and surface area (S) can be calculated as follows:

$$h = R\, tan\left(\frac{\theta^*}{2}\right) \tag{5}$$

$$V = \frac{\pi(2 + cos\,\theta^*)(1 - cos\,\theta^*)^2 R^3}{3\, sin^3\theta^*} = \frac{1}{6}\pi h\left(h^2 + 3R^2\right) \tag{6}$$

$$S = \frac{2\pi R^2}{(1 + cos\,\theta^*)} = \pi\left(h^2 + R^2\right) \tag{7}$$

2.2. The dynamics of the three-phase contact line motion

During drying of a sessile drop of solution, two major modes are proposed for the three-phase contact line motion, namely, constant contact area mode and constant contact angle mode [3]. In the constant contact area mode (**Figure 1(a)**), the three-phase contact line remains pinned and the sessile drop is assumed as a wedge-like shape, while the radius remains constant. In the constant contact angle mode (**Figure 1(b)**), the three-phase contact line recedes in a constant profile evolution as the liquid evaporates, leading to a decrease in the radius of the droplet.

In most cases of the colloidal droplet, these two modes occur successively (**Figure 1(c)**). During the first stage, the perimeter of three-phase contact line ($l = 2\pi r_0$) and the contact area ($A = \pi r_0^2$) of the sessile drop remain constant; whereas, the contact angle decreases gradually, and the three-phase contact line gets pinned. In this stage, the drying surface area (S_d) of the sessile drop is equal to its real surface area, i.e., $S_d = S$. According to Eq. (7), the drying surface area decreases with the apparent contact angle (θ^*), resulting in a slight decrease in the average evaporation rate. When the apparent contact angle reaches a receding contact angle, the

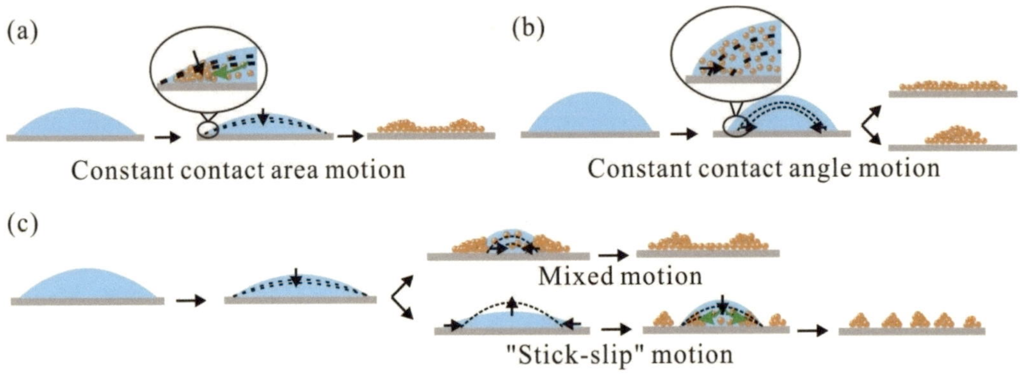

Figure 1. Motion of three-phase contact line in different modes: (a) constant contact area motion; (b) constant contact angle motion; (c) mixed motion and "stick-slip" motion.

second stage starts. In this stage, the evolution of the three-phase contact line is likely to transform to the constant contact angle mode. The three-phase contact line recedes with the propagation of evaporation ($l < 2\pi r_0$), which dominates the decrease in the drying surface area (S_d). As a result, the average evaporation rate of the droplet decreases.

Generally speaking, the local evaporation rate is also determined by the three-phase contact line motion. If the contact line is not pinned during drying, the evaporation rate along the interface of the droplet will be uniform. To the contrary, if the contact line is pinned, the local vapour pressure of the liquid above the droplet is non-uniform; the pressure in the central part is higher than that along the contact line. This makes the local evaporation rate at the contact line greater than that at the central part of the sessile drop.

Interestingly, the drying of colloidal droplets, e.g., the aqueous solution containing silica nanospheres and the ethanol suspension containing titanium dioxide (TiO_2) nanoparticles, sometimes presents a slightly different dynamics of the contact line motion from the typical evolution process [4–6]. During the second stage, if the decreasing contact angle reaches a lower critical value, the contact line will be de-pinned. At the same time, the contact angle and height of the sessile drop will increase, while the base radius will decrease ($r < r_0$). In this case, the contact line tries to get pinned again as the first stage. This motion cycle of de-pinning and re-pinning may successively repeat with further evaporation and is termed as the "stick-slip" motion of three-phase contact line (**Figure 1(c)**). This motion process is more likely to occur in the drying sessile drops of nanofluids and is affected by a number of factors: the wettability of the droplet on the substrate, the thermal conductivity of the substrate, the vapour pressure surrounding the droplet, the environmental temperature, and the relative humidity (RH) [7].

2.3. Flow fields inside the colloidal droplets during drying

Evaporation is a conversion process of volatile solvent from liquid to gas phases. It spontaneously occurs when the partial vapour pressure in the environment around the sessile drop is lower than the saturation pressure. Natural evaporation of a colloidal droplet is driven by the diffusion of molecules in the volatile solvent. Meanwhile, heat and mass convections occur within the droplet, building up a local flow field. A number of flow field modes are proposed inside colloidal droplets during their drying. Among these modes, two important flow fields are widely accepted by researchers: the capillary flow and the Marangoni flow.

The capillary flow is a compensation flow, which is induced by the differential evaporation rates along the interface of the droplet during drying [8]. When a colloidal droplet is deposited on an ideal hydrophilic surface, the contact line will be pinned during drying, and the local evaporation rate at the contact line is enhanced, as described in Section 2.2. The enhanced evaporation rate forces the liquid in the central part to flow outwards to replenish the faster liquid loss at the edge, resulting in the horizontally outward capillary flow in the droplet [9]. Within this flow field, the radius of droplet usually remains constant, while the contact angle decreases. The higher evaporation rate of the colloidal droplet leads to a faster decrease in the contact angle. It also causes the three-phase contact line to be pinned faster and the capillary flow within the sessile drop to be stronger. Accordingly, the rate of capillary flow increases along with the increase in evaporation rate of the colloidal droplet during drying [10].

Marangoni flow is a convection flow that is driven by the surface tension gradient in the droplet. This surface tension gradient can be caused by the non-uniform distribution of local temperature field and/or the differential concentration of substances, e.g., suspended particles. For example, during drying of a sessile drop of organic colloidal suspension, e.g., the mica flakes/octane and poly-methyl-methacrylate beads/octane, the liquid evaporation leads to the local variation in the concentration of solid particles and generates a surface tension gradient along the drop-air interface [11, 12]. In this situation, the liquid in the regions with lower surface tension will be pulled towards the regions with higher surface tension, producing the Marangoni convection. The features of the Marangoni flow in a colloidal droplet during drying have been presented by Hu and Larson: the liquid on the surface near the base of the sessile drop is pulled upwards to the top and then is transported downwards, causing the circular movement along the drop-air interface [12]. It is noteworthy that the local surface tension of liquid in the colloidal droplet can be decreased by increasing the local temperature ($\frac{d\gamma}{dT} < 0$), which can cause the temperature-dependent Marangoni flow [3]. This phenomenon is termed as thermo-capillary convection or Bénard-Marangoni convection. The Marangoni flow within a droplet could be described by the dimensionless Marangoni number (M_a) as follows:

$$M_a = -\frac{d\gamma}{dT}\frac{\Delta TL}{\eta\alpha} \tag{8}$$

where ΔT is the temperature difference within the colloidal droplet, L is the characteristic length, η and α are the dynamic viscosity and thermal diffusivity of the droplet, respectively.

The liquid evaporation can cause the hydrodynamic and/or thermodynamic instability within colloidal droplets during drying [13]. This may lead to the simultaneous occurrence of the capillary flow and the Marangoni flow in the droplets [14]. The inward circular Marangoni flow is assumed to counteract the outward capillary flow [9, 14]. Recently, these two flow fields have been investigated in the water droplets containing polymers on the glass substrate with different controlled temperatures [15]. When the temperature of substrate is lower than the droplet, the Marangoni flow shows stronger than the capillary flow. By contrast, on a heated substrate, the capillary flow is more significant than the Marangoni flow [15].

All of the above flow fields are discussed when the influence of the gravitational force on the droplet can be negligible, i.e., $R \leq \lambda_c$. Nevertheless, when the size of the droplet is much larger than the capillary length ($R \gg \lambda_c$), the gravitational effect should be considered. In this situation, especially for the droplets on heated substrates, another flow field, e.g., Rayleigh-Bénard convection, which is a thermo-gravitational flow, could replace the Bénard-Marangoni convection. For the sessile drop of colloidal suspension on the heated substrate, the temperature gradient induces a concentration gradient of solid particles between the top and the areas close to the three-phase contact line of the droplet [16]. The surface tension of the liquid at the top of the droplet is larger than the one of the areas close to the three-phase contact line, forming the surface tension gradient. Meanwhile, the gravitational force makes the denser liquid transport from the top to the bottom within the droplet. This is called the thermo-gravitational flow, which could be expressed by the Rayleigh number as follows [17]:

$$R_a = \frac{\Delta T \beta g L^3}{\upsilon \alpha} \tag{9}$$

where β and υ are the thermal expansion coefficient and the kinematic viscosity of the colloidal droplet, in which $\upsilon = \frac{\eta}{\rho}$. The larger Rayleigh number indicates the more significance of the gravitational effect on the thermo-gravitational flow.

2.4. Mass transportation within the drying droplets

During drying of the colloidal droplet, evaporation-induced local flow filed can give rise to the mass transport. Two ubiquitous phenomena in our daily life are related to the mass transportation within the droplets during drying. They are the "coffee ring" effect and the Marangoni effect. Also, the continuous evaporation of the volatile solvent may cause the aggregation of particles within the sessile drops.

2.4.1. The "coffee ring" effect

When a droplet of coffee is spilt onto a solid substrate, a ring-shaped deposit of coffee solids will be found around the edge of the droplet after completely dried. This phenomenon is called the "coffee ring" effect. The formation of "coffee ring" is a hydrodynamic process attributed to the capillary flow in the droplet during drying, which has been described in Section 2.3. The enhanced evaporation rate at the pinned contact line of the droplet generates the horizontally outward capillary flow; this capillary flow carries the suspended materials, e.g., coffee solids, to the edge. After evaporation, these redistributed materials are concentrated at the edge, thus forming a "coffee ring" at the perimeter of the droplet [8, 9].

http://dx.doi.org/10.5772/65301

There are two essential conditions for the "coffee ring" effect: pinning of the contact line and enhanced evaporation rate at the contact line. If the contact line motion obeys the constant contact angle mode, the suspended materials will be moved inwards by the retraction of the contact line. This process could prevent the ring formation. It is worth mentioning that the pinning of the contact line is dependent on the suspended particles rather than the nature of substrate. The substrate only provides the conditions for temporarily anchoring the contact line; this may trigger the accumulation of suspended particles there. These highly concentrated particles increase the energy barrier for the movement of the contact line, leading to the indefinite pinning and the ring formation at the edge of the droplet [18].

Many strategies are proposed to limit the "coffee ring" effect. Yunker et al. experimentally demonstrated that the control of the particle shape, e.g., ellipsoidal particles, could suppress the outward transport of spherical particles to the contact line of the droplet, ensuring the uniform deposition of particles [19]. Weon and Je studied different movements of polystyrene microspheres with two different sizes (2 and 20 μm in diameter) within the droplet and observed that the small particles have a greater tendency to move outwards to form the ring-shaped deposits, while the large particles initially moved to the edge and then reversed towards the central part of the droplet. They stated that the geometric constraint of the droplet was responsible for the formation of capillary forces on the large particles, which are in the opposite direction as the outward capillary flow, near the contact line, thus repelling the "coffee-ring" effect [20]. These capillary forces were proportional to the size of the suspended particles, while inversely proportional to the contact angle of the droplet [20, 21]. Recently, Eral et al. showed that the application of electrical potential on the sessile drops, containing DNA solution and colloidal particles of various sizes, could suppress the "coffee ring" effect. This is due to the fact that the electrical potential changes the wettability of the droplet on the substrate and prevents the pinning of the contact line, on the one hand, and produces the internal flow fields that counteract the outward capillary flow, on the other hand [22]. Furthermore, control of the Marangoni flow could counteract the capillary flow, reversing the "coffee ring" effect [14]. Therefore, temperature control of the substrate and addition of some organic mixtures were used to promote the Marangoni flow within the droplet to eliminate the ring formation. Soltman and Subramanian studied the profile of the droplet of inject-printed ink on the poly(4-vinylphenol)-coated glass substrate with controlled temperature and disclosed that a decrease in the temperature of substrate could inhibit the "coffee ring" effect [23]. Most recently, the evaporation rate-controlled regime has been revealed to be an effective way to suppress the "coffee ring" effect of a droplet containing nanoparticles [24]. The principle of this process is similar to the common natural phenomenon of the surface skinning of paint and milk, namely, particles accumulate at the interface of liquid and air. When a sessile drop of nanoparticles is drying under a high evaporation rate, the interface of the droplet shrinks much faster than the diffusion of suspended nanoparticles; in this case, the nanoparticles in the vertical evaporation flow can be captured by the shrinking interface, forming an uniform film of aggregated nanoparticles at the interface of the droplet (the film formation will be further discussed in Section 2.4.3). The "coffee ring" effect within the droplet can be prevented by this nanoparticle film, resulting in the uniform deposition of nanoparticles after complete drying [24].

The "coffee ring" effect discussed above is mainly focused on the sessile drop drying on a solid substrate. When it comes to the porous substrates, the situation could be quite different, which requires the consideration of the additional fluid penetration and removal mechanisms. Rui and Brian compared the microstructural morphologies of ink droplets containing polyethylene glycol (PEG) and zirconium dioxide (ZrO_2) nanoparticles drying on the porous substrates with different pore sizes and proposed that the ring formation is dependent on the pore size. The porous substrates play the role in draining the solvent of the droplet by capillary force. Through this mechanism, the pore size may affect the capillary force and thus influences the flow field within the droplet [25]. Most recently, Shen's group investigated the "coffee ring" effect on filter papers [26]. They found that the "coffee ring" may or may not form on filter papers. This is because of the competition between the driving force of the non-uniform evaporation and the retarding forces caused by the chromatographic and filtration effects on the transport of suspended materials in filter papers. For some filter papers, in which the chromatographic and filtration effects are the dominant retarding forces to the materials transportation, the "coffee ring" effect could be totally avoided [26].

The inhibition of the "coffee ring" effect is desirable in many applications. Hence, more significant research efforts are needed to control the "coffee ring" effect in safe, convenient and low-cost ways.

2.4.2. *The Marangoni effect*

The "tears of wine" is another interesting phenomenon widely observed in our daily life. It is also known as the "wine legs," which is manifested as a tear-like droplets running down along the side of the glass of wine, especially for the wine with high alcohol content. The "tears of wine" is the result of the Marangoni effect.

Wine is a mixture of alcohol and water, with many suspended and dissolved materials. When the wine encounters the glass, capillary force in the meniscus at the contact line between wine, air and glass walls will make the wine easily climb on the glass. During climbing, the alcohol, which has a higher equilibrium vapour pressure than water, suffers a higher evaporation rate. In the meniscus, because of the large relative surface area versus the small droplet volume, the concentration of alcohol decreases dramatically. The fast depletion of alcohol causes the different concentrations between the meniscus and bulk of wine, thus generating the surface tension gradient between these two regions. The higher surface tension in the regions with lower alcohol concentration pulls up more liquid from the regions with higher alcohol concentration (lower surface tension) to form the droplets. Once the gravitational force on the droplets overcomes the capillary force, they will creep down on the side of the glass and back to the bulk of wine. That is the "tears of wine."

2.4.3. *Aggregation of particles*

Apart from the redistribution of suspended particles driven by the local flow fields, the liquid evaporation also promotes the movement of the particles. As the evaporation propagates, the tiny menisci will be generated between particles at the interface of the droplet.

These menisci build up a negative Laplace pressure between particles, as shown in **Figure 2** [9]. This Laplace pressure (P_{cap}) is regarded as the compressive capillary force, which could be written as:

$$P_{cap} = -\frac{2\gamma^* \cos\theta_{lp}}{r_m} \tag{10}$$

where γ^* is the surface tension of the pure liquid on the drying layer of the sessile drop, θ_{lp} is the contact angle between the pure liquid and the solid particles, and r_m is the radius of the meniscus. The further evaporation will increase the curvature of the liquid menisci and decrease its radius (r_m). According to Eq. (10), the compressive capillary force will increase as the liquid evaporates; this may drain the liquid through the particle mesoscopic clusters and promote the aggregation of the adjacent clusters [27].

Meanwhile, the entropic depletion force always arises between the large particles in the colloidal suspension containing different sizes of particles [28, 29]. During drying of such colloidal droplets, the Laplace pressure forces the large particles to get closer; this may exclude the small particles from the vicinity of the large particles. This continuous local depletion of small particles produces the depletion force between the large particles and then increases in the osmotic pressure between these particles. As a result, the aggregation of these particles is promoted.

All of the evaporation-induced mass behaviours may or may not form a particle film (skin) at the drop-air interface of the colloidal droplet. It is easy to understand from the macroscopic views. If the colloidal droplet dries slowly and uniformly, the diffusion of suspended particles will be sufficient to maintain an essentially uniform particle concentration within the droplet; this does not favour the formation of the particle skin at the interface. By contrast, if the sessile drop dries fast and non-uniformly, the diffusion of suspended particles is limited, which leads to the increasing concentration of suspended particles at the drop-air interface. Once the concentration of these particles reaches to a critical value, a dense particle skin will form at the interface. The formation of particle skin is sometimes referring to the gelation process [30]. The dimensionless Péclet number (P_e) could be used to predict whether the particle film will be formed or not at the interface of the colloidal droplet during drying:

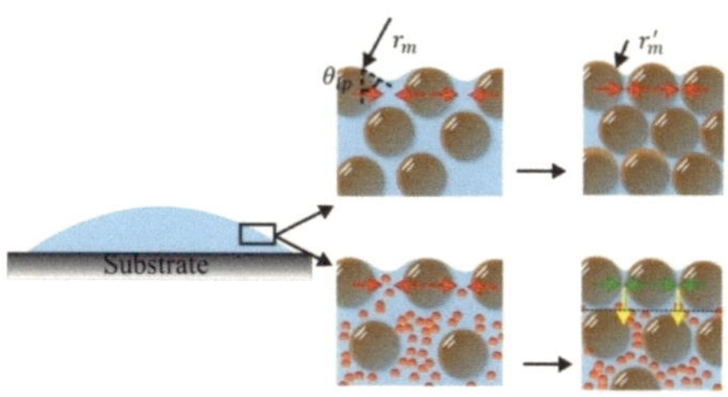

Figure 2. A schematic diagram of the Laplace pressure and the osmotic pressure in a colloidal droplet. The red and green arrows indicate the capillary and depletion forces, respectively. The yellow arrows display the different concentration of the small and red spherical particles.

$$P_e = \frac{Jh_o}{D} \tag{11}$$

where h_o is the initial height of the sessile drop, D is the diffusion coefficient of the non-volatile particles, and J is the evaporation flux of the sessile drop. The evaporation flux (J) is characterised as the mean flux of the vapour per surface area and is given by Annarelli et al. [31]:

$$J = -\frac{dV}{Sdt} \tag{12}$$

where S and V are the drying surface area and the volume of the sessile drop at the drying time t, respectively. If $P_e \ll 1$, the particle diffusion will be more significant; this will lead to a homogeneous distribution of the suspended particles within the sessile drop. If $P_e \gg 1$, the diffusion will be too weak to prevent the increase in the particle concentration at the interface, thus producing a dense skin on the surface of the colloidal droplet [32, 33].

3. Wetting and drying of colloidal droplets: pattern formation

Sessile drops of different colloidal suspensions experience different drying processes and evaporation mechanisms. This leads to the formation of various interesting patterns in the dried droplets, which is termed as the desiccation patterns. They are mainly composed of ring-stain patterns, cracking patterns, crystal patterns, and combined patterns [7, 21]. These desiccation patterns are widely observed in the sessile drops of biofluids and nanofluids. In the following sections, the desiccation patterns in these drying sessile drops will be reviewed.

3.1. Sessile drops of biofluids

3.1.1. Sessile drops of blood plasma

Blood plasma is a complicated suspension that behaves like a Newtonian fluid. It mainly contains 90% (by mass) of water, 6% of macromolecular proteins, 1% of inorganic electrolytes, and other minor components [9].

Desiccation patterns are commonly observed in a sessile drop of blood plasma drying on a solid substrate. These patterns are characterised by two distinguished regions: a peripheral region and a central part. Two major patterns generate in these regions. Cracking patterns in orthoradial and radial directions are throughout the whole sessile drop, and crystal patterns with different morphologies accumulate in the central part, as shown in **Figure 3(a)** [9].

The main components of blood plasma can be changed easily by the health condition; this leads to the different desiccation patterns. Meanwhile, these components present different evaporating behaviours and play different roles in the desiccation patterns. Moreover, they may interact with one another during drying. Because of these, it is a challenge to fully understand the underlying mechanisms for the desiccation patterns in the blood plasma.

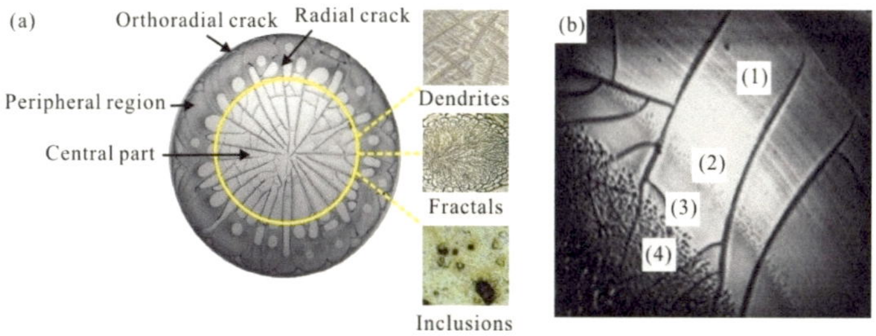

Figure 3. Desiccation patterns in the sessile drop of (a) blood plasma from a healthy adult [9]; (b) morphological details of four regional patterns in the sessile drop of BSA saline solution: (1) homogeneous protein film; (2) protein precipitates, (3) protein gel, and (4) salt crystal [34].

For the sake of better understanding, these desiccation patterns, many simple systems, e.g., bovine serum albumin (BSA) and human serum albumin (HSA) saline solutions (0.9% w/v of NaCl/water), were used to model the complex biofluids. Yakhno conducted a macroscopic study and observed that four distinguished patterns were formed in different regions: homogeneous protein film, protein precipitates, protein gel, and salt crystal, as illustrated in **Figure 3(b)** [34]. In another study, they made the observation on the sessile drop of HSA saline solution and found the similar desiccation patterns as the BSA saline solution [35]. Although these two experiments were carried out by the model macromolecular proteins and saline solution, the desiccation patterns had the similar major feature as the blood plasma: a thick outer ring of aggregated macromolecular proteins with cracking patterns and a flattened central part of accumulated inorganic salts with crystal patterns [3]. Accordingly, these desiccation patterns are referred to plasma patterns.

There are three major factors influencing on the morphologies of these plasma patterns: concentration of inorganic salts, concentration of macromolecular proteins, and the wettability of the droplet on the substrate. Inorganic salts are the main contributors to crystal patterns in the central part [36]. Without inorganic salts, crystal patterns cannot form [37, 38]. Chen and Mohamed investigated the effect of salt concentration on desiccation patterns of the sessile drops of BSA/PBS (phosphate-buffered solution) and observed different crystal patterns in the central part. They suggested that high concentration of salts promoted the aggregation of macromolecular proteins, thus changing the morphologies of crystal patterns [39]. This phenomenon was also observed in the sessile drops of HSA water and saline solutions [37]. According to the Derjaguin-Landau-Verwey-Overbeek theory (DLVO), the stability of a colloidal suspension is dependent on the inter-particle interactions via van der Waal force and Coulombic (electrostatic) force. The van der Waal force is attractive between the particles, while the electrostatic force is repulsive due to the electric double layer of counterions surrounding these particles. The ionic strength of biofluid increases with the increasing salt concentration, leads to the compression of the electric double layer around the macromolecular proteins. This process weakens the repulsive electrostatic force among these proteins, favouring the aggregation of proteins and the subsequent gelation process. Additionally, the diffusion coefficient of macromolecular proteins (D) depends on the ionic strength of colloidal suspension; it decreases proportionally to the increase in the ionic strength [38]. According to

Eq. (11), the lower diffusion coefficient (D) leads to a higher Péclet number (P_e), which means the less effect of the diffusion on the skin formation (gelation) at the interface of the droplet. Therefore, the sessile drop with higher salt concentration is more prone to gelation during drying, forming different desiccation patterns.

Concentrations of macromolecular proteins, e.g., BSA and HSA, are also responsible for the morphologies of desiccation patterns of the dried sessile drop. Tarasevich et al. observed the drying of sessile drops of BSA water and saline solutions with different BSA concentrations and indicated that the BSA concentration influenced on both of the cracking and crystal patterns [38]. The impact of protein concentration on the desiccation patterns was also reported in the sessile drops of HSA water and saline solutions [37, 40]. The effect of protein concentration on the desiccation patterns is related to the gelation because the increase in protein concentration slightly shortens the gelation time of the droplet and increases the initial spacing of radial cracks in the gelled film [31].

Different substrates also result in different morphologies of plasma patterns. Esmonde-White et al. considered that the wettability of the droplet on the substrate played the important role in the apparent contact angle of the sessile drop [41]. This may further influence on the three-phase contact line motion behaviours and the mass transportation during drying. As a consequence, different desiccation patterns form on different substrates.

The drying process of a colloidal droplet is termed as the directional drying because the propagation of evaporation-induced solidification is directed from the edge to the central part. When a droplet of blood plasma is deposited on the solid substrate, the enhanced evaporation rate at the contact line leads to the redistribution of suspended matters. These matters accumulate at the peripheral region and then start to form a film from the edge; this induces the gelation of the droplet. At first, the edge of the sessile drop dries, while the central part remains liquid [42]. In this case, the evaporation-induced shrinkage of the droplet starts from the gelled edge, building up the local stress. However, this shrinkage-caused local stress is constrained by the adhesion of sessile drop to the substrate. The competition between these two stresses results in the tensile stress on the gelled film. Once the tensile stress reaches the level of the local tensile strength, cracking will happen. Typically, the regularly spaced radial cracks occur first and propagate inwards with the retreating of the solidification front. After that, the orthoradial cracks appear parallel to the retreating solidification front [31]. This is the result of the more significance of the orthoradial tensile stress (σ_o) on the gelling film at the edge of the droplet than the radial tensile stress (σ_r), i.e., ($\frac{\sigma_o}{\sigma_r} \gg 1$) [43].

Apart from cracking patterns, crystal patterns in the central part of the droplet have attracted increasing attention of researchers. Some researchers suggested that crystal patterns were composed of the aggregation of macromolecular proteins and the crystallisation of inorganic salts; they further indicated that the increase in inorganic salts promoted the aggregation of proteins [39, 44]. However, Yakhno et al. proposed another explanation to crystal patterns and experimentally demonstrated that these patterns are the salt crystals in the gelled protein matrix; they also argued that proteins only served as seeds for the growth of salt crystals [45, 46]. With current analytical techniques, it is still difficult to clearly distinguish the major components of these crystal patterns. Therefore, the formation mechanisms of crystal patterns are still not well understood.

3.1.2. Sessile drops of whole blood

Whole human blood is a biological suspension that behaves like a non-Newtonian fluid. It is composed of plasma (55% by volume) and cellular components (45% by volume), i.e., red blood cells (RBCs), white blood cells (WBCs), and platelets; RBCs, WBCs, and platelets represent 97%, 2%, and 1% of the total volume of these cellular components, respectively [9].

Desiccation patterns in the dried sessile drop of whole human blood are significantly different from those of blood plasma without cellular components [47, 48]. Patterns in the blood droplet from healthy adults drying on the glass substrates are comprised of three distinguished zones with different characteristic cracking patterns, as displayed in **Figure 4**. They are, from outer to inner, a fine peripheral region adhering to the substrates, a coronal region with regularly ordered radial cracks and large-sized deposit plaques and a central part with disordered chaotic cracks and small-sized deposit plaques [47].

Desiccation patterns of the blood droplet are strongly affected by the external drying conditions, such as the relative humidity (RH) and the wettability of the blood droplet on the substrates [48–51]. Zeid and Brutin compared the desiccation patterns of blood sessile drop drying under different RHs and found obvious difference between these patterns [50]. The influence of the RH on the desiccation patterns of the blood droplet is related to two main aspects. On the one hand, the initial apparent contact angle of the blood droplet decreases linearly with the increase in RH; this will further influence the dynamic of contact line motion and the evaporation flux of the droplet, as well as the redistribution of suspended particles, e.g., RBCs. As a result, different desiccation patterns form under different RHs. On the other hand, a larger RH can lower the evaporation rate of the blood droplet and prolong its drying time [49]. The liquid remaining under the gelled film can release the evaporation-induced internal stresses, e.g., the capillary pressure and the tensile stress [50]. As mentioned in Section 3.1.1, cracking patterns are originated in the

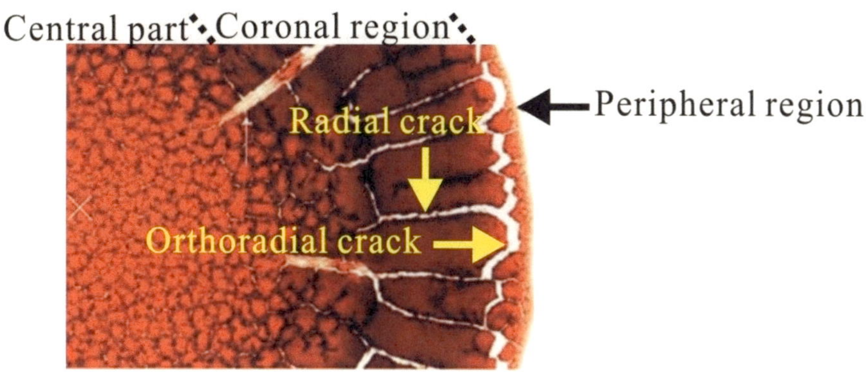

Figure 4. Desiccation patterns of a blood droplet drying on a glass substrate [47].

competition between the shrinkage-induced tensile stress within the gelled film and the adhesion of the droplet to the substrate; as a result, the cracking patterns and the deposit plaques are differing with different RHs.

Brutin et al. studied the desiccation patterns of the blood droplet drying on different substrates and found that desiccation patterns of the droplet on the metallic substrates are dramatically different from those on the glass substrate [48, 51]. The blood droplet on the metallic substrate is nearly hemispherical at the initial stage and, then, forms a uniform glassy skin at the interface; there will be no cracking pattern in the dried droplet. However, the blood droplet on the glass substrate has a lower apparent contact angle at the initial stage and forms a non-uniform solid skin at the interface; after completely dried, it exhibits complex cracking patterns. After comparing the heat and mass transfer within the blood droplet on these substrates, they suggested that the thermal diffusivity of substrates had an insignificant influence on the evaporation dynamics of blood droplet [48]. Subsequently, they attributed these obvious differences of desiccation patterns to the wettability of the blood droplets on these substrates. Because of the different surface energies of these substrates, the apparent contact angle of the blood droplet varies with the substrates. On a hydrophilic glass substrate, the apparent contact angle (θ^*) is close to 20°, while it becomes approximately 90° on the hydrophobic metallic substrates. The evaporation flux at the peripheral region of the blood droplet with low apparent contact angle ($\theta^* < 40°$) is much higher than that in the central part. Nevertheless, the difference in the evaporation flux of the blood droplet with high apparent contact angle ($\theta^* > 90°$) is small [9, 48]. The evaporation behaviours of these blood droplets would therefore be different, leading to the different desiccation patterns.

3.1.3. Sessile drops of other biological fluids

Interesting and complex desiccation patterns are also formed in the sessile drops of other biofluids, e.g., urine, saliva, and tear [52–54]. Yakhno et al. investigated the drying of the sessile drops of urine and saliva from the healthy adults and divided it into three stages: the redistribution of materials leads to the continuous flattening of the droplet; the deposited macromolecular proteins aggregate to form the gel matrix; the inorganic salts induced phase transition of macromolecular proteins to form desiccation patterns [54]. Pearce and Tomlinson investigated desiccation patterns of the tear droplet, as shown in **Figure 5** [55]. They observed a thin amorphous film in the peripheral region, while fern-like patterns in the central part. To study the microstructural morphologies of these desiccation patterns, they employed the scanning electron microscope (SEM) to further observe these regional patterns. The thin amorphous film in the peripheral region presents cracking patterns under the SEM. The fern-like patterns in the central part are composed of dendritic patterns and cubic crystals adjacent to these dendritic patterns. Considering the special components in human tear, e.g., sulphur-containing macromolecular proteins and inorganic salts, they used the energy dispersive X-ray analysis (EDXA) to measure the chemical composition of these patterns and found the present of sulphur-containing macromolecular proteins in the peripheral region. The experimental results also disclosed that dendritic patterns were predominantly made up of sodium and chloride, while cubic crystals were potassium and chloride [55]. López-Solís et al. carried out the further

http://dx.doi.org/10.5772/65301

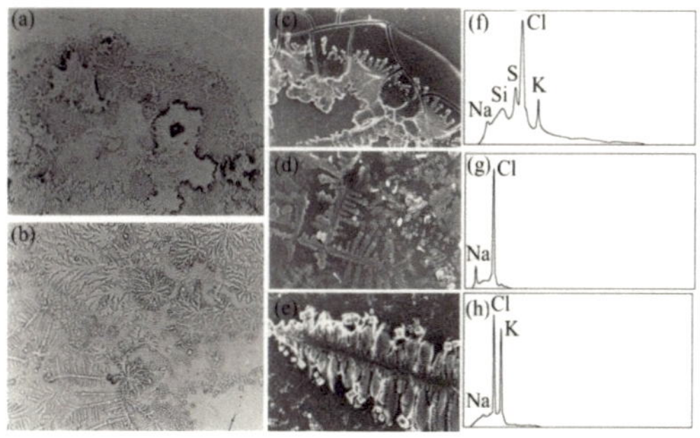

Figure 5. Desiccation patterns in a sessile drop of human tear: (a) thin amorphous film in the peripheral region; (b) fern-like patterns in the central part; (c) cracking patterns in the thin amorphous film; (d) dendritic patterns in the central part; (e) cubic crystals in the central part; (f)–(h) Energy Dispersive X-ray Analysis (EDXA) of the cracked film in (c), dendritic patterns in (d), and cubic crystals in (e), respectively [55].

experiments on desiccation patterns of tear droplets and indicated that the patterns had four distinct zones. They are an outer structured hyaline area with cracking patterns (Zone I), a band of regularly spaced and centripetally oriented crystals (Zone II), a central part comprising randomly distributed fern-like patterns (Zone III) and a circular transition between Zones I and II [52]. The desiccation patterns of tear droplets are quite similar to plasma patterns, including the accumulation of macromolecular proteins at the outer ring and the crystallisation of inorganic salts in the central part.

3.2. Sessile drops of nanofliuds

Nanofluids are the colloidal suspensions of nanoparticles (1–100 nm in size), e.g., metals, oxides, and carbon nanotubes, in a conventional base fluid, e.g., water and ethanol [56]. Various desiccation patterns have been observed in the sessile drops of nanofluids, such as "coffee ring" patterns, multi-ring patterns, uniform-deposition patterns, and cracking patterns, as illustrated in **Figure 6**. Among these patterns, the formation of "coffee ring" patterns is due to the "coffee ring" effect (explained in Section 2.4.1), the multi-ring patterns are related to the "stick-slip" motion of the contact line of the droplet during drying (mentioned in Section 2.3), the uniform-deposition patterns are the result of the inhibition of the "coffee ring" effect caused by the wide

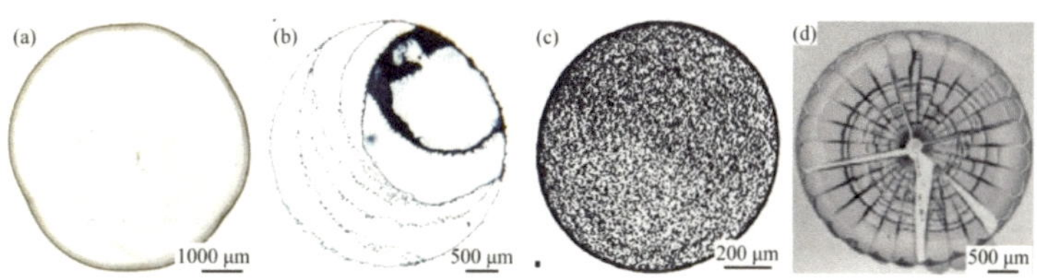

Figure 6. Desiccation patterns of sessile drops of nanofluids: (a) "coffee ring" patterns in the suspension of single-walled carbon nanotube [57]; (b) multi-ring patterns in the suspension of titanium dioxide nanoparticles in ethanol [5]; (c) uniform-deposition patterns in the suspension of graphite nanoparticles [58]; (d) cracking patterns in the suspension of silica nanoparticles [43].

coverage of nanoparticles on the surface of the droplet (referred in Section 2.4.1), and the formation of cracking patterns is connected to the competition between the shrinkage-induced tensile stress and the adhesion of the droplet to the substrates (elaborated in Section 3.1).

These characteristic desiccation patterns can be controlled by different ways. The detailed methods for the inhibition of "coffee ring" patterns have been described in Section 2.4.1. Recently, the transition of desiccation patterns from "coffee ring" patterns to cracking patterns has been found in the droplets of polystyrene nanoparticles as the concentration of nanoparticles increases. This transition is mainly attributed to the uniform distribution of the suspended nanoparticles within the droplet of high concentration of nanoparticles [59]. Most recently, Zang et al. investigated the directional drying of the sessile drops containing polytetrafluoroethylene (PTFE) nanoparticles [13]. The morphological evolution of the sessile drop of PTFE nanoparticles appears to be similar as that of the blood droplet. In the first stage, the gelation starts from the peripheral region, while the central part remains liquid; the wet front recedes as the liquid evaporates, causing the instability in the central part. When it comes to the second stage, the radial cracks occur at the peripheral region and propagate towards the central part. Interestingly, these authors found the surface wrinkling in the gelled zone of the droplet and attributed this phenomenon to the stretching effect caused by the surface tension of liquid in the central part. The surface tension of the hemispheric cap of the liquid in the central part causes the radial stresses on the gelled zone, leading to the flaws in the radial direction. With further liquid evaporation, cracks occur and develop in the radial direction [13]. To inhibit these cracking patterns, these authors added the sodium dodecylsulfate (SDS) into the droplet and suggested that the addition of SDS inhibited the formation of cracking patterns [13]. In another experiment, they studied the drying of the sessile drop containing silica nanoparticles and polyethylene oxide (PEO) additives and indicated that the PEO additives lead to the uniform deposition of silica nanoparticles during drying and to the suppression of cracking patterns after drying [60].

3.3. Applications of desiccation patterns

It is well known that some diseases can cause the changes in the composition of biofluids and the modification in the structure of macromolecular proteins; this may result in the different characteristic desiccation patterns. This implies that desiccation patterns in the sessile drops of biofluid could be used as the simple and low-cost diagnostic tools to identify the health condition of human beings [9].

The medical diagnosis technique based on desiccation patterns of biological droplets has begun with the so-called "Litos" test system in the former Soviet Union three decades ago. This system was developed after it was realised that inorganic salts in the urine droplet of patients with urolithiasis would crystallise to form the distinct desiccation patterns, which enables the diagnosis of urolithiasis at a preclinical stage [61]. After that, the medical diagnosis technique based on desiccation patterns attracted increasing attention of researchers; this is not only because of low-cost but also due to the fact that the diagnosis can be performed by a less-qualified individual [3]. Yakhno et al. presented systematic correlations and analyses of desiccation patterns in the sessile drop of blood serum from healthy individuals and patients

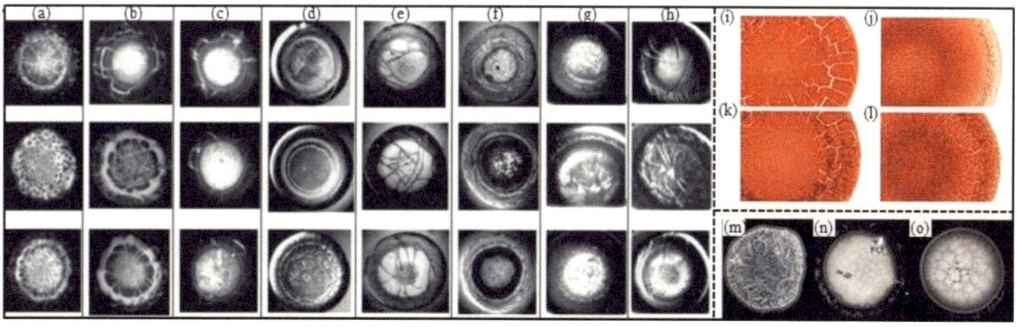

Figure 7. Desiccation patterns of sessile drops of blood serum from (a) healthy individuals and patients with different diseases or physiological states: (b) breast cancer; (c) lung cancer; (d) paraproteinemia; (e) maternal full term delivery; (f) maternal premature delivery; (g) threatened abortion (premature delivery) in different periods of gestation; (h) hepatitis [62]; blood droplets from (i) a 27-year-old healthy female; (j) a patient with anaemia; (k) a 31-year-old healthy male; (l) a patient with hyperlipidaemia [47]; tear droplets from (m) a healthy individual; (n) a 40-year-old female with moderate xerophthalmia; (o) a 72-year-old male with severe xerophthalmia [53].

with different diseases: breast cancer, lung cancer, paraproteinemia, hepatitis, as well as from females with maternal full-term delivery and premature deliveries in different periods, as displayed in **Figure 7(a)–(h)** [62]. They indicated that desiccation patterns in the blood serum droplets from healthy individuals could be easily distinguished with patients suffering from the above-mentioned diseases [62]. Muravlyvova et al. analysed desiccation patterns in the dried plasma droplets from patients with different lung-related diseases from the molecular level and implied that the distinguished patterns in the droplets from patients with interstitial pulmonary disease could be linked to the disease-induced modification of the extracellular nucleic acids and macromolecular proteins [63]. Consequently, the key information locked in desiccation patterns of the blood plasma droplet is believed to be useful for disease diagnosis.

Recently, desiccation patterns in the sessile drops of whole human blood have been reported to be connected to the diagnosis of blood-related diseases, as displayed in **Figure 7(i)–(l)** [47]. **Figure 7(i) and (k)** are the samples from healthy individuals; these two samples exhibit the similar desiccation patterns. **Figure 7(j)** presents the desiccation patterns in the blood droplet from a patient with anaemia. The peripheral region and the outer area of the coronal region are very light in colour and have more plaques compared with the samples from healthy individuals. This could be associated with the deficit of RBCs and low level of haemoglobin in these cells of anaemia sufferers. **Figure 7(l)** illustrates the desiccation patterns in the blood droplet from a patient with hyperlipidaemia. The peripheral region and the outer area of the coronal region are thick and greasy, which is most likely related to the high level of lipoproteins in the blood. Furthermore, the desiccation patterns in the blood droplet from unhealthy people do not have the radial cracking patterns in the coronal region. This could be due to the changed wettability of the blood droplet on the glass substrates caused by the disease-induced modification of blood composition [47].

Desiccation patterns left by the dried tear droplets might also give the indications of the ocular diseases. Traipe-Castro et al. compared the tear droplets from a healthy individual and those from patients with moderate and severe xerophthalmia (dry eyes) and highlighted two different points between those two samples. The tear droplets from sick people experienced much shorter drying time than that from healthy individual. After drying, the tear droplets from sick people presented the remarkably different patterns with the healthy individual, as indicated in **Figure 7 (m)–(o)** [53]. Although they did not provide the detailed mechanisms of the linkage between the desiccation patterns and the eye diseases, their experimental findings demonstrated that it could be a possible diagnosis for some eye diseases [53].

Most recently, the possible application of desiccation patterns of the biological droplets has been reported in animal detection. Yakhno et al. investigated the desiccation patterns in the sessile drops of milk, serum, and whole blood from the healthy and diseased cows (bovine leukaemia and bovine tuberculosis). They found that the desiccation patterns in the biological droplets from the healthy cows were significantly different from the diseased cows [64].

Although the interpretation of the information locked in desiccation patterns of a sessile drop of biological fluid is believed to be a low-cost and convenient diagnostic tool, there still have some critical questions need to answer. It can be seen from **Figure 7** that desiccation patterns differ from person to person under the similar conditions. Also, desiccation patterns are very sensitive to the drying conditions and the variations in the main composition in the droplets. Therefore, the first question is how to keep the repeatability of these desiccation patterns. In addition, the diagnosis based on desiccation patterns is subjective because it significantly depends on the visual perceptions of researchers. Hence, the second question is how to ensure the objectivity of the analysis on these desiccation patterns.

4. Prospective work

In summary, the physics of three-phase contact line motion, flow fields, and mass transportation play important roles in the wetting and drying of colloidal droplets as well as their pattern formation. Desiccation patterns formed during drying of blood drops find new applications in diagnosis.

Future works are expected to study on the precursor film, e.g., the length, height and main components, the relationship between the three-phase contact line motion and the evaporation flux along the surface of the sessile drop, the underlying mechanisms of evaporation-induced changes of the local flow field, as well as the control of the "coffee ring" effect. Moreover, two critical issues are required to resolve for pattern applications: the repeatability of desiccation patterns and the objectivity of the analysis on these complex patterns.

Acknowledgements

Ruoyang Chen acknowledges Monash University Institute of Graduate Research and the Faculty of Engineering for postgraduate research scholarships. Liyuan Zhang and Wei Shen acknowledge Australian Research Council Discovery Project (ARC DP1094179). Duyang Zang acknowledges the National Natural Science Foundation of China (Grant No. 51301139), the

http://dx.doi.org/10.5772/65301

Shaanxi Provincial Natural Science Foundation (Grant No. 2016JM1003) and the Fundamental Research Funds for the Central Universities (Grant No. 3102016ZY026).

Author details

Ruoyang Chen[1], Liyuan Zhang[1], Duyang Zang[2]* and Wei Shen[1]*

*Address all correspondence to: dyzang@nwpu.edu.cn and wei.shen@monash.edu

1 Department of Chemical Engineering, Monash University, Clayton, Australia

2 Department of Applied Physics, Northwestern Polytechnical University, Xi'an, China

References

[1] Kavehpour H, Ovryn B, McKinley G. Microscopic and macroscopic structure of the precursor layer in spreading viscous drops. Physical Review Letters. 2003;91:196104. DOI: 10.1103/PhysRevLett.91.196104.

[2] Bonn D, Eggers J, Indekeu J, Meunier J, Rolley E. Wetting and spreading. Reviews of Modern Physics. 2009;81:739-805. DOI: 10.1103/RevModPhys.81.739.

[3] Sefiane K. Patterns from drying drops. Advances in Colloid and Interface Science. 2014;206:372-381. DOI:10.1016/j.cis.2013.05.002.

[4] Shanahan M. Simple theory of "stick-slip" wetting hysteresis. Langmuir. 1995;11:1041-1043. DOI: 10.1021/la00003a057.

[5] Moffat J, Sefiane K, Shanahan M. Effect of TiO_2 nanoparticles on contact line stick-slip behavior of volatile drops. The Journal of Physical Chemistry B. 2009;113:8860-8866. DOI: 10.1021/jp902062z.

[6] Adachi E, Dimitrov A, Nagayama K. Stripe patterns formed on a glass surface during droplet evaporation. Langmuir. 1995;11:1057-1060. DOI: 10.1021/la00004a003.

[7] Askounis A, Sefiane K, Koutsos V, Shanahan M. Effect of particle geometry on triple line motion of nano-fluid drops and deposit nano-structuring. Advances in Colloid and Interface Science. 2015;222:44-57. DOI: 10.1016/j.cis.2014.05.003.

[8] Deegan R, Bakajin O, Dupont T, Huber G, Nagel S, Witten T. Capillary flow as the cause of ring stains from dried liquid drops. Nature. 1997;389:827-829. DOI: 10.1038/39827.

[9] Chen R, Zhang L, Zang D, Shen W. Blood drop patterns: formation and applications. Advances in Colloid and Interface Science. 2016;231:1-14. DOI:10.1016/j.cis.2016.01.008.

[10] Kajiya T, Kaneko D, Doi M. Dynamical visualization of "coffee stain phenomenon" in droplets of polymer solution via fluorescent microscopy. Langmuir. 2008;24:12369-12374. DOI: 10.1021/la8017858.

[11] Hu H, Larson R. Analysis of the effects of Marangoni stresses on the microflow in an evaporating sessile droplet. Langmuir. 2005;21:3972-3980. DOI: 10.1021/la0475270.

[12] Hu H, Larson R. Marangoni effect reverses coffee-ring depositions. The Journal of Physical Chemistry B. 2006;110:7090-7094. DOI: 10.1021/jp0609232.

[13] Zhang Y, Qian Y, Liu Z, Li Z, Zang D. Surface wrinkling and cracking dynamics in the drying of colloidal droplets. The European Physical Journal E. 2014;37:1-7. DOI: 10.1140/epje/i2014-14084-3.

[14] Parsa M, Harmand S, Sefiane K, Bigerelle M, Deltombe R. Effect of substrate temperature on pattern formation of nanoparticles from volatile drops. Langmuir. 2015;31:3354-3367. DOI: 10.1021/acs.langmuir.5b00362.

[15] Kim J, Park S, Kim J, Zin W. Polymer transports inside evaporating water droplets at various substrate temperatures. The Journal of Physical Chemistry C. 2011;115:15375-15383. DOI: 10.1021/jp202429p.

[16] Roisman I, Criscione A, Tropea C, Mandal D, Amirfazli A. Dislodging a sessile drop by a high-Reynolds-number shear flow at subfreezing temperatures. Physical Review E. 2015;92:023007. DOI: 10.1103/PhysRevE.92.023007.

[17] Carle F, Brutin D. Convection. In: Brutin D, editor. Droplet Wetting and Evaporation. Oxford: Academic Press; 2015. p. 115-228. DOI: 10.1016/B978-0-12-800722-8.00009-6.

[18] Deegan R. Pattern formation in drying drops. Physical Review E. 2000;61:475-485. DOI: 10.1103/PhysRevE.61.475.

[19] Yunker P, Still T, Lohr M, Yodh A. Suppression of the coffee-ring effect by shape-dependent capillary interactions. Nature. 2011;476:308-311. DOI: 10.1038/nature10344.

[20] Weon B, Je J. Capillary force repels coffee-ring effect. Physical Review E. 2010;82:015305. DOI: 10.1103/PhysRevE.82.015305.

[21] Zhong X, Crivoi A, Duan F. Sessile nanofluid droplet drying. Advances in Colloid and Interface Science. 2015;217:13-30. DOI:10.1016/j.cis.2014.12.003.

[22] Eral H, Augustine D, Duits M, Mugele F. Suppressing the coffee stain effect: how to control colloidal self-assembly in evaporating drops using electrowetting. Soft Matter. 2011;7:4954-4958. DOI: 10.1039/C1SM05183K.

[23] Soltman D, Subramanian V. Inkjet-printed line morphologies and temperature control of the coffee ring effect. Langmuir. 2008;24:2224-2231. DOI: 10.1021/la7026847.

[24] Li Y, Yang Q, Li M, Song Y. Rate-dependent interface capture beyond the coffee-ring effect. Scientific Reports. 2016;6:24628. DOI:10.1038/srep24628.

[25] Dou R, Derby B. Formation of coffee stains on porous surfaces. Langmuir. 2012;28:5331-5338. DOI: 10.1021/la204440w.

[26] Nilghaz A, Zhang L, Shen W. Coffee stains on paper. Chemical Engineering Science. 2015;129:34-41. DOI:10.1016/j.ces.2015.02.017.

[27] Dufresne E, Corwin E, Greenblatt N, Ashmore J, Wang D, Dinsmore A, Cheng J, Xie X, Hutchinson J, Weitzl D. Flow and fracture in drying nanoparticle suspensions. Physical Review Letters. 2003;91:224501. DOI: 10.1103/PhysRevLett.91.224501.

[28] Anachkov S, Danov K, Basheva E, Kralchevsky P, Ananthapadmanabhan K. Determination of the aggregation number and charge of ionic surfactant micelles from the stepwise thinning of foam films. Advances in Colloid and Interface Science. 2012;183–184:55-67. DOI: 10.1016/j.cis.2012.08.003.

[29] Tuinier R, Rieger J, de Kruif C. Depletion-induced phase separation in colloid-polymer mixtures. Advances in Colloid and Interface Science. 2003;103:1-31. DOI: 10.1016/S0001-8686(02)00081-7.

[30] Lu P, Zaccarelli E, Ciulla F, Schofield A, Sciortino F, Weitz D. Gelation of particles with short-range attraction. Nature. 2008;453:499-503. DOI:10.1038/nature06931.

[31] Annarelli C, Fornazero J, Bert J, Colombani J. Crack patterns in drying protein solution drops. The European Physical Journal E. 2001;5:599-603. DOI: 10.1007/s101890170043.

[32] Trueman R, Lago Domingues E, Emmett S, Murray M, Keddie J, Routh A. Autostratification in drying colloidal dispersions: experimental investigations. Langmuir. 2012;28:3420-3428. DOI: 10.1021/la203975b.

[33] Trueman R, Lago Domingues E, Emmett S, Murray M, Routh A. Auto-stratification in drying colloidal dispersions: a diffusive model. Journal of Colloid and Interface Science. 2012;377:207-212. DOI: 10.1016/j.jcis.2012.03.045.

[34] Yakhno T. Salt-induced protein phase transitions in drying drops. Journal of Colloid and Interface Science. 2008;318:225-230. DOI: 10.1016/j.jcis.2007.10.020.

[35] Yakhno T, Kazakov V, Sanin A, Shaposhnikova O, Chernov A. Dynamics of phase transitions in drying drops of human serum protein solutions. Technical Physics. 2007;52:515-520. DOI: 10.1134/S1063784207040196.

[36] Yakhno T, Kazakov V, Sanina O, Sanin A, Yakhno V. Drops of biological fluids drying on a hard substrate: variation of the morphology, weight, temperature, and mechanical properties. Technical Physics. 2010;55:929-935. DOI: 10.1134/S1063784210070030.

[37] Buzoverya M, Shcherbak Y, Shishpor I. Experimental investigation of the serum albumin fascia microstructure. Technical Physics. 2012;57:1270-1276. DOI: 10.1134/S1063784212090071.

[38] Tarasevich Y, Ayupova A. Effect of diffusion on the separation of components in a biological fluid upon wedge-shaped dehydration. Technical Physics. 2003;48:535-540. DOI: 10.1134/1.1576463.

[39] Chen G, Mohamed G. Complex protein patterns formation via salt-induced self-assembly and droplet evaporation. The European Physical Journal E. 2010;33:19-26. DOI: 10.1140/epje/i2010-10649-4

[40] Buzoverya M, Shcherbak Y, Shishpor I, Potekhina Y. Microstructural analysis of biological fluids. Technical Physics. 2012;57:1019-1024. DOI: 10.1134/S1063784212070079.

[41] Esmonde-White K, Esmonde-White F, Morris M, Roessler B. Characterization of biofluids prepared by sessile drop formation. Analyst. 2014;139:2734-2341. DOI: 10.1039/C3AN02175K.

[42] Tarasevich Y, Vodolazskaya I, Bondarenko O. Modeling of spatial-temporal distribution of the components in the drying sessile droplet of biological fluid. Colloids and Surfaces A: Physicochemical and Engineering Aspects. 2013;432:99-103. DOI: 10.1016/j.colsurfa.2013.04.069.

[43] Giorgiutti-Dauphiné F, Pauchard L. Elapsed time for crack formation during drying. The European Physical Journal E. 2014;37:1-7. DOI: 10.1140/epje/i2014-14039-8.

[44] Pauchard L, Parisse F, Allain C. Influence of salt content on crack patterns formed through colloidal suspension desiccation. Physical Review E. 1999;59:3737-3740. DOI:10.1103/PhysRevE.59.3737.

[45] Yakhno T. Complex pattern formation in sessile droplets of protein-salt solutions with low protein content. What substance fabricates these patterns? Physical Chemistry. 2011;1:10-13. DOI: 10.5923/j.pc.20110101.02.

[46] Yakhno T. Sodium chloride crystallization from drying drops of albumin-salt solutions with different albumin concentrations. Technical Physics. 2015;60:1601-1608. DOI: 10.1134/S1063784215110262.

[47] Brutin D, Sobac B, Loquet B, Sampol J. Pattern formation in drying drops of blood. Journal of Fluid Mechanics. 2011;667:85-95. DOI:10.1017/S0022112010005070.

[48] Brutin D, Sobac B, Nicloux C. Influence of substrate nature on the evaporation of a sessile drop of blood. Journal of Heat Transfer. 2012;134:061101. DOI: 10.1115/1.4006033.

[49] Bou-Zeid W, Vicente J, Brutin D. Influence of evaporation rate on cracks' formation of a drying drop of whole blood. Colloids and Surfaces A: Physicochemical and Engineering Aspects. 2013;432:139-146. DOI: 10.1016/j.colsurfa.2013.04.044.

[50] Bou-Zeid W, Brutin D. Influence of relative humidity on spreading, pattern formation and adhesion of a drying drop of whole blood. Colloids and Surfaces A: Physicochemical and Engineering Aspects. 2013;430:1-7. DOI: 10.1016/j.colsurfa.2013.03.019.

[51] Sobac B, Brutin D. Desiccation of a sessile drop of blood: cracks, folds formation and delamination. Colloids and Surfaces A: Physicochemical and Engineering Aspects. 2014;448:34-44. DOI: 10.1016/j.colsurfa.2014.01.076.

[52] López-Solís R, Traipe-Castro L, Salinas-Toro D, Srur M, Toledo-Araya H. Microdesiccates produced from normal human tears display four distinctive morphological components. Biological Research. 2013;46:299-305. DOI: 10.4067/S0716-97602013000300012.

[53] Traipe-Castro L, Salinas-Toro D, López D, Zanolli M, Srur M, Valenzuela F, Cáceres A, Toledo-Araya H, López-Solís R. Dynamics of tear fluid desiccation on a glass surface: a contribution to tear quality assessment. Biological Research. 2014;47:25. DOI: 10.1186/0717-6287-47-25.

[54] Yakhno T, Yakhno V, Sanin A, Sanina O, Pelyushenko A. Protein and salt: spatiotemporal dynamics of events in a drying drop. Technical Physics. 2004;49:1055-1063. DOI: 10.1134/1.1787668.

[55] Pearce E, Tomlinson A. Spatial location studies on the chemical composition of human tear ferns. Ophthalmic and Physiological Optics. 2000;20:306-313. DOI: 10.1046/j.1475-1313.2000.00523.x.

[56] Taylor R, Coulombe S, Otanicar T, Phelan P, Gunawan A, Lv W, Rosengarten G, Prasher R, Tyagi H. Small particles, big impacts: a review of the diverse applications of nanofluids. Journal of Applied Physics. 2013;113:011301. DOI: 10.1063/1.4754271.

[57] Li Q, Zhu Y, Kinloch I, Windle A. Self-organization of carbon nanotubes in evaporating droplets. The Journal of Physical Chemistry B. 2006;110:13926-13930. DOI: 10.1021/jp061554c.

[58] Crivoi A, Duan F. Effect of surfactant on the drying patterns of graphite nanofluid droplets. The Journal of Physical Chemistry B. 2013;117:5932-5938. DOI: 10.1021/jp401751z.

[59] Brutin D. Influence of relative humidity and nano-particle concentration on pattern formation and evaporation rate of pinned drying drops of nanofluids. Colloids and Surfaces A: Physicochemical and Engineering Aspects. 2013;429:112-120. DOI: 10.1016/j.colsurfa.2013.03.012.

[60] Zhang Y, Liu Z, Zang D, Qian Y, Lin K. Pattern transition and sluggish cracking of colloidal droplet deposition with polymer additives. Science China Physics, Mechanics and Astronomy. 2013;56:1712-1718. DOI: 10.1007/s11433-013-5280-5.

[61] Sefiane K. On the formation of regular patterns from drying droplets and their potential use for bio-medical applications. Journal of Bionic Engineering. 2010;7:S82-S93. DOI:10.1016/S1672-6529(09)60221-3.

[62] Yakhno T, Yakhno V, Sanin A, Sanina O, Pelyushenko A, Egorova N, Terentiev I, Smetanina S, Korochkina O, Yashukova E. The informative-capacity phenomenon of drying drops. Engineering in Medicine and Biology Magazine. 2005;24:96-104. DOI: 10.1109/MEMB.2005.1411354.

[63] Muravlyova L, Molotov-Luchanskiy V, Bakirova R, Zakharova Y, Klyuyev D, Bakenova P, Demidchik L, Suleimenova S. Structure-forming properties of blood plasma of patients with interstitial lung diseases. World Journal of Medical Sciences. 2014;10:478-483. DOI: 10.5829/idosi.wjms.2014.10.4.83285.

[64] Yakhno T, Sanin A, Ilyazov R, Vildanova G, Khamzin R, Astascheva N, Markovsky M, Bashirov V, Yakhno V. Drying drop technology as a possible tool for detection leukemia and tuberculosis in cattle. Journal of Biomedical Science and Engineering. 2015;8:1-23. DOI: 10.4236/jbise.2015.81001.

Chapter 2

Advances in Design and Self-Assembly of Functionalized LB Films and Supramolecular Gels

Tifeng Jiao, Ruirui Xing, Kai Ma and Lexin Zhang

Additional information is available at the end of the chapter

http://dx.doi.org/10.5772/65122

Abstract

The recent progress in functionalized LB films and supramolecular gels varies and occupies various fields. Self-assembly technique is playing an important role in preparing well-defined multilevel nanostructures and the functionalized nanomaterials with the designed and controlled properties. In this chapter, various kinds of functionalized LB films and supramolecular gels, including gold nanoparticles, inorganic-organic hybrid composites, and graphene oxide nanocomposites, have been demonstrated and analyzed. We show main research contributions in recent years in two sections: preparation and self-assembly of some functionalized LB films and preparation and self-assembly of some functionalized supramolecular gels. The above research work may give the potential perspective for the design and preparation of new self-assembly nanomaterials. Future research on preparation of LB films and supramolecular gels will depend on the novel applications and special nanostructures in order to produce novel functional nanomaterials and devices.

Keywords: LB film, supramolecular gel, self-assembly, nanomaterials, nanostructures, properties

1. Introduction

The recent progress in functionalized LB films and supramolecular gels is varied and occupies various fields. It is well known that the self-assembly techniques demonstrate important and crucial role to prepare various functionalized nanostructures and nanomaterials with the preferred properties. In comparison with the conventional self-assembly process of amphiphiles in bulk or at interfaces, the self-assembly process of inorganic-organic hybrids, nanoparticles and colloidal microspheres, and supramolecular nanostructures show special properties, especially in preparing complicated nanocomposites or nanomaterials.

In the recent reports in our groups, some systems, including LB films and supramolecular gels, have been designed and characterized. In addition, some useful analytical techniques and synthetic treatments applied in supramolecular self-assembly field have been highlighted and analyzed. In the present chapter, we summarize main work contributions in recent years in two parts: (1) preparation and self-assembly of some functionalized LB films and (2) preparation and self-assembly of some functionalized supramolecular gels.

2. Preparation and self-assembly of some functionalized LB films

The Langmuir-Blodgett technique is an important method to produce macroscopic materials organized on the molecular scale [1–5]. This approach allows amphiphilic molecules to be oriented at the air-water interface and then transferred sequentially onto a solid support [6–15]. In our research, some special amphiphiles, including bolaamphiphiles, gemini, and amphiphiles with functional substituted groups, have been designed and synthesized, and their organized supramolecular assemblies in LB films have also been investigated, which may broaden the traditional research areas about LB film and give some perspectives and clues for the relative research in the future.

Firstly, in our previous work, a bolaform and single chain Schiff base, abbreviated as BSC10 and HBOA, have been synthesized and their interfacial self-assembly in LB films and interaction with barbituric acid (BA) was demonstrated [16]. It has been found that while HBOA formed a monolayer at the air/water interface, the bolaform Schiff base formed a multilayer film with ordered layer structure on water surface. The relative interfacial morphologies were shown in **Figure 1**. The detailed investigations of the transferred multilayer films were characterized by various spectral techniques. A clear conformational change of the alkyl spacer in the bolaform Schiff base was observed during the complex formation with the barbituric acid. In the obtained self-assembly unit, the alkyl spacer showed obvious bent/gauche conformation due to the H-bond requirement. So, direct experimental data about conformation change of alkyl chains in LB films were clearly demonstrated.

In addition, the interfacial films of a series of designed gemini amphiphiles containing the Schiff base moiety have been investigated [17]. This series of gemini amphiphiles with Schiff base headgroups linked by a hydrophobic alkyl spacer (BisSBC18Cn, n = 2, 4, 6, 8, 10) could be spread to form stable monolayers and coordinated with Cu(II) ions in the monolayer. The Langmuir films of gemini amphiphiles with the spacer length of six or eight methylene groups demonstrated maximum limiting molecular area. Nanonail and tape-like morphologies were observed for amphiphile films with shorter spacers (n = 2 and 4) on the water surface. Worm-like morphologies were observed for gemini films with longer spacers of C8 and C10 when coordinated with Cu(II) ions, as illustrated in **Figure 2**. For the compounds with short spacer, the spacer showed parallel organization to the water surface. With length increment, the spacer part could show bent/gauche conformation. Therefore, the films with spacer lengths of C6 and C8 on water and Cu(II) ions demonstrated maximum limiting molecular areas, respectively.

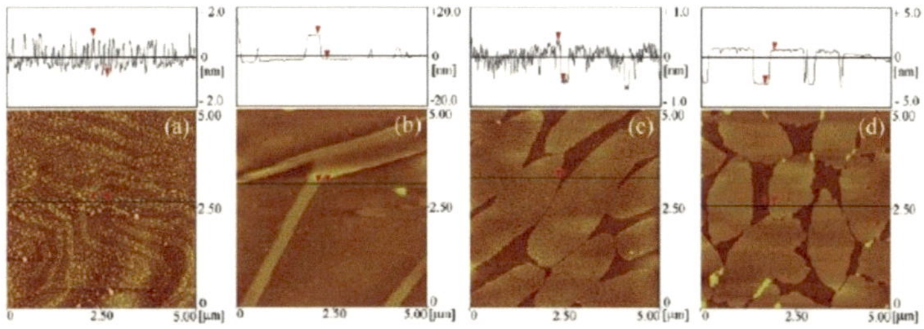

Figure 1. AFM images of monolayer LB films of bolaform Schiff base (HBOA) from pure water subphase and 1.0 mM BA subphase.

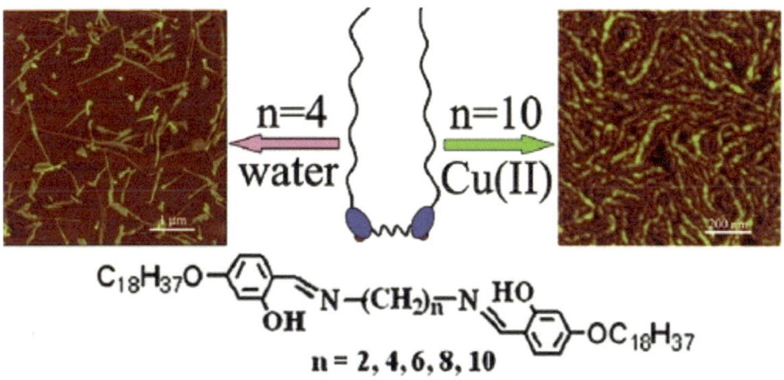

Figure 2. Schematic illustration of the packing modes of BisSBC18Cn in organized films.

In another research work, a tyrosine-based bolaamphiphile (abbreviated as C10BT) has been designed and its interfacial assemblies were investigated [18]. It was interesting to find that metal ions, such as Ag(I) and Cu(II) ions, in the subphase can greatly modulate the molecular packing of C10BT and the morphology of the subsequently deposited LB films. The X-ray diffraction and X-ray photoelectron spectra verified the orderly layer structure and the relative molar ratios compared with different metal ions. Quantitative analysis of XPS data indicated the values of 1:1.95 for C10BT:Ag(I) and 1:1.08 for C10BT:Cu(II). Considering the coordination process and spectral results, as shown in **Figure 3**, a rational self-assembly process has been proposed. For the Cu(II)-coordinated film, the molecules were connected by coordination sites with spacer in bent conformation. On the other hand, for Ag(I)-complexed film, the molecules were orderly stacked by cooperative forces.

Next, we have also investigated the interfacial self-assembly of a block-type polymer abbreviated as PEO_{45}-b-$PDMA_{69}$ by the LB technique [19]. Accompanied by a pancake-to-brush transition of PEO conformation, the worm-like surface micelles are compression induced, as shown in **Figure 4**. The micelles as the building blocks can be arrayed parallel to generate a long-range ordered structure, further bended and twisted upon compression. The obtained nanostructure change could be assigned to the special coil-semirod molecular structures and the unique capability of the PDMA part. The research work gives exploration for preparing functional polymers and nanostructures in organized films.

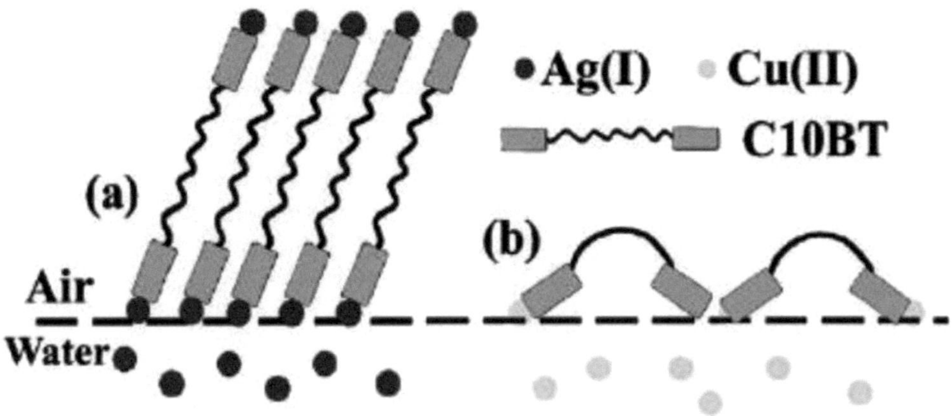

Figure 3. Cartoon illustration of the packing modes for C10BT complex films on different subphases: (a) Ag(I) ions subphase; (b) Cu(II) ions subphase.

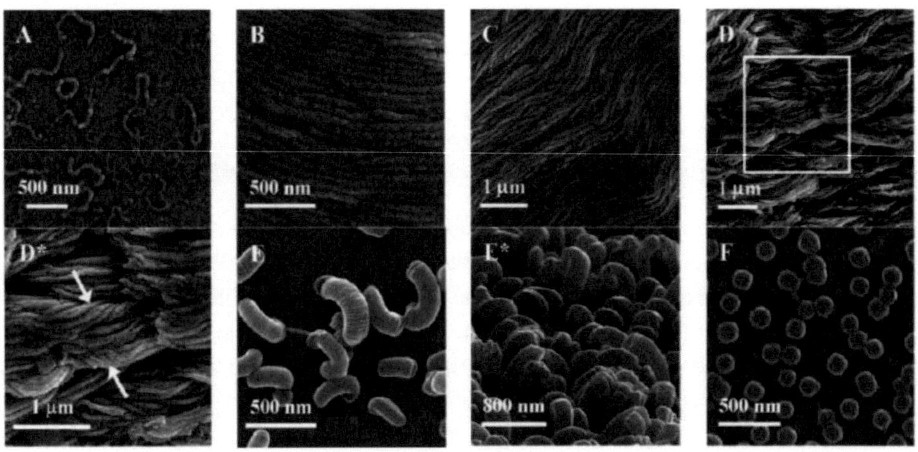

Figure 4. SEM images of the LB films at various molecular areas: (A) 28 nm^2/molecule; (B) 21 nm^2/molecule; (C) 16 nm^2/molecule; (D) 13 nm^2/molecule; (E) 7 nm^2/molecule; (F) 5 nm^2/molecule.

In another work, two coumarin derivatives were synthesized and their interfacial self-assemblies were investigated [20]. Owing to the different substituent position of the long octadecyloxy chain in the coumarin parent, the two compounds showed obviously different interfacial behaviors. The spreading films on the water surface were transferred onto solid substrates and characterized by various spectra and atomic force microscopy (AFM). Different packing of the molecules in the multilayer films was observed, as shown in **Figure 5**. Photoreaction could not occur in 4-CUMC18 film due to possible large steric hindrance, while photochemical reaction produced in 7-CUMC18 film because of the face-to-face stacking of benzene ring and steric matching.

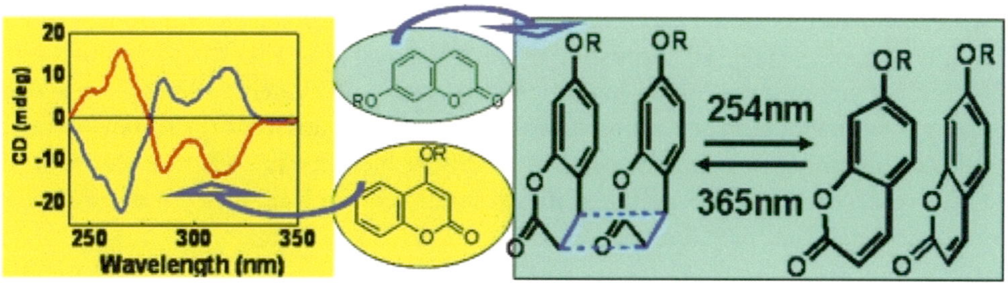

Figure 5. Illustration of the possible packing and photoreaction in the LS films.

In addition, preparation and characterization of gemini amphiphiles-gold nanostructure composite films have been demonstrated [21]. The as-formed composite monolayer films with gold nanoparticles generated by chemical or photochemical reduction in film were transferred, as shown in **Figure 6**. When UV illumination was used, gold nanoparticles can be produced. Upon subsequent reduction process in solution, the preformed nanoparticles could grow into different nanostructures as seeds and further self-assemble into larger structures in solution. This research work provided a simple way to control the aggregates and optical properties of gold nanostructures by adjusting the initial irradiation time.

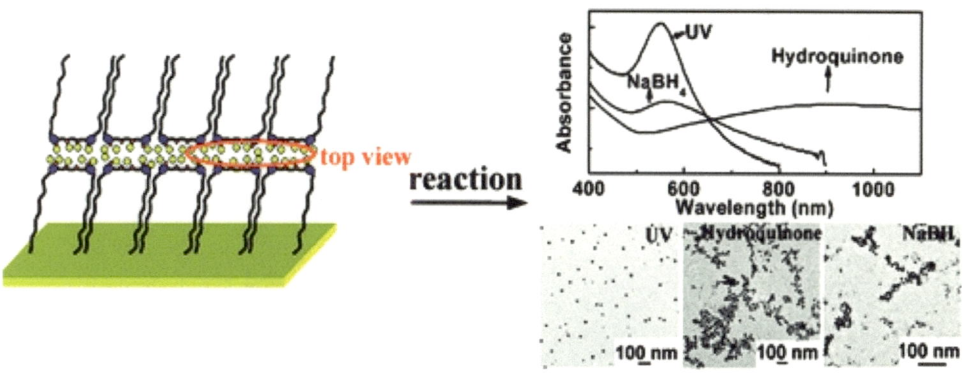

Figure 6. Schematic illustration on the generation of gold nanoparticles by photochemical and chemical reduction in GN2/$AuCl_4^-$ complex film.

In another research work, two luminol derivatives with different substituted chains (abbreviated as LC11 and TF46) were mixed with glycolipid compound GC11 to form monolayers at the air-water interface [22]. The pure and mixed interfacial Langmuir films were studied by measuring the surface pressure-molecular area isotherms, and their morphologies were characterized by Brewster angle microscopy (BAM), as shown in **Figure 7**. Dot-like domains were observed by BAM for TF46 Langmuir films, contrariwise to some strip-like aggregated domains in the case of LC11. These different morphologies may be attributed to distinct aggregation modes induced by differences in molecular structure.

In addition, in order to insert the noninhibitory anti-choline oxidase immunoglobulin (anti-ChOD IgG), mixed IgG-TF46 vesicles have been prepared and spread on a phosphate buffered saline (PBS) subphase [23]. The formation of the interfacial film, after disruption of the IgG-TF46 vesicle membranes at the air-liquid interface, has been evidenced by liposome fusion process and LB method. When acting as an electrochemiluminescent (ECL) sensor, the designed self-assembly films demonstrated capability of biomimetic sensor, as shown in **Figure 8**. The relative research work showed the exploration of application of luminol derivatives to form functional films for ECL detection.

In addition, some Schiff base compounds with trigonal molecular skeletons were prepared and the interfacial self-assembly process were characterized [24]. The Cu(II)-coordinated films could be transferred and characterized by various spectral and morphological methods. As shown in **Figure 9**, depending on different substituted headgroups and subphase solutions, various morphologies were obtained. Moreover, on different pH subphases, the interfacial assembly and nanostructures varied obviously. And reasonable explanation about the self-assembly behaviors have been proposed.

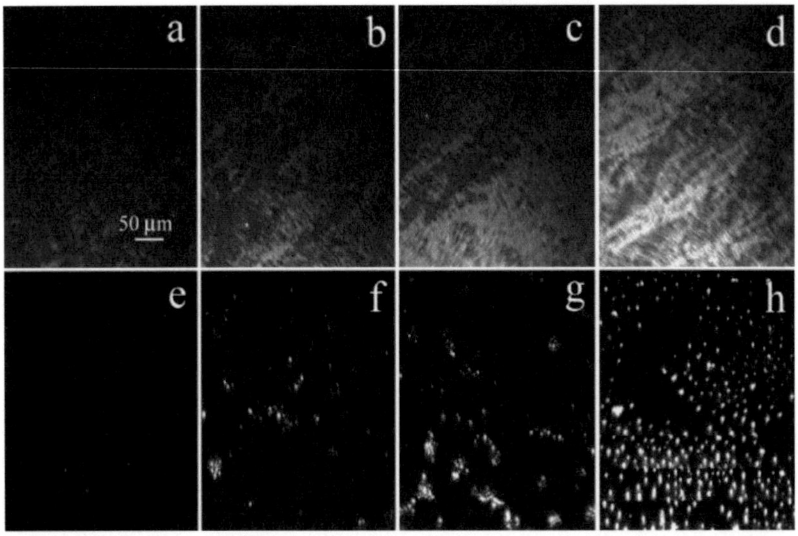

Figure 7. BAM images of LC11 and mixed LC11/GC11 monolayers at different surface pressures. For LC11 monolayer: (a) 5 mN/m; (b) 18 mN/m; (c) 24 mN/m; (d) 36 mN/m. For mixed LC11/GC11 monolayer: (e) 23 mN/m; (f) 28 mN/m; (g) 29 mN/m; (h) 31 mN/m.

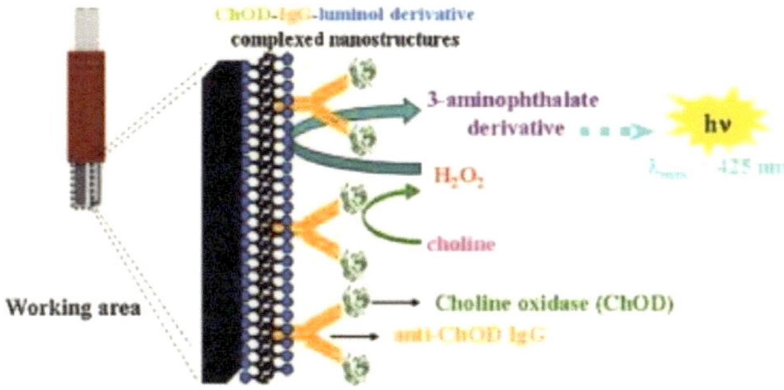

Figure 8. Reagentless detection principle of choline oxidase activity by ECL reaction triggered directly in the sensing layer.

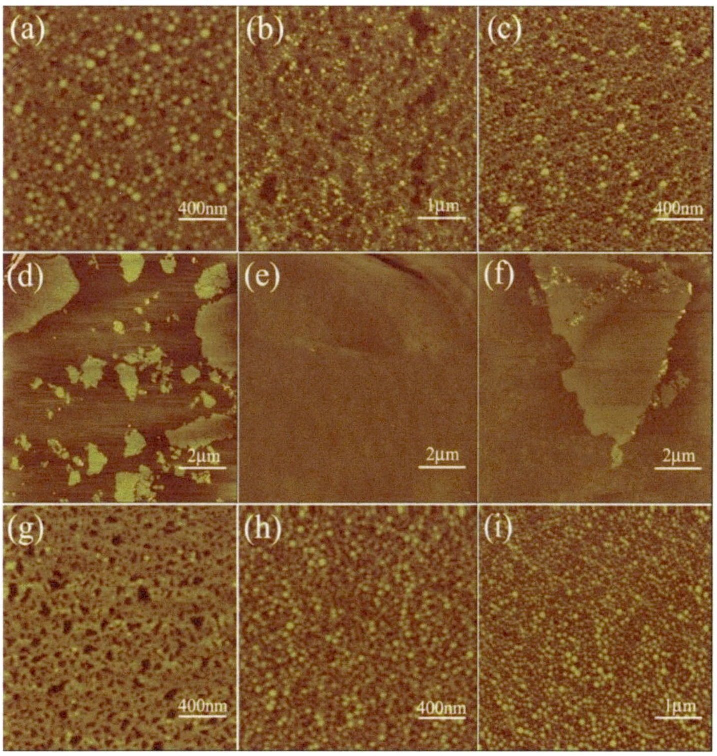

Figure 9. AFM images of one layer TSB-Sal, TSB-Np and TSB-C16 films deposited from pure water surface (a, d, g), aqueous 1.0 mM $Cu(Ac)_2$ subphase (b, e, h), and aqueous 1.0 mM $Zn(Ac)_2$ subphase (c, f, i) at 15 mN/m, respectively.

In addition, we have reported the design and preparation of LB films of an aromatic Schiff base compound and relative Cu(II)-complex [25]. We found that the synthesized compound abbreviated as m-NpSB could form stable monolayer on pure water surface with novel phase transition process. The detailed changes of nanostructures were demonstrated around the phase transition region. Many methods have been used to characterize the ligand and the corresponding complex films. A reasonable model was proposed to explain the novel phase transition process, as shown in **Figure 10**. When the ligand m-NpSB was spread on water surface, a smooth plain film was formed, which had also been verified by the AFM and XRD data. With the increment of surface pressure near the transition point, due to the spatial hindrance and lower energy level induced by the molecular rigidity and size, some molecules changed their conformational alignment and became more declined to surface, which could be well monitored by the AFM measurement at different surface pressures. After the point, the molecules completely aggregated to form fiber-like superstructure due to the high surface pressure and strong π–π stacking of aromatic moieties. Then, a change from two-dimensional (2D) to three-dimensional (3D) morphology was accomplished. The research work results provide distinct clues for the design and the fabrication of the aligned film structures at the air/water interface.

Figure 10. Schematic illustration of the phase behaviors of m-NpSB at air/water interface.

In addition, some achiral Cu(II)-coordinated Schiff base complexes containing aromatic structures were synthesized and their interfacial self-assemblies at the air/water interface were investigated [26]. The Schiff base complex molecules with naphthyl groups tended to form J-aggregate in the Langmuir-Blodgett (LB) films transferred from water surface. By investigation of atomic force microscopy shown in **Figure 11**, a multilayer film or three-dimensional structures were observed. It was interesting to note that the LB films of achiral compound Cu-NA with naphthyl segment and without methyl groups transferred from water surface showed chirality, which could be due to a cooperative stereoregular π–π stacking of the functional groups in a helical sense.

In another research work, a naphthyl-containing Schiff base derivative was synthesized, and its coordination with various metal ions at the air/water interface has been investigated [27]. Different nanostructures and interfacial coordination were obtained in the monolayers with metal ion subphase. And only the Cu(II)-coordinated composite film demonstrated supramolecular chirality. A possible organization mechanism at the air/water interface was suggested, as shown in **Figure 12**. When spreading on the subphase containing Cu(II) ion, an in situ coordination occurred between Cu(II) ion and the two hydrophilic groups. Due to this coordination and spatial hindrance, the long side of the triangle molecule would contact with the water surface, and the adjacent molecules are suggested to align cooperatively in a helical sense. Although both the left- and right-handed helical sense will occur due to the confinement of a two-dimensional platform, one kind of the helical sense could possibly be predominant, and thus we got a macroscopic chirality of the complex films.

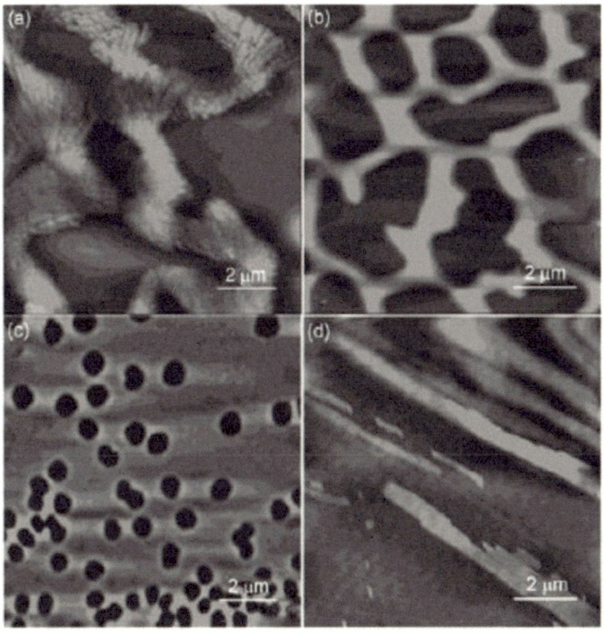

Figure 11. AFM images of one-layer LB films on pure water surface at surface pressure of 15 mN m^{-1}. (a) Cu-SA; (b) Cu-SAM; (c) Cu-NA; and (d) Cu-NAM.

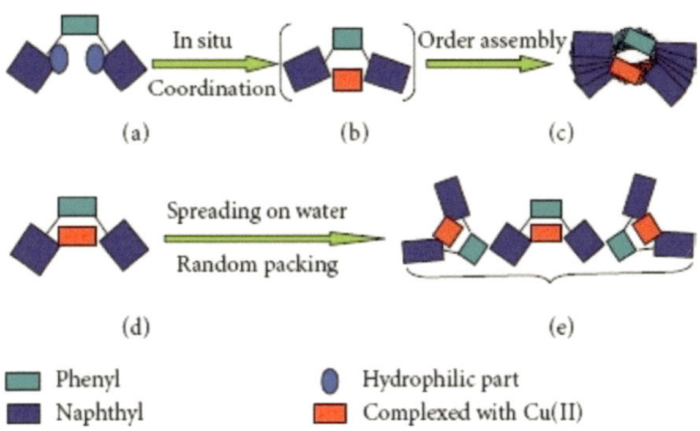

Figure 12. A possible schematic illustration on the formation of the chiral assemblies: (a) ligand molecule; (b) in situ Cu(II)-coordinated complex; (c) stacked in a helical sense to form chiral assembly; (d) preformed Cu(o-NpSB); (e) stacked in a random way.

3. Preparation and self-assembly of some functionalized supramolecular gels

In recent years, supramolecular gels have demonstrated more attentions due to the organized three-dimensional nanostructures and potential wide applications on functionalized drug delivery substrates and wastewater treatment [28–32]. The main driving forces in gel formations are cooperative noncovalent interactions, such as hydrogen bonding, π–π stacking, host-guest interaction, and so on [33–37]. The designed properties need the preparation of various functional gels with organized and controllable nanostructures [38–41]. In this section, we have showed some typical gel systems, including amide amphiphiles, binary gelators, graphene oxide composites, and AuNPs composites.

Firstly, two cholesterol amide derivatives with azobenzene substituent groups have been synthesized, and their gelation behaviors have also been investigated [42]. The experimental data revealed that the compound with headgroups of hydrogen units could only form gel in DMF, whereas the other compound with headgroups of methyl units cannot gelate any used solvent. Under UV light irradiation, the change of trans-cis isomerization in azobenzene segment appeared, which resulted in the gel-sol transition, as shown in **Figure 13**. In addition, the gel could be recovered by the reverse cis-trans isomerization process after visible light exposure. At the same time, the bulk solution with the macroscopic gel to sol transition became transparent.

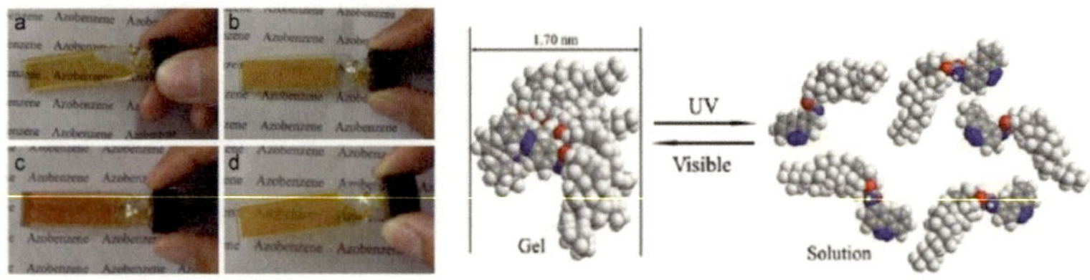

Figure 13. Left, Photographs of Ch-azo in hot DMF solution (a), formed gel in room temperature (b), after UV irradiation for 170 min (c), and subsequent visible irradiation for 40min (d). Right, Schematic representation of photo-induced gel-sol transition of DMF gel of Ch-azo.

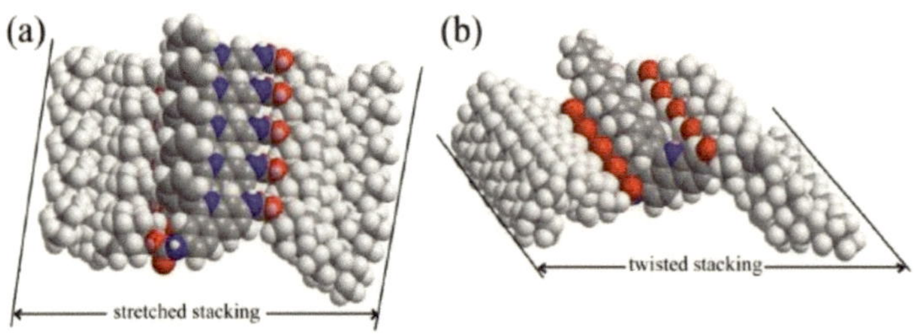

Figure 14. Rational assembly modes of CH-PY organogels in stretched stacking (a) and twisted stacking (b), respectively.

In addition, we have characterized the gelation behaviors of some bolaform cholesteryl amide compounds with large conjugated spacer [43, 44]. We found that the formed nanostructures and self-assembly process in organogels could be regulated by solvent change. Morphological characterization showed different aggregates, including wrinkle, belt, and fiber. We proposed possible self-assembly modes in gels, as shown in **Figure 14**. Different solvents could adjust molecular conformation to self-assemble and form twisted stretched stacking nanostructures.

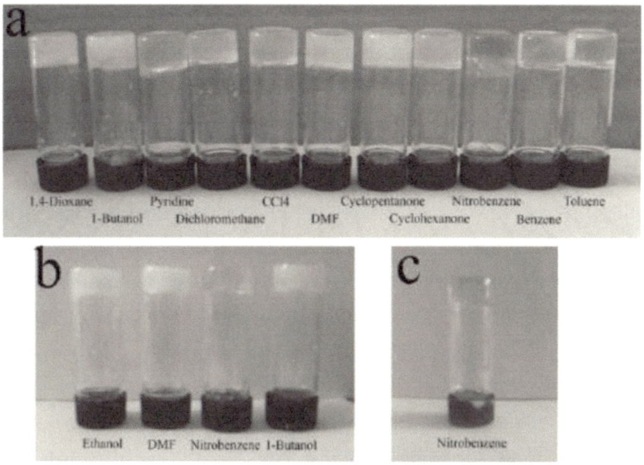

Figure 15. Photographs of organogels of GC16 (a), GC14 (b), and GC12 (c) in different solvents.

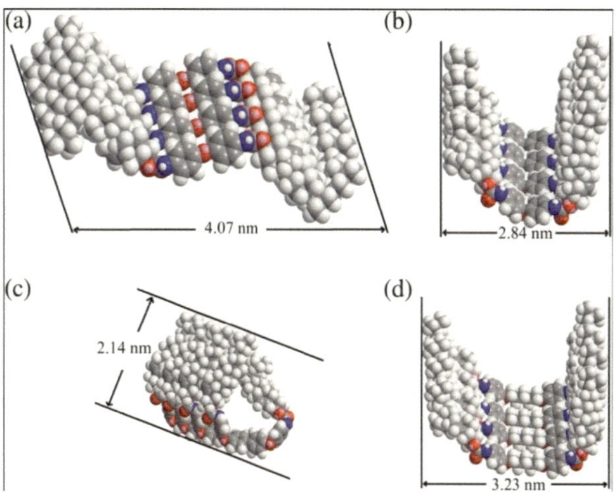

Figure 16. Rational assembly modes of CH-C1, CH-C3, and CH-C4 in gels. Experimental values of CH-C1 in 1,4-dioxane and nitrobenzene (a, b), CH-C3 in nitrobenzene (c), and CH-C4 in nitrobenzene (d).

In another work, we have demonstrated gelation behaviors of some glutamic acid diethyl ester amide compounds [45]. The obtained data indicated that the length of substituted alkyl chains in gelator skeletons showed obvious regulation in gel formation. The photographs of all organogels in different solvents are shown in **Figure 15**. The reasonable explanation for the strong gelation behaviors for GC16 can be due to the enhanced hydrophobic force and organized spatial conformation.

Next, the gelation behaviors of some bolaform cholesteryl amide compounds with different kinds of spacers were demonstrated [46–50]. The obtained experimental results showed that these suitable flexible/rigid segments in spacers were helpful to form gels. Considering the all obtained data, some possible packing modes of these gelators were proposed and schematically shown in **Figure 16**. Due to the flexibility of spacers in the molecular skeleton and different intermolecular forces with solvents, after the intermolecular hydrogen bonding and orderly stacking in different solvents, various repeating units with different lengths were obtained.

In another research work, some binary organogels based on glutamic acid derivatives and acids with different molecular skeletons were designed and prepared [51]. The obtained results demonstrated that the suitable solvents or volume ratios in binary solvents were favorable for gel formation due to cooperative intermolecular interactions. In addition, the gelation behaviors in ethanol/water mixed solvents were also investigated, as shown in **Figure 17**. Interestingly, the results indicated the present mixtures can also form opaque organogels in mixed solvents with different volume ratios of mixed solvents. For example, GC2 can form gels in ethanol/water mixed solvents from the range of 2:1–1:5. The present experimental results suggested that some factors, such as solvents as well as the volume ratios of mixed solvents, aromatic/alkyl cores, and molecular skeletons in acid derivatives, could efficiently change and regulate the gelling abilities of these binary mixtures. Rational assembly modes in organogels were proposed and discussed.

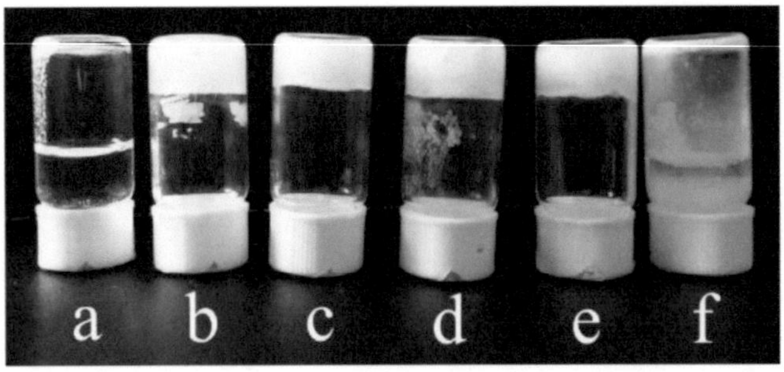

Figure 17. Photographs of GC2 organogels from ethanol/water mixed solvent with the volume ratios of 5:1, 2:1, 1:1, 1:2, 1:5, and 1:10 (a, b, c, d, e, and f, respectively).

In another continuous work, the gelation behaviors of binary organogels composed of azobenzene amino derivatives and fatty acids with different alkyl chains in various organic solvents were designed and investigated [52]. Their gelation behaviors in 20 solvents were tested as new binary organic gelators. Longer alkyl chains in molecular skeletons in present gelators are favorable for the gelation of organic solvents. Morphological studies revealed that the gelator molecules self-assemble into different aggregates from lamella, wrinkle, to belt with change of solvents, as shown in **Figure 18**. In addition, it is interesting to note that these belt aggregates showed a tendency to aggregate together due to highly directional intermolecular interactions and/or solvent evaporation. The difference of morphologies can be mainly due to the different strengths of the intermolecular hydrophobic force between alkyl chains of fatty acids, which have played an important role in regulating the intermolecular orderly stacking and formation of special aggregates.

In another system, we have investigated the gelation behaviors of binary gelators, including glutamic acid amino derivative/azobenzene amino compounds and fatty acids [53, 54]. Considering the obtained XRD data, two possible assembly modes of Glu-C18 were proposed and schematically shown in **Figure 19**. For the Glu-C18 gels in some solvents, such as toluene, the layer distance in gel structures appeared at 3.2 nm due to the penetration of alkyl chains of the neighboring gel molecules. In another kind of gel, repeating unit with 2 nm length was obtained because of parallel chains.

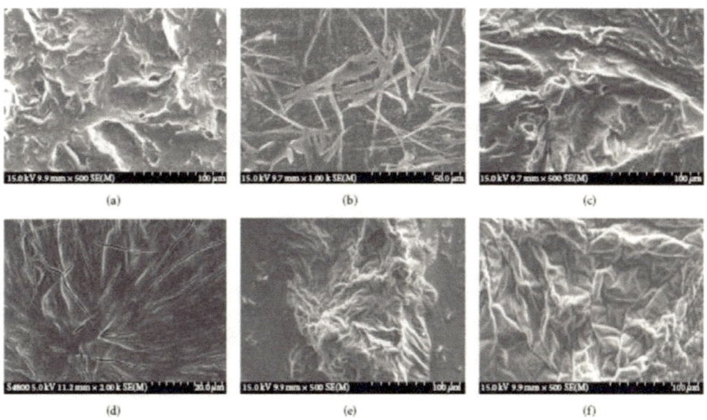

Figure 18. SEM images of xerogels: (a and c–f): C18-Azo, C16-Azo, C14-Azo, C18-Azo-Me, and C16-Azo-Me in ethanolamine, respectively; (b) C16-Azo in nitrobenzene.

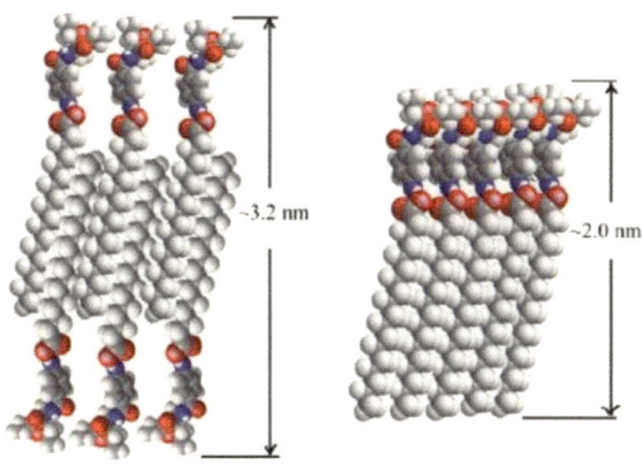

Figure 19. Two possible assembly modes for Glu-C18 organogels in different solvents.

In recent years, graphene oxide (GO)-based nanomaterials have drawn abundant attentions. In recent report, we have prepared organogels through cationic gemini compounds-GO composites [55]. The obtained data demonstrated that the gelation behaviors could be regulated by change of substituted headgroups in compounds. In addition, different self-assembly nanostructures were obtained, as shown in **Figure 20**, indicating special assembly modes in gels. Moreover, it is well known that the thickness of GO sheet (about 0.5–1.0 nm) is larger than the theoretical value of graphene layer (0.34 nm). This is mainly due to the abundant oxygen-containing groups (hydroxyl and epoxy groups) remaining on the surface of the GO sheets. A possible mechanism for headgroup effects on self-assembly and as-prepared nanostructures is proposed.

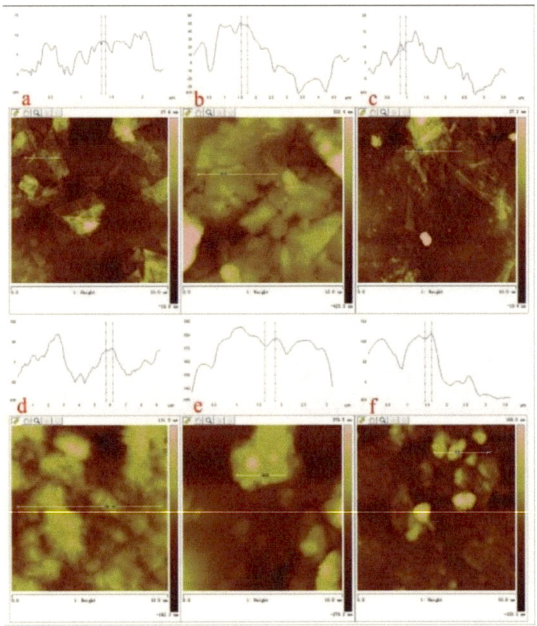

Figure 20. AFM images with section analysis of xerogels. C16Py-GO gels, BPy-GO gels, and CTAB-GO gels in DMF (a, b, and c) and in THF (d, e, and f).

In another continuous work, we have reported the preparation of composite supramolecular organogels by self-assembling cationic functional gemini compound-GO nanocomposites [56]. The gel formation properties of as-prepared nanocomposites present in different organic solvents could be controlled by regulating symmetry in skeletons of amphiphilic compounds. In combination with the spectral and morphological data of as-prepared organogels, some possible and reasonable self-assembly modes for present gemini compound-GO nanocomposite gels are given and demonstrated in **Figure 21**. As for C18-6-0/GO composite and C18-6-6/GO gel with shorter substituent alkyl chain in gemini molecules, due to the intermolecular weak van der Waals interaction of substituent alkyl chains caused by asymmetric molecular skeletons, disorderly stacking appeared in the interlayer of self-assembled nanocomposite

http://dx.doi.org/10.5772/65122

units, which seemed to be difficult to connect each other as fundamental building blocks to fabricate three-dimensional organized net to form gel state. So C18-6-6/GO composite can only fabricate one kind of gel in present 20 organic solvents. With the increment of alkyl chain to carbon 12 and 18, for the case of C18-6-12/GO and C18-6-18/GO composites, longer alkyl chains in molecular skeletons helped to increase hydrophobic force and flexibility in self-assembly process. After integration with GO, organized stacking units appeared in various solvents with the strong van der Waals force of gemini compounds with functional oxygen-containing chemical groups on surface or at edge.

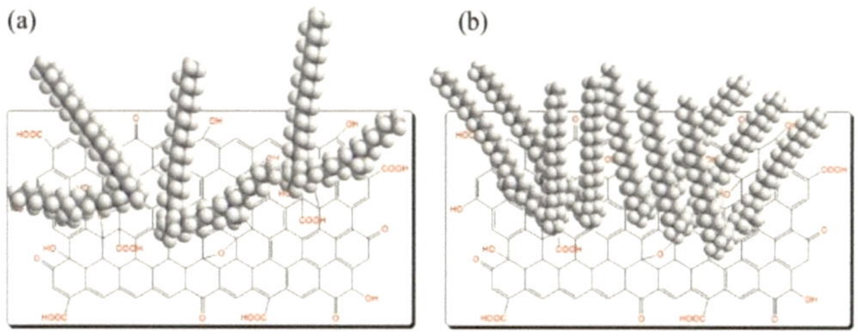

Figure 21. Schematic pictures of different assembly modes in C18-6-6/GO gels with asymmetric skeleton (a) and C18-6-18/GO gels with symmetric skeleton (b).

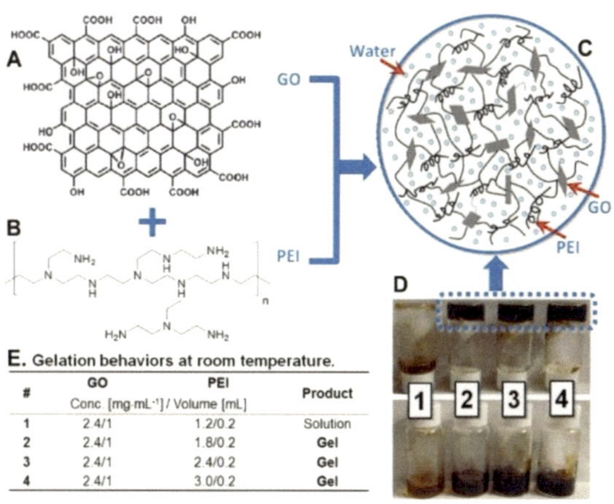

#	GO	PEI	Product
	Conc. [mg·mL⁻¹] / Volume [mL]		
1	2.4/1	1.2/0.2	Solution
2	2.4/1	1.8/0.2	**Gel**
3	2.4/1	2.4/0.2	**Gel**
4	2.4/1	3.0/0.2	**Gel**

Figure 22. Schematic depiction of the formation of GO/PEI gels. (A) GO and (B) amine-rich PEI was combined to give (C) GO/PEI hydrogels. (D and E) Gelation pictures.

In another work, the preparation of GO/polyethylenimine (PEI) and GO/chitosan (CS) hydrogels as dye adsorbents for wastewater treatment was reported [57, 58]. The GO/PEI hydrogels were obtained through both hydrogen bonding and electrostatic interactions between amine-rich PEI and GO sheets. **Figure 22** depicts the complete preparation process of GO/PEI hydrogels by combining the GO suspension and the PEI aqueous solution. The as-prepared GO/PEI hydrogels exhibited good removal rates for both MB and RhB in accordance with the pseudo-second-order model. More importantly, the dye-adsorbed hydrogels can be conveniently separated from an aqueous environment, suggesting potential large-scale applications of the GO-based hydrogels for organic dye removal and wastewater treatment.

In a recent report, some composite hydrogels through GO and multiamine molecules have been designed and prepared from the self-assembly of GO in the presence of multiamine molecules, including diethylenetriamine (DETA) and triethylenetetramine (TETA) [59]. The micro/nanostructures in nanocomposite hydrogel were characterized by morphological investigation. The characteristic bands of graphene samples in Raman spectra appeared, as shown in **Figure 23**. This change can be mainly attributed to the self-assembly of GO in the net-like composite nanostructures. The next adsorption properties demonstrate that these designed and synthesized GO-based composite hydrogels can act as efficient absorbents for dye removal from wastewater in well accordance with the pseudo-second order model.

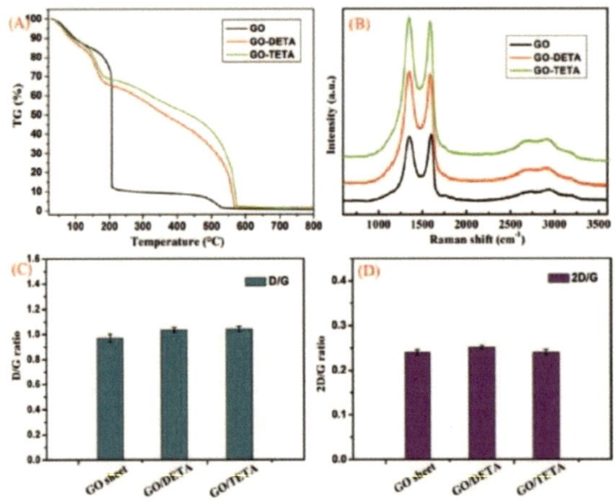

Figure 23. TG curves (A) and Raman spectra (B) of lyophilized GO sheet, GO-DETA hydrogel, and GO-TETA hydrogel at concentration of 0.25 wt%, respectively. (C) and (D) are D/G and 2D/G ratios of the Raman spectra shown in B, respectively.

In addition, we have reported the synthesis of RGO/PEI/Ag and RGO/CS/Ag composite gel materials and evaluated its dye degradation capacity [60, 61]. The CS molecule was chosen for its functional amine segments in the molecular skeleton that can form porous gel nanostructures through interactions such as hydrogen bonding. The photographs of the GO aqueous solution, GO/CS gel, RGO/CS gel, and RGO/CS/Ag composite gels are demonstrated in **Figure 24**. The formed composite gels can provide enough space among its 3D nanostructure for the adsorption and degradation of organic dyes. In addition, the in situ formed silver nanoparticles were homogeneously anchored on RGO surface to form a ternary nanocomposite material. The data of photocatalytic capacity experiments suggest that the prepared 3D GO-based hydrogels can efficiently remove dyes and exhibit good photocatalytic performance for presently used RhB and MB single or mixed solutions in accordance with the pseudo-second order model.

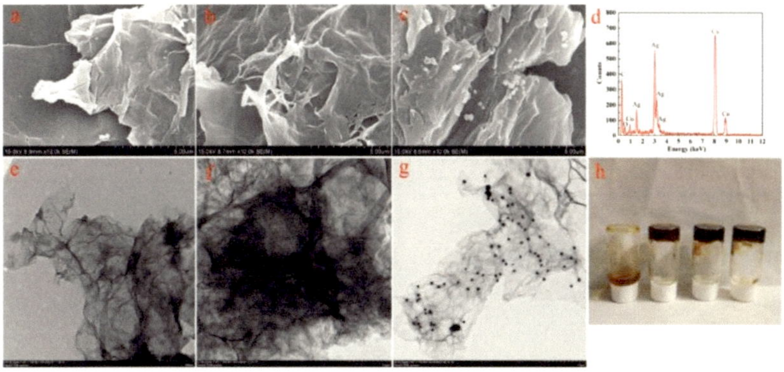

Figure 24. SEM and TEM images for the lyophilized GO/CS hydrogel (a, e), RGO/CS hydrogel (b, f), and RGO/CS/Ag hydrogel (c, g). (d) EDXS taken on the RGO/CS/Ag hydrogel shown in part (g). (h) Photographs of the following: GO aqueous solution, GO/CS, RGO/CS, and RGO/CS/Ag composite hydrogels (from left to right).

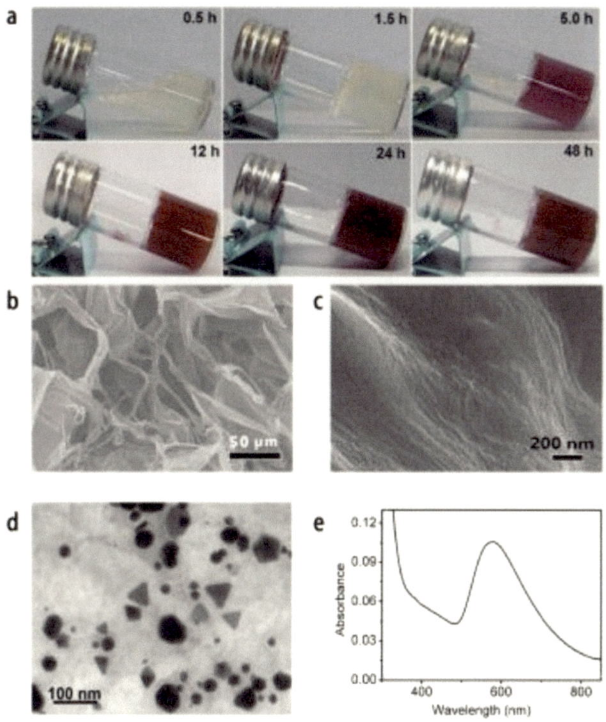

Figure 25. Photographs and morphological characterization of the obtained collagen-AuNPs composite hydrogel: (a) photographs of the formation and color change of the collagen hydrogel versus time when a collagen aqueous acidic solution was mixed with an aliquot of HAuCl4 solution under ambient conditions; (b) and (c) SEM images at low and high magnification of a typical collagen-based hydrogel; (d) TEM image of the AuNP spreading throughout the hydrogel network; (e) UV–vis absorption spectrum of the collagen-based hydrogel containing AuNPs, showing the characteristic SPR absorbance.

In a recent work, in order to enhance the mechanical behaviors of hydrogels, we have designed and prepared an injectable and self-healing collagen-protein-based hydrogel by a gold-biomineralization-triggered self-assembly [62]. The locally synthesized gold nanoparticles are demonstrated to tune the mechanical properties of the collagen-based hydrogels, in which reversible weak interactions between collagen chains and gold nanoparticles endow the hydrogels with shear-thinning and self-healing functions. The photographs and morphological characterization were shown in **Figure 25**. Such biocompatible collagen-based hydrogels have been developed as a novel tool for localized delivery and sustained release of therapeutic drugs, with the advantages to reduce the drug dosage, to lower the toxicity, and to improve the bioavailability.

4. Conclusion and perspective

We are working on the design, preparation, and self-assembly of functionalized LB films and supramolecular gels. In this chapter, various kinds of LB films including bolaform amphiphiles, gemini-type compounds, inorganic-organic hybrid composites, and supramolecular gels through functionalized amide amphiphiles/binary gelators/graphene oxide nanocomposites have all been demonstrated and investigated. The above research work may give the potential perspective for the design and preparation of new self-assembly systems and nanomaterials. In closing, LB films and supramolecular gels can be regarded as good research platforms for various assembly systems. In addition, supramolecular self-assembly demonstrate charming applications, such as physics, biosensors, catalysis, nanomaterials, environmental treatment, and so on. It can be expected that future research systems of LB films and supramolecular gels will be relative to the novel applications and special nanostructures to obtain new functional nanocomposites and nanostructures. The development of preparation of functionalized nanomaterials with self-assembled nanostructures has been fascinating in future.

Acknowledgements

The authors like to extend their thanks to the National Natural Science Foundation of China (Project No. 21473153), the Support Program for the Top Young Talents of Hebei Province, the China Postdoctoral Science Foundation (No. 2015M580214), the Science Foundation for the Excellent Youth Scholars from Universities and Colleges of Hebei Province (No. YQ2013026), the Post-graduate's Innovation Fund Project of Hebei Province (No. 2016SJBS009), the Scientific and Technological Research and Development Program of Qinhuangdao City (No. 201502A006), and the Open Funding Project of the National Key Laboratory of Biochemical Engineering (No. 2013KF-02).

http://dx.doi.org/10.5772/65122

Author details

Tifeng Jiao[1,2*], Ruirui Xing[1,2], Kai Ma[1] and Lexin Zhang[1]

*Address all correspondence to: tfjiao@ysu.edu.cn

1 Hebei Key Laboratory of Applied Chemistry, School of Environmental and Chemical Engineering, Yanshan University, Qinhuangdao, P. R. China

2 National Key Laboratory of Biochemical Engineering, Institute of Process Engineering, Chinese Academy of Sciences, Beijing, P. R. China

References

[1] De Barros A, Constantino CJL, Bortoleto JRR, Da Cruz NC, Ferreira M. Incorporation of gold nanoparticles into Langmuir-Blodgett films of polyaniline and montmorillonite for enhanced detection of metallic ions. Sensor Actuat B-Chem 2016;236:408–417.

[2] Debnath P, Chakraborty S, Deb S, Nath J, Dey B, Bhattacharjee D, Hussain SA. Stability of J-aggregated species in an indocarbocyanine dye in Langmuir–Blodgett films. J Lumin 2016;179:287–296.

[3] Grauby-Heywang C, Moroté F, Mathelié-Guinlet M, Gammoudi I, Faye NR, Cohen-Bouhacina T. Influence of oxidized lipids on palmitoyl-oleoyl-phosphatidylcholine organization, contribution of Langmuir monolayers and Langmuir–Blodgett films. Chem Phys Lipids 2016;200:74–82.

[4] Wang L, Li Y, Wang Q, Zou L, Ye B. The construction of well-aligned MWCNTs-PANI Langmuir–Blodgett film modified glassy carbon electrode and its analytical application. Sensor Actuat B-Chem 2016;228:214–220.

[5] Wang KH, Wu JY, Chen LH, Lee YL. Architecture effects of glucose oxidase/Au nanoparticle composite Langmuir-Blodgett films on glucose sensing performance. Appl Surf Sci 2016;366:202–209.

[6] Saha S, Ghosh M, Dutta B, Chowdhury J. Silver coated gold nanocolloids entrapped in organized Langmuir–Blodgett film of stearic acid: Potential evidence of a new SERS active substrate. Appl Surf Sci 2016;362:364–373.

[7] Harish TS, Viswanath P. Annealing assisted structural and surface morphological changes in Langmuir–Blodgett films of nickel octabutoxy phthalocyanine. Thin Solid Films 2016;598:170–176.

[8] Wang F, Chi C, Yu B, Ye B. Simultaneous voltammetric determination of dopamine and uric acid based on Langmuir–Blodgett film of calixarene modified glassy carbon electrode. Sensor Actuat B-Chem 2015;221:1586–1593.

[9] de Araújo FT, Caseli L. Rhodanese incorporated in Langmuir and Langmuir–Blodgett films of dimyristoylphosphatidic acid: Physical chemical properties and improvement of the enzyme activity. Colloids Surf B Biointerfaces 2016;141:59–64.

[10] Miura YF, Matsui H, Inoue K, Hoshino J, Ikegami K. Structure and properties of the highly conductive Langmuir–Blodgett films based on ditetradecyldimethylammonium-Au(dmit)$_2$ salt. Synthetic Met 2015;207:54–64.

[11] Pandit P, Banerjee M, Pandey KK, Sharma SM, Gupta A. Role of substrate in melting behavior of Langmuir–Blodgett films. Colloids Surface A 2015;471:159–163.

[12] Pazinato J, Hoffmeister DM, Naidek KP, Westphal E, Gallardo H, Winnischofer H. Amphiphilic ruthenium bipyridine complex containing long-chain azopyridine group and the mechanism of electron transfer in Langmuir-Blodgett films. Electrochim Acta 2015;153:574–582.

[13] Zou L, Li Y, Cao S, Ye B. A new voltammetric sensor for sensitive and selective determination of xanthine based on DNA and polyaniline composite Langmuir–Blodgett film. Talanta 2014;129:346–351.

[14] Ferreira GC, Caseli L, Péres LO. Block copolymers of o-PPV organized at the molecular scale as Langmuir and Langmuir–Blodgett films. Synthetic Met 2014;194:65–70.

[15] Karthik C, Manjuladevi V, Gupta RK, Kumar S. Pattern formation in Langmuir–Blodgett films of tricycloquinazoline based discotic liquid crystal molecules. J Mol Struct 2014;1070:52–57.

[16] Jiao T, Liu M. Supramolecular assemblies and molecular recognition of amphiphilic Schiff bases with barbituric acid in organized molecular films. J Phys Chem B 2005;109(7):2532–2539.

[17] Jiao T, Liu M. Supramolecular assemblies of a new series of gemini-type Schiff base amphiphiles at the air/water interface: interfacial nanoarchitectures, in situ coordination and spacer effect. Langmuir 2006;22(11):5005–5012.

[18] Jiao T, Cheng C, Xi F, Liu M. Metal ion modulated organization and nanostructure construction of amphiphilic tyrosine-based bolaamphiphile at the air/water interface. Thin Solid Films 2006;503(1-2):230–235.

[19] Cheng C, Jiao T, Tang R, Chen E, Liu M, Xi F. Compression-induced hierarchical nanostructures of a poly(ethylene oxide)-block-dendronized polymethacrylate copolymer at the air/water interface. Macromolecules 2006;39(19):6327–6330.

[20] Guo Z, Jiao T, Liu M. Effect of substituent position in coumarin derivatives on the interfacial assembly: reversible photodimerization and supramolecular chirality. Langmuir 2007;23(4):1824–1829.

[21] Zhong L, Jiao T, Liu M. Synthesis and assembly of gold nanoparticles in organized molecular films of gemini amphiphiles. Langmuir 2008;24(20):11677–11683.

[22] Jiao T, Leca-Bouvier BD, Boullanger P, Blum LJ, Girard-Egrot AP. Phase behavior and optical investigation of two synthetic luminol derivatives and glycolipid mixed monolayers at the air-water interface. Colloid Surf A 2008;321(1-3):137–142.

[23] Jiao T, Leca-Bouvier BD, Boullanger P, Blum LJ, Girard-Egrot AP. A chemiluminescent Langmuir–Blodgett membrane as the sensing layer for the reagentless monitoring of an immobilized enzyme activity. Colloids Surf A 2010;354(1-3):284–290.

[24] Jiao T, Li X, Zhang Q, Duan P, Zhang L, Liu M, Luo X, Li Q, Gao F. Interfacial assembly of a series of trigonal Schiff base amphiphiles in organized molecular films. Colloids Surf A 2012;407(5):108–115.

[25] Jiao T, Liu M. Phase behaviors and 2D-3D morphological transition of aromatic Schiff base derivatives in organized molecular films. Acta Phys-Chim Sin 2012;28(6):1418–1424.

[26] Jiao T, Li X, Zhang Q, Li Q, Zhou J, Gao F. Interfacial assembly of a series of Cu(II)-coordinated Schiff bases complexes: orderly nanostructures and supramolecular chirality. Sci China Technol Sci 2013;56(1):20–24.

[27] Jiao T, Xing R, Zhang Q, Lv Y, Zhou J, Gao F. Self-assembly, interfacial nanostructure, and supramolecular chirality of the Langmuir-Blodgett films of some Schiff base derivatives without alkyl chain. J Nanomater 2013;2013:297564.

[28] Kowalczuk J, Rachocki A, Bielejewski M, Tritt-Goc J. Effect of gel matrix confinement on the solvent dynamics in supramolecular gels. J Colloid Interf Sci 2016;472:60–68.

[29] Marcos X, Pérez-Casas S, Llovo J, Concheiro A, Alvarez-Lorenzo C. Poloxamer-hydroxyethyl cellulose-α-cyclodextrin supramolecular gels for sustained release of griseofulvin. Int J Pharm 2016;500(1-2):11–19.

[30] Zeng X, Sun Z, Wang H, Wang Q, Yang Y. Supramolecular gel composites reinforced by using halloysite nanotubes loading with in-situ formed Fe_3O_4 nanoparticles and used for dye adsorption. Compos Sci Technol 2016;122:149–154.

[31] Tanaka T, Gotanda R, Tsutsui A, Sasayama S, Yamamoto K, Kimura Y, Kadokawa J. Synthesis and gel formation of hyperbranched supramolecular polymer by vine-twining polymerization using branched primer–guest conjugate. Polymer 2015;73:9–16.

[32] Chu X, Xing P, Li S, Ma M, Hao A. Inorganic salt-tuned multiple self-assemblies of supramolecular β-cyclodextrin gel. Colloid Surface A 2014;461:11–17.

[33] Zhu F, Lin XY, Wu ZL, Cheng L, Yin J, Song Y, Qian J, Zheng Q. Processing tough supramolecular hydrogels with tunable strength of polyion complex. Polymer 2016;95:9–17.

[34] Simões SMN, Veiga F, Ribeiro ACF, Figueiras AR, Taboada P, Concheiro A, Alvarez-Lorenzo C. Supramolecular gels of poly-α-cyclodextrin and PEO-based copolymers for controlled drug release. Eur J Pharm Biopharm 2014;87(3):579–588.

[35] Cheng X, Jin Y, Sun T, Qi R, Li H, Fan W. An injectable, dual pH and oxidation-responsive supramolecular hydrogel for controlled dual drug delivery. Colloids Surf B 2016;141:44–52.

[36] Klaewklod A, Tantishaiyakul V, Hirun N, Sangfai T, Li L. Characterization of supramolecular gels based on β-cyclodextrin and polyethyleneglycol and their potential use for topical drug delivery. Mater Sci Eng-C 2015;50:242–250.

[37] Lin Y, Li L, Li G. A new supramolecular gel via host–guest complexation with cucurbit[8]uril and N-(4-diethylaminobenzyl)chitosan. Carbohydr Polym 2013;92(1):429–434.

[38] Yang Y, Chen X, Huang D, Ye W. Supramolecular gel based on the cyclodextrin inclusion assembly of Ag-Fe_3O_4 nanodimers and Pluronic F127. Colloids Surf A 2013;436:467–473.

[39] Delbecq F. Supramolecular gels from lipopeptide gelators: Template improvement and strategies for the in-situ preparation of inorganic nanomaterials and for the dispersion of carbon nanomaterials. Adv Colloid Interface Sci 2014;209:98–108.

[40] Zhang YM, Zhang WQ, Li JQ, Dang JP, Wei TB. Cadmium-induced supramolecular hydrogels: Reversible sol–gel transitions induced by EDTA and acid/base exchange. Mater Lett 2012;82:227–229.

[41] Jin H, Dai XH, Wu C, Pan JM, Wang XH, Yan YS, Liu DM, Sun L. Rational design of shear-thinning supramolecular hydrogels with porphyrin for controlled chemotherapeutics release and photodynamic therapy. Eur Polym J 2015;66:149–159.

[42] Jiao T, Wang Y, Gao FQ, Zhou J, Gao FM. Photoresponsive organogel and organized nanostructures of cholesterol imide derivatives with azobenzene substituent groups. Prog Nat Sci: Mater Int 2012;22(1):64–70.

[43] Jiao T, Gao FQ, Wang Y, Zhou J, Gao FM, Luo X. Supramolecular gel and nanostructures of bolaform and trigonal cholesteryl derivatives with different aromatic spacers. Curr Nanosci 2012;8(1):111–116.

[44] Jiao TF, Gao FQ, Shen XH, Zhang QR, Zhang XF, Zhou JX, Gao FM. Self-assembly and nanostructures in organogels based on a bolaform cholesteryl imide compound with conjugated aromatic spacer. Materials 2013;6(12):5893–5906.

[45] Jiao T, Wang R, Zhang Q, Yan X, Zhou J, Gao F. Nanostructures and substituent alkyl chains effect on assembly of organogels based on some glutamic acid diethyl ester imide derivatives. Curr Nanosci 2013;9(4):536–542.

[46] Jiao T, Gao F, Zhang Q, Zhou J, Gao F. Spacer effect on nanostructures and self-assembly in organogels via some bolaform cholesteryl imide derivatives with different spacers. Nanoscale Res Lett 2013;8:406.

[47] Jiao T, Huang Q, Zhang Q, Xiao D, Zhou J, Gao F. Self-assembly of organogels via new luminol imide derivatives: diverse nanostructures and substituent chain effect. Nanoscale Res Lett 2013;8:278.

[48] Jiao T, Wang Y, Zhang Q, Zhou J, Gao F. Regulation of substituent groups on morphologies and self-assembly of organogels based on some azobenzene imide derivatives. Nanoscale Res Lett 2013;8:160.

[49] Jiao T, Ma K, Shen X, Zhang Q, Li X, Zhou J, Gao F. Self-assembly and soft material preparation of binary organogels via aminobenzimidazole/benzothiazole and acids with different alkyl substituent chains. J Nanomater 2013;2013:762732.

[50] Shen X, Jiao T, Zhang Q, Guo H, Lv Y, Zhou J, Gao F. Nanostructures and self-assembly of organogels via benzimidazole/benzothiazole imide derivatives with different alkyl substituent chains. J Nanomater 2013;2013:409087.

[51] Guo H, Jiao T, Shen X, Zhang Q, Li A, Zhou J, Gao F. Binary organogels based on glutamic acid derivatives and different acids: solvent effect and molecular skeletons on self-assembly and nanostructures. Colloids Surf A 2014;447:88–96.

[52] Guo H, Jiao T, Shen X, Zhang Q, Li A, Gao F. Preparation and characterization of binary organogels via some azobenzene amino derivatives and different fatty acids: self-assembly and nanostructures. J Spectrosc 2014;2014:758765.

[53] Jiao T, Xing R, Shen X, Zhang Q, Zhou J, Gao F. Investigation of orderly nanostructures and assembly modes of binary organogels via glutamic acid amino derivative and different fatty acids. Integr Ferroelect: Int J 2014;151(1):31–41.

[54] Hu Y, Li Q, Hong W, Jiao T, Xing G, Jiang Q. Characterization of binary organogels based on some azobenzene compounds and alkyloxybenzoic acids with different chain lengths. J Spectrosc 2014;2014:970827.

[55] Jiao T, Wang Y, Zhang Q, Yan X, Zhao X, Zhou J, Gao F. Self-assembly and headgroup effect in nanostructured organogels via cationic amphiphile-graphene oxide composites. PLoS One 2014;9(7):e101620.

[56] Jiao T, Wang Y, Zhang Q, Yan X, Zhao X, Huo Q, Zhou J, Gao F. Organogels via gemini amphiphile-graphene oxide composites: self-assembly and symmetry effect. Sci Adv Mater 2015;7(9):1677–1685.

[57] Guo H, Jiao T, Zhang Q, Guo W, Peng Q, Yan X. Preparation of graphene oxide-based hydrogels as efficient dye adsorbents for wastewater treatment. Nanoscale Res Lett 2015;10:272.

[58] Zhao H, Jiao T, Zhang L, Zhou J, Zhang Q, Peng Q, Yan X. Preparation and adsorption capacity evaluation of graphene oxide-chitosan composite hydrogels. Sci China Mater 2015;58(10):811–818.

[59] Zhu K, Jiao T, Zhang L, Xing R, Guo R, Zhou J, Hou C, Zhang Q, Peng Q, Li X. Preparation and absorption capacity evaluation of composite hydrogels via graphene oxide and multi-amine molecules. Sci Adv Mater 2016;8(7):in press.

[60] Jiao T, Guo H, Zhang Q, Peng Q, Tang Y, Yan X, Li B. Reduced graphene oxide-based silver nanoparticle-containing composite hydrogel as highly efficient dye catalysts for wastewater treatment. Sci Rep-UK 2015;5:11873.

[61] Jiao T, Zhao H, Zhou J, Zhang Q, Luo X, Hu J, Peng Q, Yan X. Self-assembly reduced graphene oxide nanosheet hydrogel fabrication by anchorage of chitosan/silver and its potential efficient application toward dyes degradation for wastewater treatments. ACS Sustain Chem Eng 2015;3(12):3130–3139.

[62] Xing R, Liu K, Jiao T, Zhang N, Ma K, Zhang R, Zou Q, Ma G, Yan X. An injectable self-assembling collagen-gold hybrid hydrogel for combinatorial antitumor photothermal/photodynamic therapy. Adv Mater 2016;28(19):3669–3676.

Chapter 3

Mechanism of Aggregation Colloid Centers on Surface Ionic Crystals

Utkirjon Sharopov, Bakhtiyar Atabaev,
Ruzmat Djabbarganov and Muzaffar Qurbanov

Additional information is available at the end of the chapter

http://dx.doi.org/10.5772/65517

Abstract

In this chapter, we analyze the kinetics of changes in the intensities of peaks of these aggregate defects as a function of the substrate temperature, as well as study the degradation of these aggregate F centers in order to understand the mechanisms of their formation and transformation to other types of defects, to colloids. The results obtained using methods of total current (TC) spectroscopy and secondary-ion mass spectroscopy (SIMS) under ion bombardment of LiF crystals are analyzed. The temperature dependence of the generation kinetics of F centers and their aggregates in a LiF/Si(1 1 1) thin-film system after irradiation with low-energy (80 eV) electrons was studied by TC spectroscopy. It was shown that, in all cases, low-temperature annealing results in the degradation of the formed centers followed by their coalescence. By SIMS, it is shown that the majority of the products of crystal sputtering contain point defects. A procedure for determining defects in sputtered clusters of ionic crystals is developed.

Keywords: defects, colloids, coalescence, surface, temperature, ion electron bombardment, annealing, ionic crystals.

1. Introduction

It is known that under ionizing irradiation of ionic crystals the anion vacancies at which electrons are trapped are first formed and then the F centers emerge. Their migration over the lattice results in coalescence of point defects and in the formation of large stable aggregates of defects. The migration of defects depends on crystal temperature [1].

In the 1970s, the studies of damage created under electron and ion irradiation were started with the first accelerators [1–5]. For the ions having energy of 1–10 MeV/u the main interaction with diverse solids is Coulomb one, which leads to an extremely high level of electron excitation and makes the processes similar to those in dielectrics and metals [5]. New radiation phenomena in diverse solids such as local phase transitions, volume change, internal stress, etc. [6–8] are due to the high electronic excitation level.

Such features of damage created under fast heavy ion irradiation were observed in LiF and some other ionic crystals (NaCl, NaF, CaF_2). In these crystals the two different zones of damage are induced: in a cylindrical region of the radius of 1–2 nm the defect aggregates are dominating (aggregate damage zone), while in the more extended lateral track region ($r > 2$ nm, track halo), the point defects prevailed [8–10]. It should be noted that the LiF crystals occupy a unique place among the alkali and other ionic crystals owing to their large band gap ($E_g = 14.6$ eV) and corresponding high energy of lattice. As a result, the main radiation defects (F- and V-centers) in the LiF crystals are stable up to 200°C and defect formation in the LiF crystals can be studied for the different values of temperature and within the wide range of excitation energy density.

In alkali halides, the radiation damage was studied in-depth for several decades [1–4, 12–14], which led to the discovery of an exciton mechanism of defect formation. The first V_K-centers and self-trapped excitons were observed in the LiF crystals [4].

The damage induced under irradiation strongly depends on absorbed dose, dose rate, and irradiation [11] temperature. The effect of high dose radiation on the LiF crystals was studied by thermal neutron irradiation. LiF is very sensitive to thermal neutrons due to the reaction of 6Li (n, α)3 T, where 4.78 MeV energy is released. Under irradiation up to the dose of ~1 MGy the single point defects such as F-, F_n-, V-, and V_n-centers ($n \leq 4$) are dominating. For higher dose and higher irradiation temperature ($T \geq 100$–200°C), the single point defects are aggregated and larger aggregates of F- and V-centers are formed. The Li colloids and fluorine molecular clusters (nF_2) are produced within the dose range of 10–100 MGy. The fluorine molecules F_2 are accumulated in the crystal bulk and only a small part of the fluorine molecules goes out of the crystal. For high irradiation dose, the dislocations are also formed and the mechanical stress and swelling effects were observed [12–14].

LiF thin films containing different optically active centers are used in adjustable lasers [15, 16], fiber optics, and molecular spectroscopy. Investigations of the formation kinetics of defects in LiF crystals were performed in Refs. [17–19].

In our study, the formation and aggregation of various point defects, such as F, F_2, F_3, and X centers and colloid centers, were observed upon thermal annealing. The formation and degradation kinetics of aggregate centers and the time-dependent change in the peak intensities of aggregate centers in the total current spectrum upon heating from room temperature to 150°C were also studied.

Upon the irradiation of halide ionic crystals, halogen vacancies, at which electrons are trapped, emerge first to form F centers. It has been shown that F_2^+, Li_4, F_2, F_3, and X centers are produced upon low-dose electron irradiation of LiF films [29].

In the present chapter, we analyze the kinetics of changes in the intensities of peaks of these aggregate defects as a function of the substrate temperature, as well as study the degradation of these aggregate F centers in order to understand the mechanisms of their formation and transformation to other types of defects, to colloids. And the obtained results using methods of SIMS under ion bombardment of LiF crystals as well as procedure for determining defects in sputtered clusters are analyzed. The problem of this paper lies in determining the composition of secondary clusters sputtered from the surface of LiF.

2. Materials and methods

2.1. Total current spectroscopy

The method of TC spectroscopy is some kind of low-energy threshold secondary-ion spectroscopy. For experimental realization of this method a measuring scheme presented in **Figure 1** is necessary. A monokinetic beam of electrons focused with an electronic optical system is directed to the sample surface. In the space between the focused system and plane sample the electrons move in the homogeneous decelerating field and come to the sample with energy defined by the displacement potential. A primary electron current I_1 is defined by a value of the current formed between the electron gun cathode and the sample under study. When the primary electrons interact with a target, some part of the electrons is reflected and creates a current of the secondary electrons I_2 and another one remains in the target and makes a contribution to the current I running through the sample. The current balance is written as follows

$$I_1 = I + I_2 \quad (1)$$

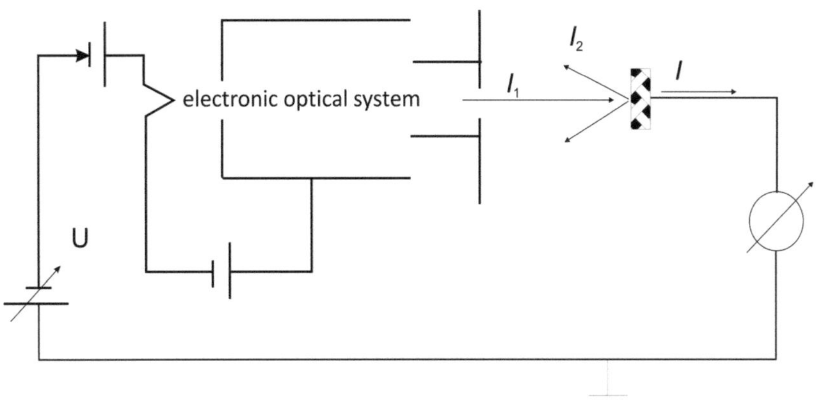

Figure 1. An electric scheme of the total current spectrometer.

Since the relative potentials of the electrostatic system remains the same when the retarding field over the sample changes, the primary electron beam does not lose its focus and the current I_1 remains unchanged in value [20].

Taking that fact into account, it is possible to define the value of the secondary electron current I_2 by measuring the target circuit current I

$$I = I_1 - I_2, \quad I_1 = \text{const} \tag{2}$$

Applying a small (0.1–0.2 V) sinusoidal voltage of frequency ω to the cathode unit, one can modulate a primary beam in energy and select the first derivative of the sample circuit current with a system of synchronous detection. Then the expression for the current balance can be written in the following form

$$\frac{dI}{dE_p} = \frac{d(I_1 - I_2)}{dE_p} = -\frac{dI_2}{dE_p} = S(E_p) \tag{3}$$

Thus, a TC spectrum $S(E_p)$ is a derivative of the target circuit current in energy of incident electrons.

Several works devoted to dielectrics, semiconductors, and metals studied by the TC spectroscopy method showed that the TC spectra have a fine structure that is characteristic of each substance within the low-energy range (0–15 eV) [20, 21]. In Refs. [21, 22], on the basis of the analysis of the energy dependences of elastic and inelastic reflection of electrons from solid surface and taking into account the effect of inelastic interaction on the intensity of elastic reflection, the model describes about TC spectroscopy signal formation, which unambiguously relates the TC spectra structure with the special features of the electronic state density of the valent and free zones of the sample under study.

Colloid formation on the LiF film surface under electron and ion irradiation is described below.

In the high vacuum chamber, an electronic gun is mounted to produce an electron beam with energy $1 < E \leq 80$ eV. Spectrum registration and electron irradiation of film were made with one gun to get information from an irradiated part of the film.

An ion gun was also mounted to irradiate the samples by Cs^+ and Cl^- ions with energy $200 < E \leq 2000$ eV.

An experimental technique for studying the defect and colloid formation on the surface of the growing LiF film according to the following algorithm was developed:

1. Cleaning of the single crystal Si(1 1 1) surface by ion etching and subsequent annealing.
2. Registration of a TC spectrum of the cleaned Si(1 1 1).
3. Deposition of a LiF film on the Si(1 1 1) substrate and further irradiation by electrons with energy $5 < E \leq 80$ eV or Cs^+ ions with energy $200 < E \leq 2000$ eV.

http://dx.doi.org/10.5772/65517

4. TC spectrum registration every 1 min for 30–40 min.

This algorithm was used for different energy irradiation and substrate temperature (25 ≤ *T* ≤ 800°C).

The advantage of our total current spectrometer is that it provides much more information as compared with others. For example, in Ref. [23] the total current from the LiF films was measured lower than the threshold of electron-stimulated desorption (up to 12 eV). **Figure 2** illustrates the total current spectra of the LiF films of 17 monolayer in thickness. The authors associated the minimums at 3 and 7 eV with deceleration voltage on the surface. Deceleration voltage manifests itself on the surface when the thickness is over 2 monolayers. There are special features near zero energy of the spectrum for thickness of ~10 monolayers.

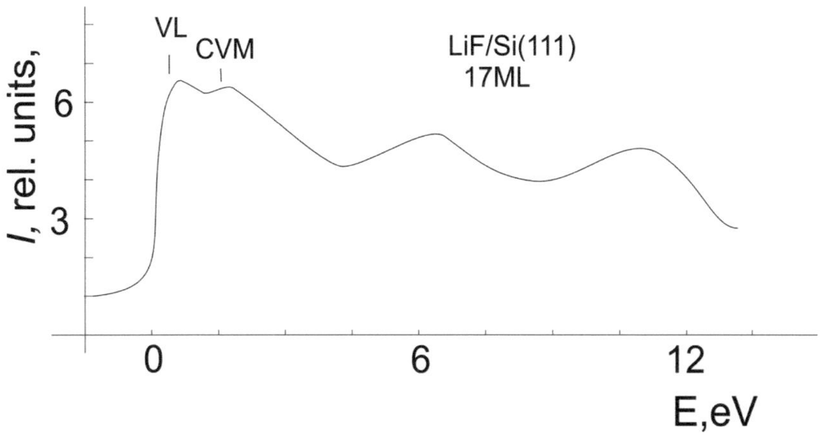

Figure 2. The total current spectrum of the LiF films of 17 monolayers in thickness [23].

From the energy position of the first minimum and decrease in total current, it was established that the electron affinity of the LiF film is 1.2 eV. Also, it was shown that the low-energy part of the spectra demonstrates the basic properties of the density of the electronic states in the conductivity zone. As seen, in this case, the total current has little information, the peaks of the local levels are almost unresolved, and the peak of exciton absorption is not seen since in this technique the current in the sample circuit is not integrated by a registering device. Nevertheless, some of these data can be used for the interpretation of our results.

Also in Refs. [1, 20–23] it is shown that the changes in the intensity of the TC spectra can show defect formation when the whole spectrum of total current is shifted, and also the work function changes, as well as bending of the energy zones takes place, islands are formed, and changes in the surface potential are defined by doubling of the primary peak.

Thus, at present, the TC spectroscopy has become an efficient tool for studying the surface properties of solid. This technique has a whole series of attractive features: it is simple for experimental realization and at the same time can provide much information, it has high surface sensitivity (an order of 10 Åi, and no destruction of the surface takes place under study.

However, for the studies of the surface one should not seek a "better" method but apply, as possible, a combination of the independent methods providing information about the features that manifest themselves more pronouncedly. For example, the characteristics of some atoms of substance are defined by secondary-ion mass-spectrometry, which makes it possible for an elemental analysis of the surface according to the mass spectra. Each technique provides information on some properties of the surface area and only application of several independent techniques allows more exactly and unambiguously solving the study problems in case of correct comparison of the results of independent measurements.

A design of the total current spectrometer allows effective solution of surface cleaning and ordering, and a mass spectrometer can define an elemental composition of a sample under study.

2.2. Vacuum system and experimental setup elements

The measurements were performed with an all-metal chamber having three-stepped oil-free pumping. Preliminary pumping was made by a zeolite pump up to pressure 10^{-3} Torr, then the zeolite pump was blocked with a high-vacuum valve and further pumping was made by an ion pump NORD-160 allowing working pressure of up to 10^{-9} Torr. The experimental setup had additional devices allowing the complex effect on a sample under study with no disturbing of vacuum conditions.

A graduated system of sample heating is necessary for thermal cleaning and measurements within a wide range of temperature. The system consists of a tantalum cylinder with a tungsten spiral of indirect heating inside it. On the end face of the cylinder a sample to be studied is mounted with special holders. A chromel-alumel thermocouple for heating control within the wide range of temperature (20–1500°C) is mounted on the side part of the sample. A special spiral winding was used to prevent the effect of the magnetic field of the heater on the sample under study. To exclude noises of thermoelectron current (when a useful signal was measured in the sample circuit) a positive potential relative to the sample was applied to the heater spiral; that potential did not allow thermoelectrons to go to the sample.

An ion gun operating according to the ionization mechanism was used to clean the sample surface by ion etching. The gun provides the ion current of an order of 1–10 nA and the beam diameter of 0.4 cm. Ion bombardment was performed at an angle of 25° relative to the normal sample surface.

The study objects were thin LiF films (~150 Å50re thin LiF fil 1 1) side of the silicon substrate by thermal evaporation in ultrahigh vacuum. The films of ion crystals were chosen as model objects since they are usually used to develop diverse technologies. LiF is a very simple crystal so the studies with it are very easy. In practice, monocrystals and films of LiF are used to produce high-effective lasers (80% in efficiency).

A LiF evaporator was used to carry on experiments with film systems. The evaporator is a quartz pipe; on its external side a nichrome wire is reeled up. Special glasses located at some distance from the evaporator were used to define the thickness of the deposited layers. After the end of the experiments, these glasses from the chamber were removed, and by measuring the thickness of the deposited layer (for example, with an interference microscope), one can judge the thickness of the deposited layer on the sample by taking into account the distances of the evaporator-sample and evaporator-glass. Preliminarily, a mode for evaporator operation was selected and the total current spectra were not registered in the process of film growth for different rates of deposition. A film with fewer defects was that spectrum which had less structure, and such conditions occurred at the deposition rate of 3–4 Åf deposition rate fects of film growth for different rates of deposition. A filmsample surface.

2.3. Additional methods

For an element composition of the alkali-haloid crystal surface to be defined by SIMS, comparative studies of the mass composition of the LiF sputtering products were conducted.

It is known that dielectric sputtering under ion bombardment can increase if there are a lot of defects on the target surface. To study the role of defects and colloids in the sputtering processes, some experiments were carried on with a setup described below.

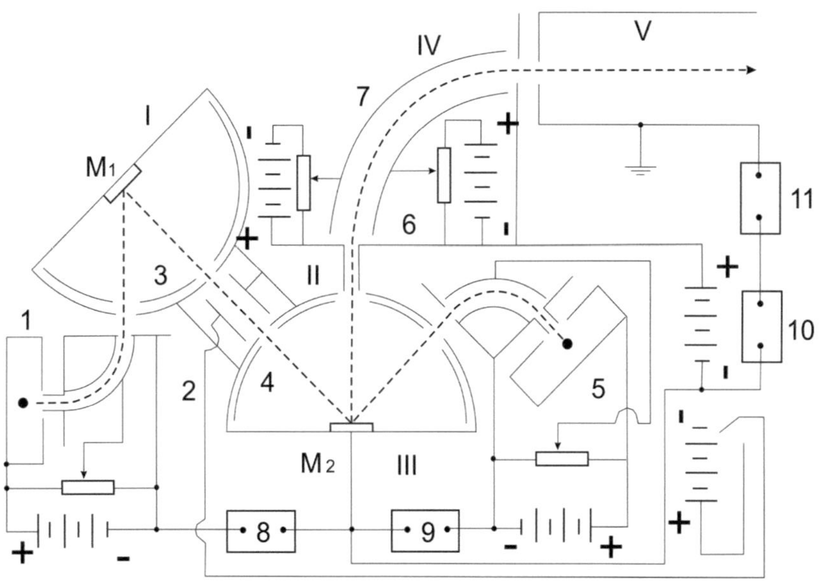

Figure 3. A mass-spectrometer of negative ions.

A setup consisted of the following main parts (**Figure 3**): a sputtering chamber I with an evaporator of primary ions I, a target to be studied M_1, an electrode system to draw out and accelerate the primary ions and to focus them on the target M_1; a cylindrical condenser (in its electrical field a beam of charged particles are separated from neutrals); a chamber to deviate an ion component of sputtering (II); a chamber to ionize the sputtered particles III with a target-ionizer M_2 and a source of primary ions 5; an electrode system to draw out and accelerate the primary ions and to focus them on the target M_2; a cylindrical condenser to separate a beam of charged particles from neutrals; an electrode 6 to draw out secondary negative ions; a 90°

energy analyzer IV and a mass-analyzer V with a mass-spectrometer to analyze the sputtered ions in mass.

The targets to be studied were mounted in the chamber III at the place of the target M_2. In this case, the surface of the studied target M_2 is bombarded by a positive ion beam produced and formed from the ion source 5. Some part of the secondary negative ions sputtered from the target M_2 to the whole front semisphere is taken by a sagged field of the electrode 6 and passing through the cylindrical energy analyzer 7 goes through a gap to the mass-analyzer operating in a mode of energy modulation.

Before entering the mass-analyzer, the negative ions passed through the exit gap of the energy analyzer III and are accelerated by an electric field created by series sources of constant voltage (BC-23) and generator of saw-tooth voltage. The ions that passed through the mass-analyzer enter a receiver of the secondary ions. For rapid involute of the mass spectra, the modulation of voltage accelerating the secondary ions with the help of a generator of saw-tooth pulses was used, now secondary-electron multipliers provide sufficient rate of rapid registration of ion current pulses. Therefore, an electronic multiplier of an open type was chosen as an amplifying device. The output of the secondary-electron multiplier is connected with a vertical input of the electronic oscillograph, and its horizontal involute is synchronized with the modulation of secondary ion energy. For some value of the magnetic field on the oscillograph monitor will be a motionless picture of some area of the mass-spectrogram, its size is defined by the depth of ion energy modulation by the saw-tooth pulse generator. The different regions of the mass spectrum were observed and measured by magnetic field variation.

The mass-spectrometric setup was manufactured from stainless nonmagnetic steel. Vacuum was created by pumps NORD-250 and maintained at a level of ~10^{-8} Torr.

Thus, it is possible to simultaneously make a quantitative analysis of the composition and state of the surface and accurately define how the work function of the sample changes in electron volts.

3. Aggregation colloid centers on surface ionic crystals

Figure 4 shows the kinetics of the TC spectra upon annealing of the LiF film for 120 min at 25°C after irradiation with 80 eV electrons. The TC spectrum exhibits the formation of the peaks of F_2^+, F_2, F_3, and F centers. The peak intensities of the F_2 and F_3 centers increase for 30 min due to the coalescence of F_2^+ and F centers at the surface, after which the peaks disappear slowly due to the formation of X centers in the film.

Upon annealing of the irradiated LiF film at 50°C, the peak intensities of the F_2 and F_3 centers increase for 10 min and disappear after 90 min due to the formation of X centers whose concentration becomes high.

http://dx.doi.org/10.5772/65517

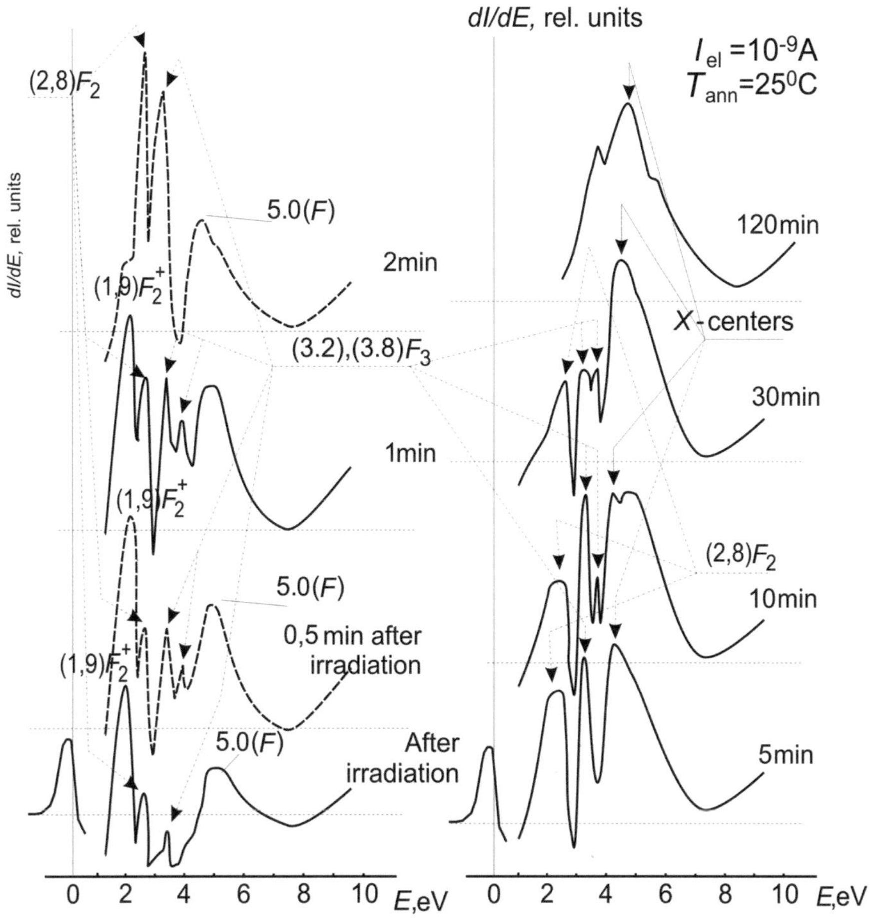

Figure 4. Kinetics of the TC spectra at T = 25°C after irradiation of the LiF film with 80 eV electrons.

When the annealing temperature of the irradiated LiF film is 75°C, the peak intensities of the F_2 and F_3 centers increase for 5 min, the F_2 centers disappear after 15 min, and the F_3 centers disappear after 20 min. A high-intensity peak with an energy of 2 eV appears in the TC spectrum, which belongs to small-size colloid centers. This means that colloid centers can form upon annealing at 75°C, but this process depends on the concentration of the preliminarily generated F centers. Note that 15 min after annealing at 75°C, the peak with an energy of 2.7 eV increases. Since the concentrations of F_2 and F_3 centers decrease to minimum values upon 10-min annealing, the peak with an energy of 2.7 eV belongs only to large colloid centers [18]. Subsequent annealing for 20 min leads to an increase in the concentration of the colloid centers and the appearance of a new peak with an energy of 3.6 eV, which, probably, belongs to halide complexes. Upon intense colloid formation, the concentration of F centers decreases greatly (~70% of F centers coagulate into colloid centers) [18]. The concentration of colloid centers increases due to electron abstraction from F centers. With an increase in the annealing time to ~70–170 min, the intensity of the colloid peak increases significantly.

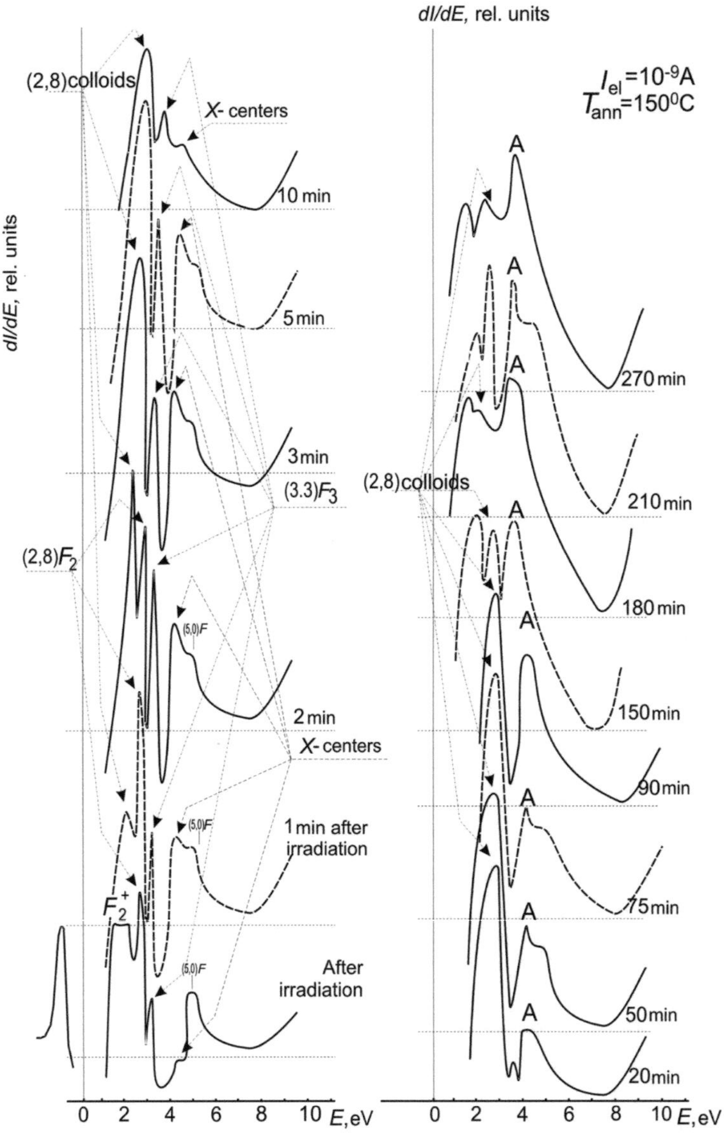

Figure 5. TC spectra at $T = 150°C$ at different time points after irradiation of the LiF film with 80 eV electrons.

After 10-min annealing of the irradiated LiF film at 100°C, the peak intensities of the F_2 and F_3 centers decrease fast and, at the same time, the peak intensities of the X centers ($E = 4.5$ eV) and colloid centers ($E = 2.0$ eV) increase greatly. After 20-min annealing, peaks of the large colloids ($E = 2.7$ eV) and anionic complexes ($E = 3.6$ eV) appear. Further annealing for 210 min leads to an increase in the concentrations of the colloid centers and anionic complexes.

Annealing of the irradiated LiF film at $T = 150°C$ results in the disappearance of F_2 centers (3 min) and F_3 centers (10 min) (**Figure 5**). Immediately after 1-min annealing, peaks of the X and

http://dx.doi.org/10.5772/65517

colloid centers appear. After 20-min annealing at 150°C, large colloid centers and halide cluster complexes form in the film. TC spectral analysis of the LiF films annealed at different temperatures (25, 50, 75, 100, and 150°C) shows that the F_2 and F_3 centers are in existence for a long time. With an increase in the temperature by 50°C, they exist in the LiF film for just 10–20 min. Annealing of the LiF film at 75°C results in a large accumulation of X centers and colloid centers. With increasing annealing time at different temperatures (75, 100, and 150°C), anionic complexes form along with the colloid centers, presumably, due to the migration of cations to the surface upon metallization. Lattice deformation results in the accumulation of halide atoms and H centers and the formation of anionic complexes.

Figure 6 shows the degradation kinetics of F_2 centers at different temperatures after deposition of the LiF film (F_2 centers form due to the localization of two electrons at two anion vacancies). The model of an F_2 center has been described in Ref. [10]. The point symmetry of F_2 centers is identical in all crystals; this means that an F_2 center has only one absorption band in the spectrum (2.76 eV). It follows from the analysis of **Figure 6** that low-temperature (50–150°C) annealing leads to an increase in the peak intensity of the F_2 centers in the TC spectrum due to the migration and coalescence of F centers and their components from the bulk to the surface. A sharp decrease in the F_2 center peak at 75–150°C indicates the formation of colloid and X centers at the surface. Room-temperature annealing leads to an increase in the peak intensity of F_2 centers due to the coalescence of F centers and their components (vacancies, F_2^+, and F). With increasing temperature to 50°C, the kinetics of the peak intensity of F_2 centers increases two-fold. With a further increase in the annealing temperature to 75 and 100°C, the peak intensity of the F_2 centers decreases appreciably, but their survival time remains unchanged. At 150°C, the peak intensity of the F_2 center increases quickly and then drops to zero in 3 min. As seen, the concentration of F_2 centers always increases upon annealing. The increase in the concentration of a certain defect type always results in the degradation of this defect due to the formation of another large defect upon their fusion with one another. Therefore, the concentration of F_2 centers drops to zero after several minutes. It is seen that the degradation of F_2 centers occurs faster when the annealing temperature is higher.

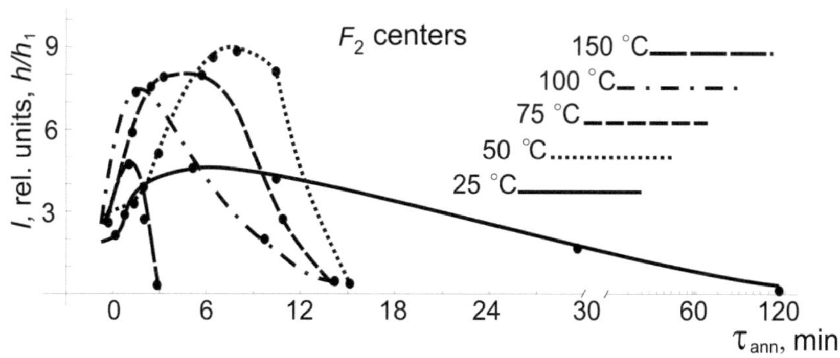

Figure 6. Kinetics of the peak intensity of F_2 centers at different annealing temperatures of the LiF film.

Figure 7 shows the kinetics of the peak intensity of F_3 centers at an annealing temperature of 25–150°C. It is seen that the degradation kinetics of F_3 centers differs from that of F_2 centers: at low temperatures, F_3 centers exist longer. An F_3 center corresponds to three F centers, which can be represented as three anionic vacancies forming an equilateral triangle in the (1 1 1) plane of the LiF lattice and as three electrons moving in the positive field of vacancies. Particularly, an F_3 center has several absorption bands in alkali halide crystals and there are two such bands in LiF (E = 3.3 and 3.9 eV) [10].

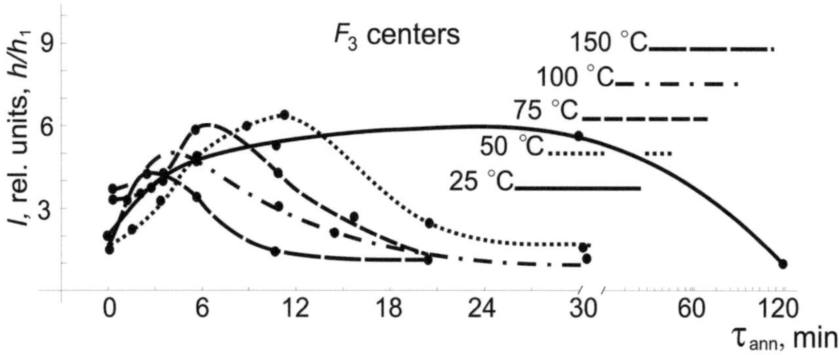

Figure 7. Kinetics of the peak intensity of F_3 centers at different annealing temperatures of the LiF film.

With increasing temperature, the concentration of F_3 centers increases fast and then decreases to zero. One can note that the higher the annealing temperature, the faster the increase in the peak intensity of these aggregate centers. This can be caused by the fact that at high annealing temperatures the coagulation and diffusion of large aggregate centers proceeds in less time. The data in **Figure 4** suggest that the peak intensity of F_3 centers at T = 25°C increases slightly and drops to zero after 2 h of observation due to the formation of the X centers. At 50, 75, and 100°C, the concentration of F_3 centers increases by four-, three-, and two-fold, respectively. Such an increase can be explained by the fact that annealing of the film accelerates the migration and diffusion of color centers from the bulk to the near-surface layer, but with increasing annealing temperature, the formation of X centers and colloid centers proceeds faster, which leads to a decrease in the concentration of aggregate centers. With a further increase in the temperature, the surface is enriched with colloid centers and X centers, which prevents growth in the peak intensity of F_3 centers due to their coalescence into large aggregate centers.

As seen, all aggregate centers coalesce after thermal annealing to form large centers, such as colloid and X centers, which are more stable than other aggregate centers. In our experiments, the X center peak appears in the TC spectrum at an energy of 4.5 eV. Their creation requires a corresponding dose rate and temperature [22]. **Figure 8** shows the kinetics of the peak intensity of the X centers at different substrate temperatures after deposition of the LiF film. It is seen that the X centers do not degrade and coalesce into colloid centers at low temperatures. At an annealing temperature of 50°C, the peak intensity of the X centers decreases appreciably, but at this temperature, the X centers also do not coalesce into colloid centers. With increasing temperature to 75°C, the X centers in the LiF film degrade in 30 min. Only at T = 150°C does the peak intensity of the X centers decrease quickly (in 15 min). The concentration of X centers depends on the concentration of precreated F centers and their aggregates. The increase in annealing temperature results in their degradation and transition to the state of a colloid center. The increase in the concentration of F centers results in the accumulation of cations surrounded by F centers. After trapping a sufficient amount of electrons, such a center stabilizes into the colloid aggregate (negatively charged center), which acquires a structure untypical of a macroscopic metal. With increasing size of such particles, aggregates with a true metallic bond form [17, 22]. Colloid centers emerge at high radiation doses, but, for their formation, thermal annealing can also be used, wherein conditions that accelerate the coalescence and migration of F centers can be created.

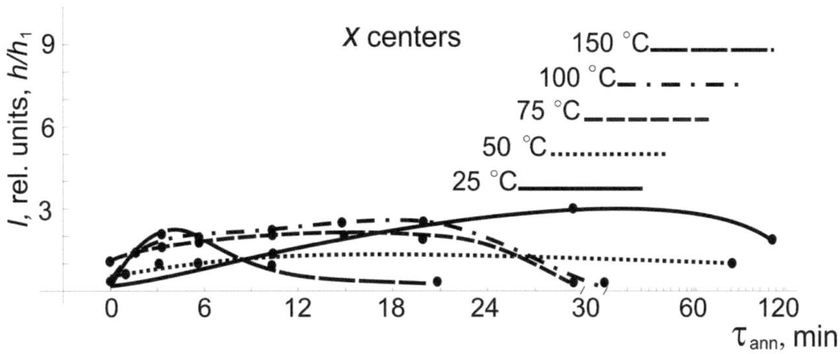

Figure 8. Kinetics of the peak intensity of X centers at different annealing temperatures of the LiF film.

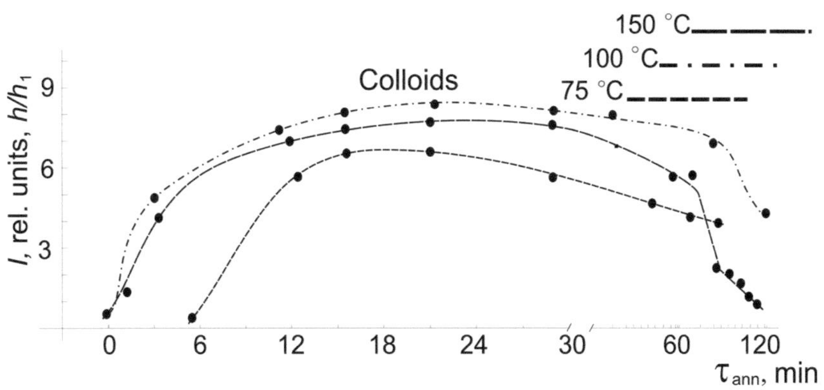

Figure 9. Time change in the peak intensity of colloid centers at different annealing temperatures of the LiF film.

Figure 9 shows the time dependence of the peak intensity of colloid centers upon annealing of the LiF film after deposition. It is seen that with increasing temperature, the formation time of colloid centers decreases. At room temperature, the migration of F centers proceeds slower than at high temperatures. At 25–50°C, no colloid centers form and, with increasing temperature to 75°C, they emerge even 5 min after annealing. At $T = 100°C$, the colloid centers form after 1 min. At T = 150°C, the concentration of colloid centers increases fast, and after 1-h annealing, begins decreasing due to evaporation of the colloid center components. At higher temperatures, evaporation occurs faster.

4. Procedure for determining defects in sputtered clusters of ionic crystals

Secondary ions, which are recorded in the spectra of SIMS, can be products of recombination [24, 25] or direct emission [26] from the lattice and can also be produced as a result of defect formation on the crystal surface [27–30]. The authors of almost all papers concerning sputtering have asserted that defects stimulate the formation of secondary clusters [24, 27–37]. Defect formation during crystal sputtering was considered in Refs. [31–34] as a result of the formation of self-localized excitons such as electron-hole pairs, which is accompanied by excitation of the valence band. In some insulator materials (alkali-halide crystals, SiO_2, Al_2O_3), defects can be formed under bombardment with electrons (electron-stimulated desorption) and photons (photon-stimulated desorption) [35–37]. In this case, as a result of the decay of self-localized excitons, neutralized anions (halides in alkali-halide crystals and oxygen in SiO_2 and Al_2O_3) are desorbed and cations are neutralized in the case of alkali-halide crystals, which lead to desorption in the case of small transfer of the bombarding ion momentum.

An analysis of published papers indicates that, at present, there is no procedure for recording point defects in secondary clusters neither in terms of experiment nor theory. Upon collision between primary particles and a crystal, defects are formed on its surface; they not only stimulate sputtering, but also can be emitted together with secondary ions.

Bombardment of the LiF crystals with 0.6 keV Cs^+ ions led to the appearance of a series of peaks, which corresponds to H_N, F_2^-, and F centers, in the TC spectra (**Figure 10**). (An F center is an anion vacancy with a captured electron [22], and H_N centers are attributed to interstitial crystal defects and appear in the absorption energy region of 1.7 eV.) In our case, H_N centers are formed in the target during Cs-ion implantation. These centers are very stable and did not disappear upon annealing up to a temperature of 200°C. Their concentration increased with increasing dose. An increase in the dose of bombarding ions up to 4.5×10^{13} ion/cm^2 led to an increase in the F, F_2, F_3, and F_4 peak intensities (**Figure 10**, curve 3) and also to the appearance of a peak of X centers, which formed in the case of cation accumulation around the F centers. The larger the concentration of F centers, the more stable the X centers [22].

The change in the position of the primary peak indicates the presence of positive charge on the sample surface [20]. A further increase in dose of up to 9×10^{13} ion/cm^2 led to a decrease in the intensities of all peaks (**Figure 10**, curve 5), which indicates surface amorphization. In this case, a wide peak at an energy of 3.5–4.1 eV appeared in the TC spectrum; it can be identified as a complex of anions (V_k centers) in LiF. This can mean that, under Cs^+-ion bombardment, aggregates of F centers lost weakly bound electrons during colloid formation, which affected the formation of anion complexes because of a decrease in the concentration of F centers and their aggregates. As a result, the concentration of colloids and halide complexes in LiF increased. A further increase in the dose of Cs ions up to 10^{14} ion/cm^2 led to the formation of

http://dx.doi.org/10.5772/65517

metal islands on the substrate surface. Based on the data in **Figure 10**, we can state that, if anion complexes are formed under bombardment, then the lattice contains interstitial F atoms, which are recorded during sputtering in the form of positive and negative F_2, F_3, and F_4 ions. If colloid centers are formed, then positive and negative ions of the Li_2, Li_3, and Li_4 types should be recorded. This can be verified using SIMS.

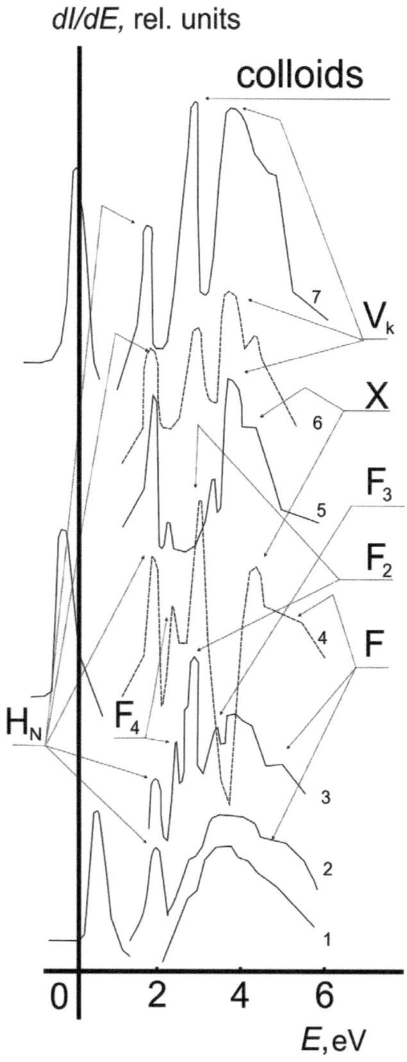

Figure 10. Change in the spectrum of the total current from the LiF crystal under Cs^+ bombardment. Cs^+ doses (ion/cm^2): (1) 0, (2) 2×10^{13}, (3) 4.5×10^{13}, (4) 5.4×10^{13}, (5) 9×10^{13}, (6) 1.4×10^{13}, and (7) 10^{14}.

Figure 11 shows data on secondary ion emission from the LiF crystal under bombardment with positive Ar^+ ions. When measuring the yield of positive ions from the cleaned target surface in this case, it was established that the SIMS spectrum of sputtered particles consists mainly of Li^+ cations, F^- anions, and molecular Li^+, $Li^+(LiF)n$ ions, where $n = 1–5$. Li^+ cations in the complex with neutral clusters $(LiF)n$ of the compositions $Li^+(LiF)$, $Li^+(LiF)_2$, and so on, dominate in the SIMS spectra of positive ions.

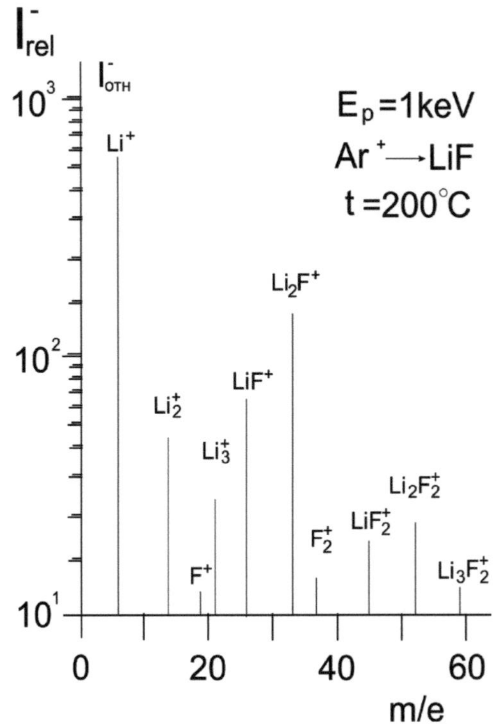

Figure 11. SIMS spectra of positive secondary-ion emission from the LiF crystal under Ar^+ ion bombardment.

Secondary particles			
Li^+	Ion from the lattice	Li^-	Ion from the colloid center
Li_2^+	Ion from the colloid center	F^-	Ion from the lattice
Li_3^+	Ion from the colloid center	F_2^-	F ion with a *H* center
F^+	Ion from the lattice	F_3^-	F ion with a *H* center
F_2^+	*H* center	LiF^-	F ion with a *H* center
LiF^+	Li ion with a *H* center	LiF_2^-	Ion from the lattice
LiF_2^+	Li ion with a *H* center	Li_2F^-	Cluster from the colloid center
$Li_2F_2^+$	Cluster with a *H* center	LiF_3^-	Cluster with a *H* center
$Li_3F_2^+$	Cluster from the lattice	LiF_4^-	Cluster with a *H* center

Table 1. Positive and negative secondary ions that can be sputtered from the LiF crystal surface together with possible defects.

http://dx.doi.org/10.5772/65517

Table 1 shows positive secondary ions emitted from the LiF crystal, which can be sputtered from a surface with possible defects. Analysis of the positive secondary clusters showed that every three clusters in five contain defects formed on the ion crystal surface during transfer of the energy of the bombarding ion.

For example, the secondary Li^+ ion could be knocked out from the colloid center, because two Li atoms cannot be located at neighboring lattice sites. The secondary $Li_2F_2^+$ cluster consists of separate atoms in the form (LiF)(F)Li^+:(LiF), which is a part of the lattice; Li^+ can also be a part of the lattice, but neutral F does not bind with any of the constituents of this cluster (**Figure 12a**).

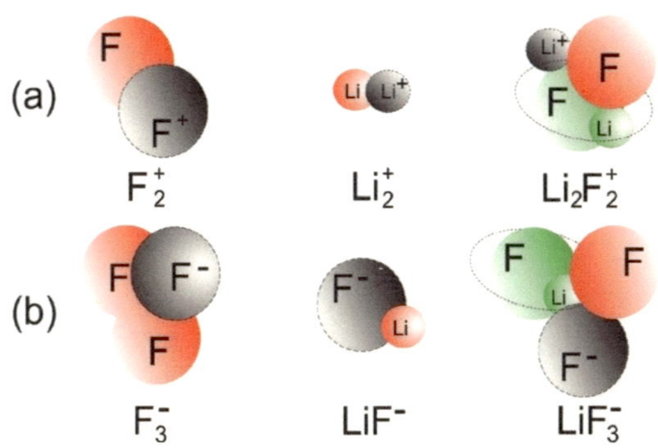

Figure 12. Possible variants of defect structures and orientations in secondary clusters produced under ion bombardment of the LiF crystal: (a) positive and (b) negative clusters.

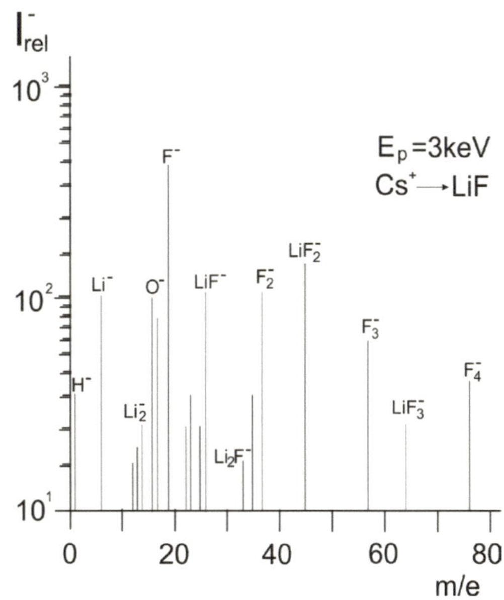

Figure 13. SIMS spectra of negative secondary ion emission from the LiF crystal under Cs^+-ion bombardment.

The study of secondary ion emission from an uncleaned LiF crystal under Cs^+-ion bombardment showed that negative ions F ~ (n = 1–4), Li_m (m = 1–3), $(Li_mF_n)^-$, where n = 1–4, m = 1, 2, $(LiO)^-$, $(CF)^-$, $(FO)^-$, and $(LiFO)^-$ are contained in the SIMS spectrum of LiF sputtering products. The yields of heavy ions observed in the SIMS spectra can be associated with an increased content of defect structures on the uncleaned surface. In this case, the rate of defect formation under bombardment and the intensity of surface-layer sputtering are very large.

Figure 13 shows the SIMS spectrum of LiF crystal sputtering products after long-duration cleaning with Cs^+ ions. Because fluorine is an electronegative element, the number of negative ions is larger, which is seen from this spectrum.

The table contains negative secondary ions from the LiF crystal, which can form on the surface during defect formation. It can be seen that every four negatively charged clusters in five contain surface defects produced by bombardment of the ionic crystal and attached to them. For example, the secondary F_2^-, ion was, probably, ejected together with the H center, because two F atoms cannot be located at neighboring lattice sites, and, especially as one of them is neutral.

The secondary LiF_3^- cluster consists of separate atoms in the form $(LiF)(F)F^-$, where (LiF) is a part of the lattice, F^- can also be a part of the lattice, but neutral fluorine does not bind with any of the components of this cluster. The likely structure of these ions is shown in **Figure 12(b)**. Here, the secondary ion is $F_3^- = F^-(FF)$. This means that the cluster F_3^- consists of one F^- ion and two interstitial defects (FF). The cluster F_3^- cannot be a crystal-lattice product, because only F^- and Li^+ ions together can form a part of the ionic crystal (**Figure 12b**, ellipse). In all other cases, secondary negative clusters are formed by interstitial, colloid, and aggregate V_k centers. Thus, it is very probable that the main part of the charged products of LiF-ionic-crystal sputtering consist of point defects.

It was established that, at low temperatures of alkali-halide crystals, the main sputtering mechanism includes elastic collisions between the impinging primary particles and the target components and, as the temperature increases, the increase in sputtering can be associated with the generation of defects in the form of F—H pairs in the surface layer [22]. The F—H pair is an anion vacancy with a captured electron (the F center) and an interstitial halide atom (the H center), which are formed under bombardment with charged particles. Approximately 60 eV is spent on the formation of an F—H pair in LiF crystals. The formation of primary F—H pairs is most important for sputtering. In accordance with Refs. [22, 31, 35], the increase in the sputtering of alkali-halide crystals with increasing temperature was related to the diffusion of these defects to the surface, which results in the neutralization of surface cations and anions. The latter lacking bonds can participate in both sputtering and evaporation of the target material. If the fact that, in the case of the given energy E_0, the ranges of heavy Cs^+ ions in the LiF crystal, which are smaller than that of Ar^+ ions, is taken into account, then the effectiveness of the formation of defects (including F—H pairs in the surface layer) is relatively high. In this case, defect diffusion to the surface is facilitated as a result of an increase in the bombarded-target temperature, which was observed in our experiment (**Figure 11**). The secondary-emission characteristics of insulators of the alkali-halide type are usually changed if a charge is produced on the surface bombarded with ions. Surface charging causes the deterioration of beam focusing, a shift of the energy spectrum of secondary ions, a decrease in its intensity, and the migration of primary ions over the sample surface. The effect can be enhanced as the density of the bombarding-ion beam increases. To remove the charge, a thermal method is traditionally used. Our studies showed (**Figure 11**) that, if beams of Ar^+ ions with an intensity of 10^{-9} A are used, the charge on the surface will be suppressed in the temperature range of 100 to 100–180°C, where the LiF crystal does not undergo evaporation.

At low bombardment doses, the concentrations of vacancies and implanted defects in crystals are usually small. Therefore, an insignificant number of cluster ions are observed in the case of LiF-surface bombardment with ions with small energies (300 eV) and at low doses. The increase in the yield of cluster ions with increasing primary-ion energy and dose is due to the fact that, in this case, the number of vacancies and implanted defects increases considerably and more and more F and Li ions penetrate into positions of implanted defects or vacancies near other Li and F ions, which facilitates the process of cluster formation. Thus, as the dose, the bombardment energy, and the temperature increase, a large number of aggregate defects, colloids, and neutralized anion complexes form and are sputtered from the surface together with the clusters.

Analysis of the SIMS spectra of ionic crystals showed that neutralized atoms in the compositions of secondary ions can mainly indicate the presence of defects in them. This procedure can now be used only for ionic crystals. It follows from the foregoing that defects formed on the surfaces of alkali-halide crystals stimulate the sputtering process and are emitted from the surface together with secondary ions in this case.

5. Conclusion

In all cases, low-temperature annealing of the LiF/Si(1 1 1) film results in the degradation of the formed centers followed by their coalescence into large aggregates. Long-term annealing results in evaporation of the halogen atoms and, after electron trapping by cations at the film surface, the formation of colloid centers begins followed by metallization of the surface. The optimal temperatures for the formation of colloid centers on the surface are 75–150°C: at these temperatures, they remain in the film for 5 h. At high temperature (75–500°C), long-term annealing of the LiF film, along with colloid centers (E = 2.7 eV), anionic complexes (E = 3.6 eV) form on the surface.

Point defects formed on the surface of a LiF crystal under ion bombardment not only stimulate the process of cluster formation, but also can be ejected from the surface together with secondary ions. Colloids stimulate Li-cluster sputtering and aggregates of V_k centers, whose compositions contain H centers, increase the yield of F cluster ions. Analysis of the SIMS spectra of LiF ionic crystals subjected to ion bombardment showed that the neutralized atoms in the compositions of secondary ions can mainly indicate the existence of defects in the sputtered clusters. The developed procedure can be useful for the development of theoretical models of structures and orientations of defects in clusters, which are products of the ion bombardment of alkali-halide crystals.

Author details

Utkirjon Sharopov*, Bakhtiyar Atabaev, Ruzmat Djabbarganov and Muzaffar Qurbanov

*Address all correspondence to: utkirstar@gmail.com

Tashkent State Technical University, Tashkent, Uzbekistan

References

[1] Hersch H.H. Proposed excitonic mehcanism of colour center formation in alkali halides. Phys. Rev. 1966;148(2):928–932. DOI:10.1103/PhysRev.148.928

[2] Itoh N. Formation of lattice defects by ionizing radiation in alkali halides. J. Phys. Colloques. 1976;37(C7-27-C7):37. DOI: 10.1051/jphyscol:1976703

[3] Itoh N. Bond scission induced by electronic excitation in solids: A tool for nanomanipulation. Nucl. Instr. Meth. B. 1997;(122) :405.

[4] Lushchik C.B. Physics of radiation effects in crystals. R.A. Johnson and A.N. Orlov (ed.). Elsevier, Amsterdam. 1986; pp. 473–525.

[5] Vetter J., Scholz R., Angert N. Investigation of latent tracks from heavy ions in GeS crystals by high resolution TEM. Nucl. Instr. Meth. B. 1994;(91):129. DOI: 10.1016/0168-583X(94)96202-2

[6] Wiesner J., Traeholt C., Wen J.-G., Zandenbergen H.W., Wirth G., Fuess H. HREM of heavy ion induced defects in superconducting Bi-2212 thin films in relation to their effect on Jc. Physica C. 1996;(268):161.

[7] Schwartz K. Electronic excitations and defect creation in LiF crystals. Nucl. Instr. Meth. B. 1996;(107):128. DOI: 10.1016/0168-583X(95)00846-2

[8] Schwartz K., Geiss O. Chemical etching of ion tracks in LiF crystals Journal of applied physics. J. Appl. Phys. 1998;83(7):3560–3564. DOI: 10.1063/1.366572

[9] Schwartz K., Trautmann C., Steckenreiter T., Geiss O. Damage and track morphology in LiF crystals irradiated with GeV ions. Phys. Rev. 1998;58(17):11232 . DOI: 11232 10.1103/PhysRevB.58.11232

[10] Agullo-Lopez F., Catlow C.R.A., Townsend P.D. Point defects in materials. Academic Press (ed.). London: Academic Press; 1988. 125 p .

[11] Gilman J.J., Johnston W.G. Dislocations, point-defect clusters, and cavities in neutron irradiated LiF crystals. J. Appl. Phys. 1958;(29):877. DOI: 10.1063/1.1723322

[12] Peisl H., Balzer R., Waidelich W. Schottky and Frenkel disorder in KCl with color centers. Phys. Rev. Lett. 1966;(17):1129. DOI: 10.1103/PhysRevLett.17.1129

[13] Montereali R.M. Miniaturized structures based on color centers in lithium fluoride: Optical properties and applications. Rad. Eff. Defects Solids. 2002;(157):545. DOI: 10.1080/10420150215835

[14] Weber M.J. (ed.). Handbook of Lasers. Washington, DC: CRC Press; 2001. 235 p.

[15] Scaglione S., Montereali R.M., Mussi V., Nichelatti E. F-color centers and lithium nanoclusters in ionbeam assisted LiF thin films. Optoelectron. Adv. Mater. 2005;(7):207.

[16] Roberts J.G., van Hove M.A., Somorjai G.A. Surface structural analysis of LiF(1 0 0) thin films grown on Pt(1 1 1). Surf. Sci. 2002;(518):49. .

[17] Ostrovskii I., Ostrovskaya N., Korotchenkov O. Radiation defects manipulation by ultrasound in ionic crystals. IEEE Trans. Nucl. Sci. 2005;52(6):3068–3073. DOI: 10.1109/TNS.2005.861476

[18] Komolov S.A. Integral secondary electron spectroscopy of surface. Leningr. Gos. Univ. (ed.). Leningrad: Leningr. Gos. Univ.; 1986. 146 p.

[19] Henrich V.E. Fast, accurate secondary-electron yield measurements at low primary energies. Rew. Sci. Instr. 1973;44:456–462. DOI: 10.1063/1.1686155

[20] Komolov S.A., Chadderton L.T. Interaction of slow electrons with surface: Desorbtion and adsorption of gas on (1 0 0) vanadium. Rad. Eff. 1976;31:1–6.

[21] Alidzanov E.K., Atabaev B.G., Gaipov C.G., et al. Target current spectroscopy of the alkali halides KCl, CsCl and KBr. Thin Solid Films. 1994;270:268 .

[22] Shvarts K.K., Ekmanis Y.A. Radiation Processes and Radiation Hardening of Dielectric Materials. Riga: Zinatne [in Russian]; 1989. 168 p.

[23] Golek F., Sobolewski W.J. Transmission of low-energy electrons through polycrystalline LiF films. Phys. Status Solidi. 1999;(1):210–211.

[24] Sigmund P. Fundamental processes in sputtering of atoms and molecules. Mat. Fys. Medd. 1993;(43):1.

[25] Dzhemilev N.K. Statistical mechanism of the formation of the energy spectra of sputtered molecular clusters. Surf. Invest. 2012;(6):654.

[26] Wittmaack K. On the mechanism of cluster emission in sputtering. Phys. Lett. 1979;(69): 322.

[27] Seifert N., Yan Q., Barnes A.V., et al. Beam interactions with materials and atoms. Nucl. Instrum. Methods Phys. Res. 1995;(101):131.

[28] Reichling M. Electron stimulated desorption from CaF_2: Penetration depth of electrons and sample charging. Nucl. Instr. Meth. Phys. Res. 1995;101:118.

[29] Atabaev B.G., Gaipov S., Sharopov U.B. Investigation of point defects and their clusters formation on a growing film LiF/Si (1 1 1) surface under electrons irradiation. Poverkhnost. 2007;(10):52.

[30] Sharopov U.B., Atabaev B.G., Djabbarganov R., Kurbanov M.K. Kinetics of aggregations of F2, F3, X, and colloid centers in LiF/Si(1 1 1) films upon low-temperature annealing. J. Surf. Invest. X-ray, Synchrotron Neutron Tech. 2013;7:195. DOI: 10.1134/S1027451012120117

[31] Neidhart T., Pichler F., Aumayr F., et al. Potential sputtering of lithium fluoride by slow multicharged ions. Phys. Rev. Lett. 1995;(74):5280. DOI: 10.1103/PhysRevLett.74.5280

[32] Sporn M., Libiseller G., Neidhart T., et al. Potential sputtering of clean SiO2 by slow highly charged ions. Phys. Rev. Lett. 1997;945. DOI: 10.1103/PhysRevLett.79.945

[33] Aumayr F., Burgdorfer J., Varga P., Winter H.P. Sputtering of insulator surfaces by slow highly charged ions: Coulomb explosion or defect mediated desorption?. Comm. At. Mol. Phys. 1999;(34):201.

[34] Hayderer G., Schmid M., Varga P., et al. Threshold for potential sputtering of LiF. Phys. Rev. Lett. 1999;(83):3948. DOI: 10.1103/PhysRevLett.83.3948

[35] Green T.A., Loubriel G.M., Richards P.M., et al. Time dependence of desorbed ground-state lithium atoms following pulsed-electron-beam irradiation of lithium fluoride. Phys. Rev. 1987;(35):781. DOI: 10.1103/PhysRevB.35.781

[36] Walkup R.E., Avouris P., Ghosh A. Positive-ion production by electron bombardment of alkali halides. Phys. Rev. B. 1987;(36):4577. DOI: 10.1103/PhysRevB.36.4577

[37] Szymonski M., Poradisz A., Czuba P., et al. Electron stimulated desorption of neutral species from (1 0 0) KCl surfaces. Surf. Sci. 1992;(260):295. DOI: 10.1016/0039-6028(92)90044-ZnO

Section 2

State-of-the-Art Nanomaterials

Chapter 4

Colloidal Photonic Crystals Containing Copper-Oxide and Silver Nanoparticles with Tunable Structural Colors

Chun-Feng Lai

Additional information is available at the end of the chapter

http://dx.doi.org/10.5772/65007

Abstract

In this chapter, we investigated polystyrene (PS) colloidal photonic crystal (CPhC) color films containing copper-oxide (CuO) nanoparticles (NPs) and silver (Ag) NPs and exhibiting tunable structural colors. PS CPhC color films containing CuO-NPs and Ag-NPs were prepared through thermal-assisted self-assembly by using a gravitational sedimentation method. Doped CuO-NPs and Ag-NPs deposited on the bottom of the substrate and acted as black materials that absorb background and scattering light. Experimental results showed that brilliant structural colors were enhanced because of the absorption of incoherently scattered light, and color saturation was increased by the distribution of metal NPs on PS CPhC surfaces. The brilliant structural colors of CuO-NPs/PS and Ag-NPs/PS hybrid CPhC color films were based on the scattering absorption and Bragg diffraction theory. The reflection peaks of metal-NPs/PS hybrid CPhCs and pure PS CPhCs were measured by UV-Visible reflection spectrometry and theoretically calculated based on the Bragg diffraction law. Additionally, the structural colors of metal-NPs/PS hybrid CPhC color films were assessed through color measurements based on the Commission International d'Eclairage 1931 standard colorimetric system. Finally, this chapter exhibits a simple method to generate tunable structural color of functional materials for numerous applications, such as in textile fabrics, bionic colors, catalysis, and paint.

Keywords: colloidal photonic crystals, copper-oxide, silver, nanoparticles, structural colors

1. Introduction

Structural colors generated by nanostructures are generally classified as iridescent [1–3] and non-iridescent colors [4–7]. The brilliant iridescent character of nanostructures has attracted considerable attention for a variety of applications [8–10]. Iridescent colors produced by interference, diffraction, or light scattering are generated by periodic lattice crystals. Colloidal photonic crystals (CPhCs) of periodic crystals, which can reflect light at a specific photonic band gap (PBG) caused by periodic variation of refractive index in crystal structures [11, 12], are artificially fabricated. Polystyrene (PS), SiO_2, or polymethyl methacrylate (PMMA) nanospheres of 3D CPhCs were generally fabricated by a self-assembly method [13–15]. 3D CPhCs still have several problems which limit the applications, containing low color visibility (opalescent appearance) and poor crystal quality.

The structural color of CPhCs is extremely dull because of the interferences of scattering and background light (**Figure 1(a)**). An effective method to address low color visibility is to mix highly absorbent black materials in the visible light region into colloidal systems to enhance color saturation, because black materials can reduce incoherent scattering to enhance color [16–19] (**Figure 1(b)**). Carbon black (CB) [20–23], carbon [24, 25], carbon-modified colloidal nanospheres [26, 27], and metal nanoparticles (NPs) [28–32] of black materials as additives have been recently used to fabricate CPhCs that exhibit enhanced structural colors.

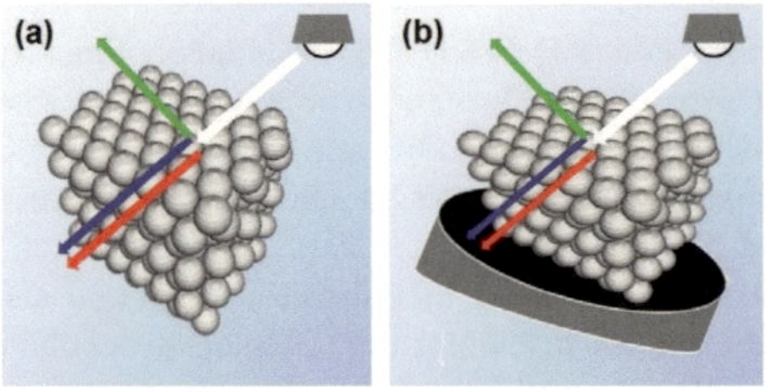

Figure 1. (a) A highly crystalline CPhC transmits light of certain wavelengths and reflects the other wavelengths. This response results in an opalescent appearance. (b) CPhCs are doped on black materials; transmitted light is absorbed, and the reflected light became more visible.

Incorporating a variety of NPs in CPhCs produces brilliant iridescent colors [20–32] because NPs can absorb and scatter visible light. In this chapter, CPhCs were incorporated directly into copper-oxide (CuO)-NPs and Ag-NPs of metal NPs by using a simple method [30, 33]. The metal NPs in the CPhC structures absorb the visible light and reduce scattering light that enhanced the color under natural lighting conditions. The method is mainly used to distribute metal NPs on PS nanosphere surfaces due to electrostatic interactions. Additionally, the sediment of metal NPs on the bottom of substrate could absorb scattering light because Cu and Ag materials are denser than PS, resulting in production of vivid structural colors. The reflection wavelengths of PS CPhC films with and without CuO-NPs and Ag-NPs were predicted theoretically and compared with the corresponding experimental results. The structural colors of the CuO-NPs/PS and Ag-NPs/PS hybrid CPhC color films were not only substantially enhanced but also varied with observation angles. CuO-NPs/PS and Ag-NPs/PS hybrid CPhCs exhibiting enhanced structural colors were obtained using a simple and inexpensive method. In addition, Cu-NPs and Ag-NPs are generally used as antimicrobial agents [34, 35] and catalysts [36, 37] in the biomedical field. Therefore, CPhCs incorporated

http://dx.doi.org/10.5772/65007

with CuO-NPs or Ag-NPs could provide additional advantages, such as antibacterial activity, in textile fabrics.

2. Materials and methods

2.1. Materials

Styrene (St) was distilled prior to use. Sodium dodecyl sulfate (SDS), potassium persulfate (KPS), sodiumbicarbonate ($NaHCO_3$), Cu nanopowder (40–60 nm particle size; 99.5% trace metal basis), and Ag nanopowder (<100 nm particle size; 99.5% trace metal basis) were used as received. Deionized (DI) water (18.2 MΩ cm resistivity) was purified using a PURELAB purification system. Cover glasses (Marienfeld) with hydrophilic surface were prepared using O_2 plasma.

2.2. Synthesis of monodisperse PS nanospheres

Negatively charged PS nanospheres were synthesized via an emulsion polymerization method as described previously [28] by using St as monomer, SDS as emulsifier, KPS as initiator, and $NaHCO_3$ as buffer. SDS was used to render negative charge to the surface of PS nanospheres [28]. Different sizes of colloidal PS nanospheres were synthesized using the same method by varying the quantity of SDS and St. The diameter of PS nanospheres (D_{PS}) was approximately linearly dependent on the amount of SDS and St (**Figure 2(a)** and **(b)**), respectively. **Table 1** shows the structural parameters for six PS CPhCs used in this chapter. For example, the sample B synthesis process is as follows: First, 200 mL of DI water, 125 mg of $NaHCO_3$, and 150 mg of SDS were mixed in a 500-mL three-necked flask, which was placed in a water bath under nitrogen protection. Then, 15 mL of St and 250 mg of KPS were added into the mixture. After stirring for 24 h, PS nanospheres with an average D_{PS} of 230 nm were obtained. Using the same synthesis process (sample E), we used 150 mg of SDS, 125 mg of $NaHCO_3$, 200 mL of DI water, 250 mg of KPS dissolved in 50 mL of DI water, and 8 g of styrene to obtain a D_{PS} of 215 nm. PS nanosphere powders were purified through dialysis, collected via centrifugation at 15,000 rpm for 1 h, and purified five times by washing with DI water before drying in a vacuum oven.

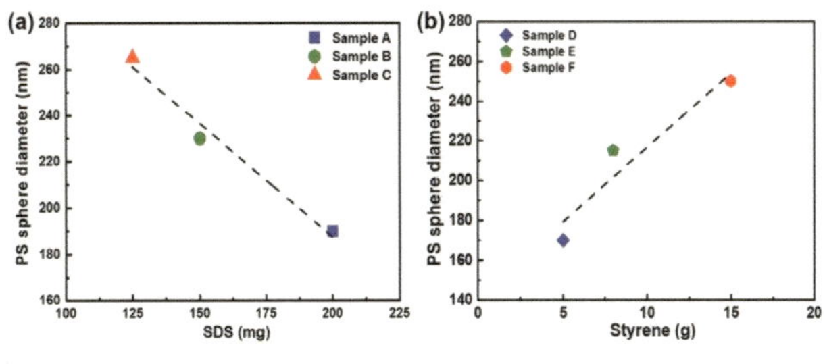

Figure 2. Diameter of colloidal PS nanospheres was dependent on the amount of (a) SDS and (b) styrene. Black dashed line indicates linear fitting.

Sample	Diameter (D_{PS}, nm)	Normal reflection peak (λ_R, nm)
A	190	452
B	230	547
C	265	630
D	170	417
E	215	515
F	250	601

Table 1. Structural parameters for the six PS CPhCs used in this chapter.

2.3. Attachment of CuO-NPs and Ag-NPs onto colloidal PS latex

Cu and Ag nanopowder with 99.5% metal basis were purchased from Sigma-Aldrich. Cu-NPs are easily oxidized upon exposure to an ambient laboratory atmosphere at room temperature. We utilized the random adsorption of CuO-NPs and Ag-NPs on negatively charged PS nanospheres. PS latex suspensions of six PS CPhCs with different D_{PS} (samples A–F, **Table 1**) at a concentration of approximately 4.8 weight percent (wt%) were prepared through ultrasonication. Varying PS latex suspensions (4.8 wt%) were added with 0.05, 0.10, and 0.50 wt% Cu nanopowder, and 1.0 and 4.5 wt% Ag nanopowder. Random distribution of metal NPs on PS surfaces through electrostatic interactions was achieved by ultrasonication for 8 h; the unabsorbed metal NPs settled at the bottom of the bottle.

2.4. Preparation of CuO-NPs/PS and Ag-NPs/PS hybrid CPhC color films

CuO-NPs/PS and Ag-NPs/PS hybrid CPhC color films were fabricated by the thermal-assisted self-assembly process and the gravitational sedimentation method (**Figure 3**). 200 μL of metal NPs/PS mixture suspensions were dropped on a cover glass. The suspension was spread to the glass surface. Then, the glass was placed in an oven at a constant temperature of 50°C for 2 h. The oven temperature was subsequently raised to 80°C for 30 min to improve the physical rigidity of the metal NPs/PS hybrid CPhC color films. Field-emission scanning electron microscopy (FESEM) was used to study the crystalline structure of the metal NPs/PS hybrid CPhC color films (**Figure 4(a)–(f)**).

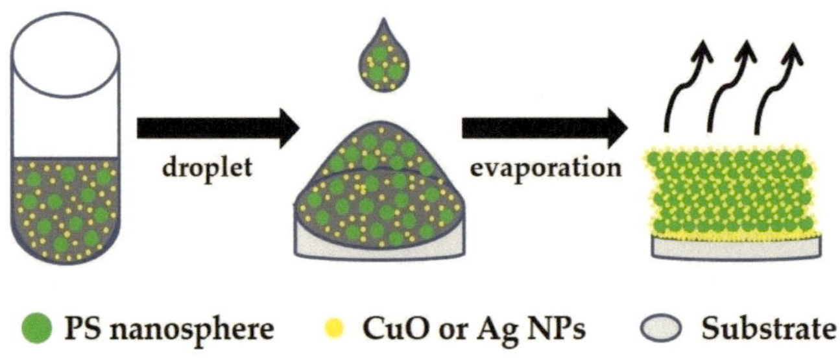

Figure 3. Schematic of the thermal-assisted self-assembly via gravitational sedimentation method.

http://dx.doi.org/10.5772/65007

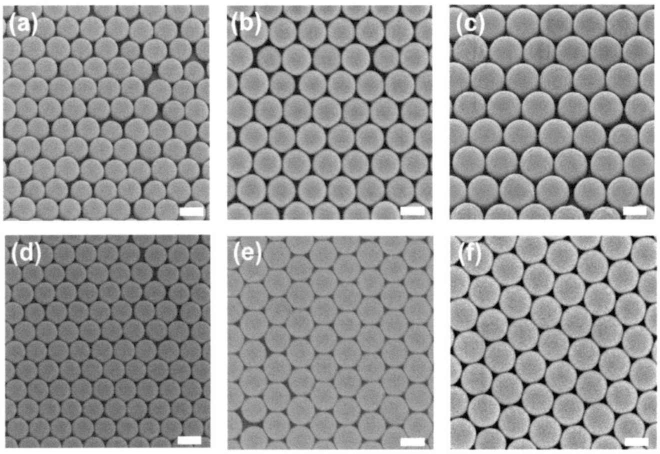

Figure 4. FESEM images of PS CPhC structures: (a) sample A, (b) sample B, (c) sample C, (d) sample D, (e) sample E, and (f) sample F. All scale bars represent 200 nm.

2.5. Characterization

The morphology of CuO-NPs/PS and Ag-NPs/PS hybrid CPhC color films was observed using FESEM (S-4800, Hitachi) and field-emission transmission electron microscope (FETEM, JEM-2100F, JEOL). Energy dispersive spectrometer (EDS, EMAX400, Horiba) was used to verify the presence of CuO-NPs and Ag-NPs on PS nanosphere surface. UV-Visible extinction spectra were obtained in standard transmittance mode, and the reflectance and color of metal NPs/PS hybrid CPhC color films were measured on a HR2000 spectrometer (Ocean Optics) with unpolarized white light provided by a Xe light source. Fiber optic Y-cables were used in reflection measurement. The color of the metal NPs/PS hybrid CPhCs color films was measured using SpectraSuite software (Ocean Optics) according to the Commission International d'Eclairage (CIE) 1931 standard colorimetric system. Raman spectra were analyzed from the reference database KnowItAll Information System (Bio-Rad Laboratories). Cover glasses (diameter, 10 mm; Marienfeld) were treated with plasma (Zepto, Diener) to obtain a hydrophilic surface before use.

3. Results

3.1. Optical properties of the CuO-NPs

Cu is the most commonly used metal in various applications in electronics due to its conductivity and low cost. As Cu is prone to oxidation, this could significantly affect the optical properties of CuO-NPs. Therefore, the optical characteristics of CuO-NPs are initially discussed in this chapter. **Figure 5** exhibits the surface adsorption of CuO-NPs on PS nanospheres which could cause visible light absorption. Furthermore, **Figure 5** demonstrates the absorbance degree of light which was dependent on the CuO-NPs content. **Figure 5(a)** shows the black-brown color of diluted solution because visible light was absorbed by CuO-NPs. In addition, **Figure 5(c)** shows the transmittance spectra of different CuO-NP concentrations, in which the CuO-NPs were strongly absorbed in the visible light. The transmittance was approximately 6% or less depending on CuO-NP concentrations. Moreover, scattering is typically very sensitive to the aggregation state of the NPs, with the scattering contribution increases as particles further aggregate [21, 38]. **Figure 5(d)** demonstrates the extinction spectra depending on the aggregation state of CuO-NPs, which are nearly entirely caused by photon absorption. The extinction spectra in **Figure 5(b and d)** clearly show that CuO-NPs agglomerated.

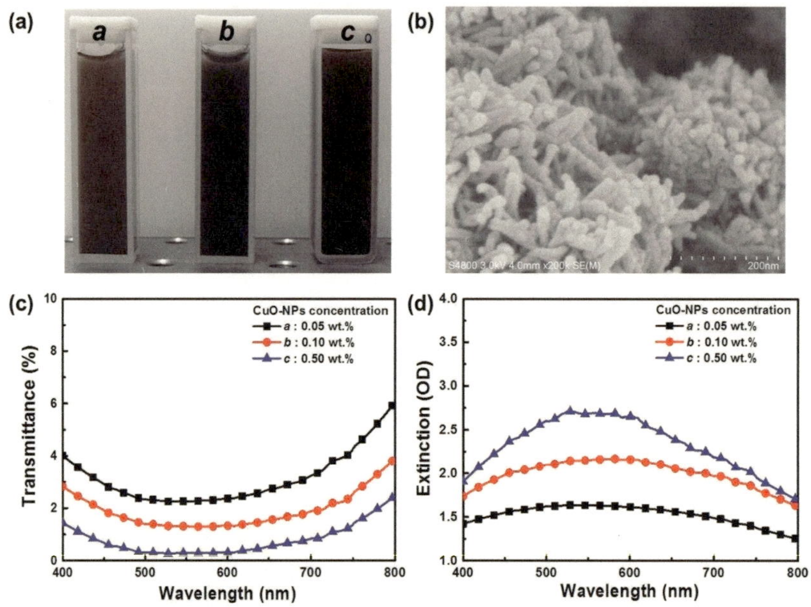

Figure 5. (a) Diluted solution appears black-brown because light is absorbed across the entire visible spectrum. (b) FESEM image of the agglomerates of 0.5 wt% CuO-NPs. (c) Transmittance and (d) extinction spectra of CuO-NP solutions resulted nearly entirely by light absorption with negligible scattering.

3.2. Optical properties of concentrated Ag-NPs

Ag-NPs are generally used for medical purposes due to its antibacterial, antimicrobial, and anti-inflammatory properties [39, 40]. In this chapter, dispersion in water could significantly affect the optical properties of concentrated Ag-NPs. Therefore, we discussed the optical characteristics of diluted Ag-NP solutions. **Figure 6(a)** shows the 1.0 and 4.5 wt% black-brown diluted solutions in a standard quartz cuvette because visible light was absorbed by Ag-NPs. FESEM images in **Figure 6(b)** demonstrated the agglomeration of Ag-NPs. The optical properties of Ag-NPs upon aggregation are affected, and conduction electrons located near each NP surface become delocalized and are shared among neighboring NPs. The extinction spectra in **Figure 6(d)** demonstrated the LSPR of Ag-NPs when they aggregated, therefore localized surface plasmon resonance (LSPR) was not observed in this study. In addition, **Figure 6(c)** illustrates the transmittance spectra of different Ag-NP concentrations, in which strong absorption was observed in the visible spectrum. The transmittance dropped to approximately 5% or less depending on Ag-NP concentrations. **Figure 6(d)** shows the extinction spectra depending on the size, shape, and aggregation state of Ag-NPs, which are nearly entirely caused by photon absorption.

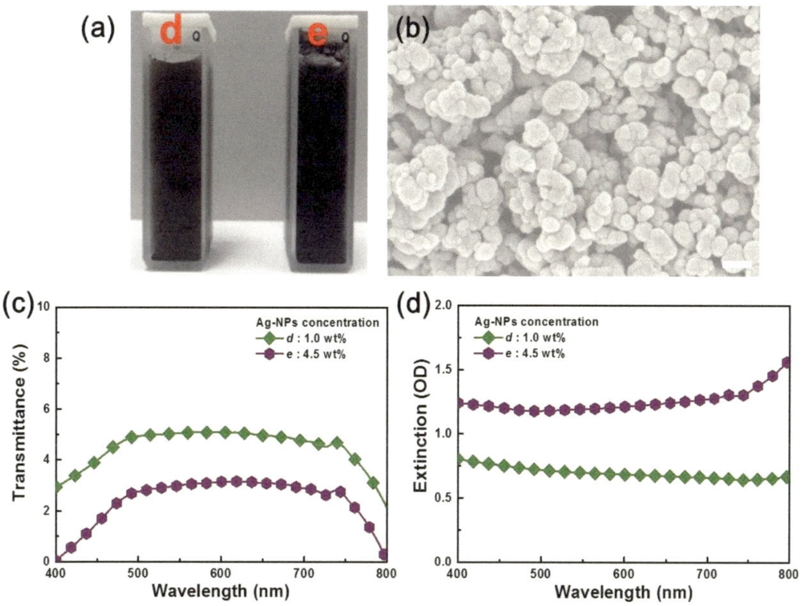

Figure 6. (a) Diluted solution appears black-brown because light is absorbed across the entire visible spectrum. (b) FESEM image of Ag-NP agglomerates at 4.5 wt% concentration. Scale bar in the inset represents 200 nm. (c) Transmittance and (d) extinction spectra of diluted Ag-NP solutions resulting nearly entirely from the absorption of all wavelengths of visible light with negligible scattering.

3.3. Crystal structures of the prepared CuO-NPs/PS hybrid CPhC color films

These studied samples contain PS nanospheres with different diameters a corresponding D_{PS} of 190, 230, and 265 nm, and different CuO-NP concentrations (samples A–C, **Table 1**). The CuO-NPs/PS hybrid CPhC color films were fabricated by mixing of CuO-NPs via a thermal-assisted self-assembly process by using a gravitational sedimentation method. FESEM was used to observe the crystalline arrangement of CuO-NPs/PS hybrid CPhC color films (**Figure 4(a)–(c)**). The CuO-NPs/PS hybrid CPhC color films exhibiting face-centered cubic (*fcc*) structures were assembled through gravitational sedimentation, and their top surfaces were parallel to the (1 1 1) crystallographic plane. **Figure 4(a)–(c)** demonstrated that the CuO-NP dopant did not disrupt the structures of CPhCs. The EDS mapping images of the CuO-NPs/PS hybrid CPhC films showed the randomly adsorbed CuO-NPs on PS nanospheres, and they appeared more often at the nanospheres compared with Cu atoms and O atoms (inset of

Figure 7); this finding indicated that surface oxidization occurred. In addition, the CuO-NPs/PS hybrid CPhC color film structures were investigated by obtaining their Raman spectra (**Figure 7**). The PS CPhC and CuO structures were measured based on their Raman spectra [41, 42] (**Figure 8**). All Raman spectra were analyzed by a commercial library search (KnowItAll Information System) and based on the literature [41, 42]. We confirmed that the doping of low CuO-NP concentrations did not affect the structural quality of the PS CPhC films.

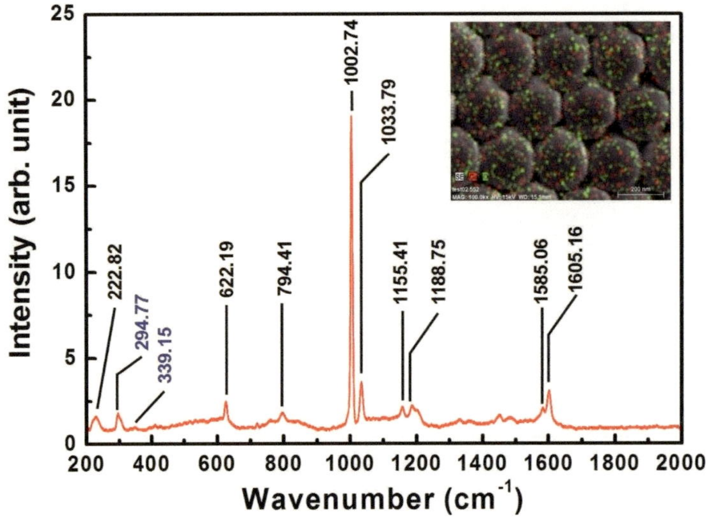

Figure 7. Raman spectrum of CuO-NPs/PS hybrid CPhC color films, in which black and blue indicate PS and CuO signals, respectively. Inset shows the EDS compositional mapping (Cu/O overlay).

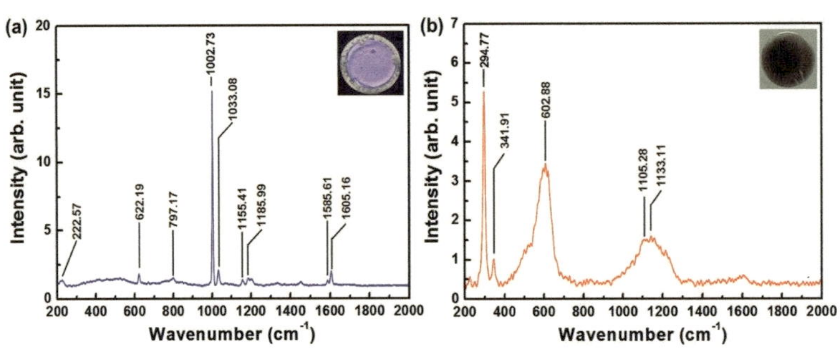

Figure 8. Raman spectrum of (a) PS CPhC and (b) CuO films. Inset shows the optical image of PS CPhC and CuO films.

3.4. Crystal structures of the prepared Ag-NPs/PS hybrid CPhC color films

Samples containing PS nanosphere with sizes (D_{PS}) of 170, 215, and 250 nm and containing various concentrations of Ag-NPs were prepared in this study (samples D–F, **Table 1**). Ag-NPs/PS hybrid CPhC color films were also fabricated by mixing Ag-NPs via a thermal-assisted self-assembly process by using a gravitational sedimentation method. The FESEM images indicated that the Ag-NPs dopant did not disrupt the structures of CPhCs (**Figure 4(d)–(f)**). The FETEM and EDS mapping images of the Ag-NPs/PS hybrid CPhC color films showed that the Ag-NPs were randomly adsorbed on the PS nanosphere surface (**Figure 9**). **Figure 9** shows the EDS mapping images of a sample consisting of nanospheres with adsorbed Ag and O atoms. Thus, Ag-NPs easily adsorbed O atoms, indicating its antibacterial effect [43, 44]. These results suggested that introduction of high Ag-NP concentrations did not affect the structural quality of the PS CPhC films.

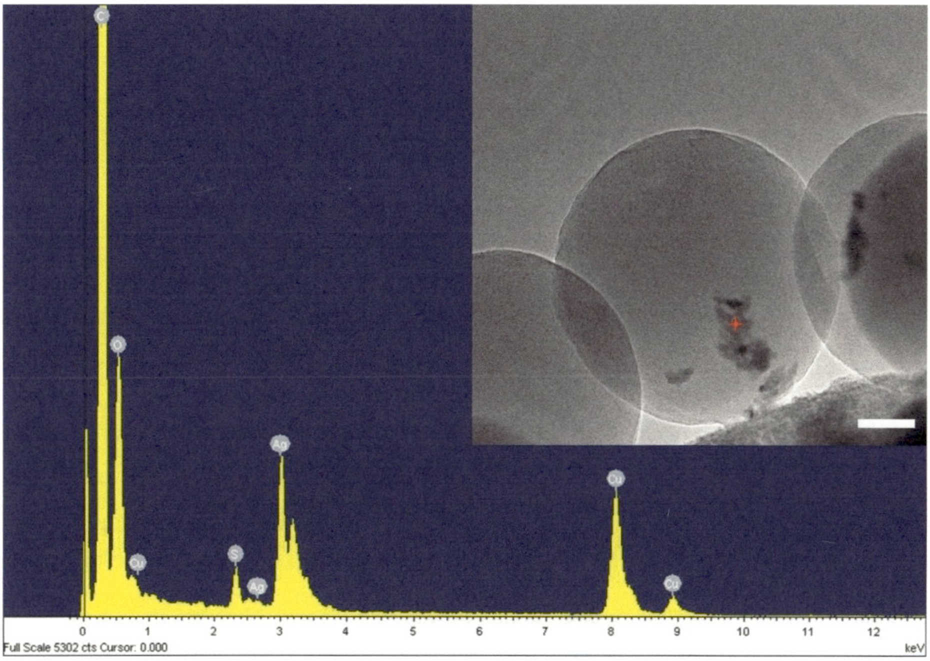

Figure 9. EDS compositional mapping (Ag) scans from the red cross point of inset FETEM image. Scale bars represent 20 nm.

3.5. Effects of CuO-NPs content on PS CPhC color films

Figure 10 shows the photographs of PS CPhCs without and with CuO-NPs. Pure PS CPhC films are usually milky white with extremely faint structural colors (**Figure 10(a)**, standard). As shown in **Figure 10(a)**, the visual appearance of metal NPs/PS hybrid CPhC color films from exhibiting dull color to vivid iridescent color clearly changed by doping CuO-NPs into PS CPhCs. These pictures were observed under a fluorescent lamp and exhibited a highly angular-dependent Bragg diffraction law. These pictures show a tunable color mechanism, which contributed to the absorbance of scattering light by doped CuO-NPs. For example, in this study was obtained by doping only 0.50 wt% CuO-NPs into PS CPhCs, and the visual appearance of hybrid CPhC color films changed evidently (**Figure 10**). Additionally, **Figure 10(b)** exhibits the back surface photographs of CuO-NPs/PS hybrid CPhC color films. The visual appearance of the CPhC color films changes remarkably from milky white to deep brown by doping CuO-NPs into the *fcc* structure (**Figure 10(b)**, sample B). Consequently, CuO-NPs may absorb scattering light and increase color saturation, producing bright structural colors of CPhCs.

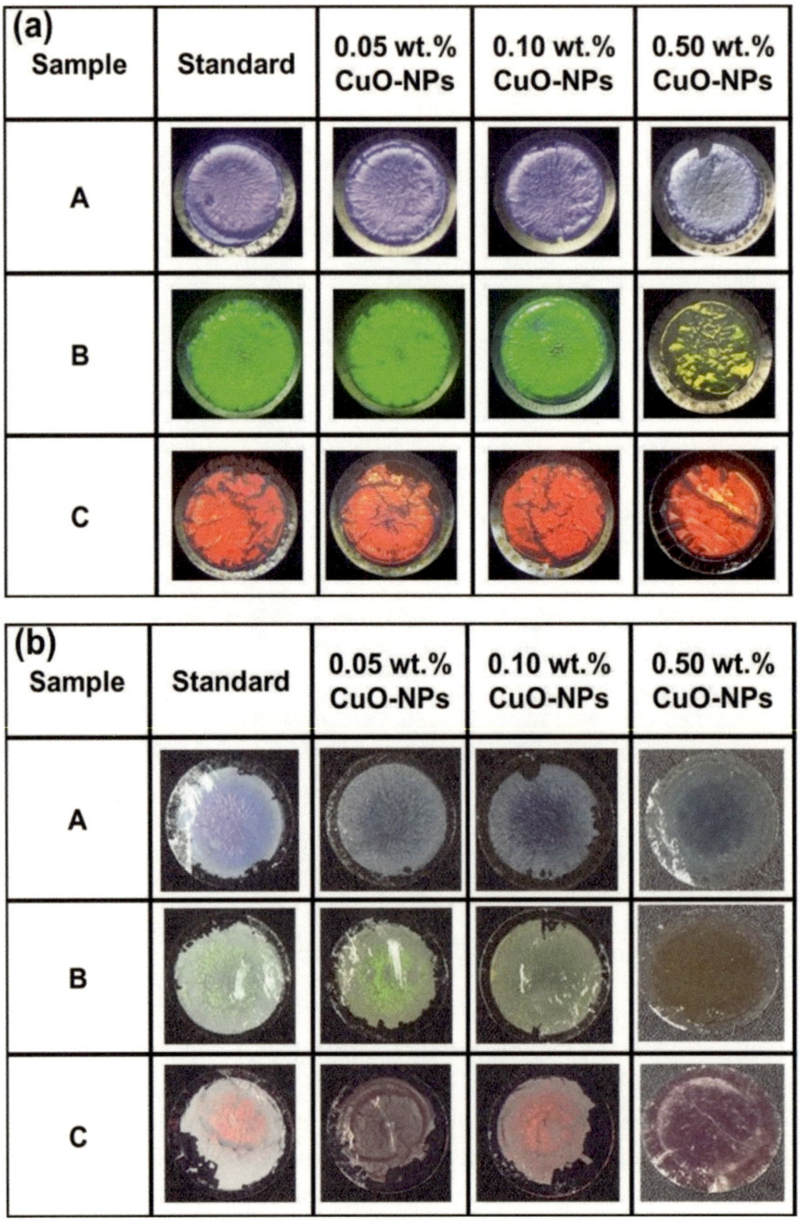

Figure 10. (a) Top surface and (b) back surface of CuO-NPs/PS hybrid CPhC color films (1) without CuO-NPs (standard), and (2) with 0.05 wt% and (3) 0.10 wt% and (4) with 0.50 wt% CuO-NPs.

3.6. Effects of Ag-NPs content on PS CPhC color films

Figure 11 shows the photographs of Ag-NPs/PS hybrid CPhC color films. Pure PS CPhC color films are generally milky white with faint structural colors (**Figure 11(a)**, standard). The visual appearance of hybrid CPhCs clearly changed from dull color to vivid iridescent color by doping Ag-NPs into PS CPhCs (**Figure 11(a)**). The pictures indicated a tunable color mechanism, which contributed to the absorbance of scattering light by doping Ag-NPs. After doping the Ag-NPs into the PS CPhCs, the visual appearance of hybrid CPhC color films changed evidently (**Figure 11(a)**). **Figure 11(b)** exhibits the back surface photographs of Ag-NPs/PS hybrid CPhC color films. The visual appearance of the hybrid CPhC color films changes quite clearly from milky white to deep green after doping Ag-NPs into the *fcc* structure (**Figure 11(b)**, sample E). Consequently, Ag-NPs may absorb scattered light and increase color saturation, producing bright structural colors of CPhCs. In addition, we observed the coffee ring of CPhC color films (**Figure 11(a)**).

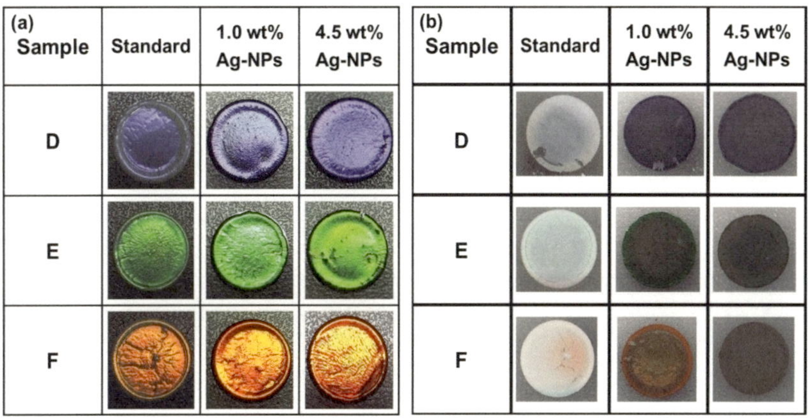

Figure 11. (a) Top surface and (b) back surface of AgNPs/PS hybrid CPhC color films (1) without AgNPs (standard) and with (2) 1.0 wt% and (3) 4.5 wt% AgNPs.

4. Discussion

The normal reflectance spectra of all samples were measured on a UV-Visible spectrometer equipped with a Xe lamp. Collimated broadband white light from a UV-enhanced Xe lamp was incident of normal surface of the sample, in which the Ag mirror as reference for reflectance measurement was the incident light of the Xe light source. The light spot size on the sample was approximately 4.0 mm. Fiber optic Y-cables were used for reflectance measurement. The theoretical reflection wavelengths of the Bragg diffraction law of metal NPs/PS hybrid CPhC color films were calculated by using the combined form of Bragg law and Snell law. According to the Bragg-Snell law Eq. (1), the relationship between the reflection peak wavelength (λ_R) and the incident angle (α) of an *fcc* structure is as follows:

$$\lambda_R = \sqrt{8/3} D_{PS} \sqrt{n_{eff}^2 - \sin^2 \alpha} \quad (1)$$

where D_{PS} is the diameter of PS nanospheres. To obtain n_{eff}, which is the effective refractive index of the medium. In this case, n_{NPs_air} was obtained using Eq. (2), as follows:

$$n_{eff} = [n_{PS}^2 f_{PS} + n_{NPs\_air}^2 (1 - f_{PS})]^{1/2} \quad (2)$$

where $n_{PS} = 1.59236$ is the refractive index of PS nanospheres and $f_{PS} = 0.74$ is the volume fraction of the PS nanosphere. Therefore, we can obtain n_{NPs_air}, which is the refractive index of NPs.

4.1. Tunable structural color of the CuO-NPs/PS CPhC color films

Figure 12(a)–(c) shows the reflectance spectra of PS CPhC color films with and without CuO-NPs. Pure PS CPhC color films exhibited the reflectance peaks at 452 (sample A), 547 (sample B), and 630 nm (sample C), corresponding to blue, green, and red, respectively. After introducing 0.50 wt% of CuO-NPs into the three kinds of PS nanospheres, the hybrid CPhC color films showed a low reflection intensity and broadband reflectance peaks at 459, 557, and 638 nm, corresponding to cyan blue, yellow, and deep red. The reflectance spectra of PS CPhCs

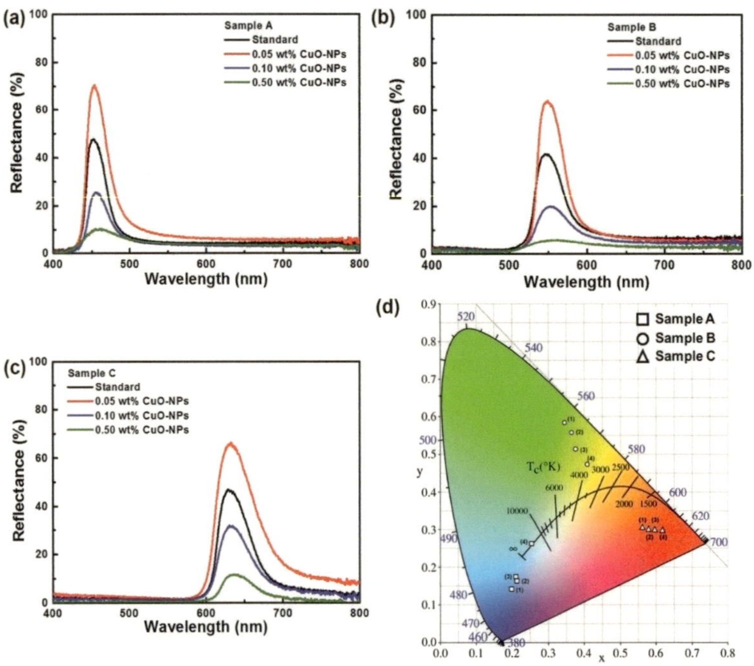

Figure 12. Reflection spectra of PS CPhC films with and without CuO-NPs: (a) sample A (□), (b) sample B (○), and (c) sample C (△). (d) CIE chromaticity diagram of CuO-NPs/PS hybrid CPhC color films for different samples (A–C) and CuO-NP contents.

containing various amounts of CuO-NPs showed different reflection peak positions, consistent with the Bragg law. The color of CuO-NPs/PS hybrid CPhC color films was calculated by using the corresponding chromaticity coordinates of the CIE 1931 standard colorimetric system. The visible light (380–700 nm) of the CIE 1931 chromaticity diagram visually calculated the changes in hybrid CPhC structural colors. The color of the CuO-NPs/PS hybrid CPhC color films was measured by irradiation on a D65 light source. Then, the reflection spectra obtained from CuO-NPs/PS hybrid CPhC color films were used to calculate the CIE 1931 chromaticity coordinates by using SpectraSuite software. The structural color of CPhCs was visually presented in the CIE 1931 chromaticity diagram, in which the color coordinates indicated the corresponding color. The PS nanospheres with a D_{PS} of 190 (sample A, □), 230 (sample B, O), and 265 nm (sample C, Δ), the corresponding structural colors were blue, green, and red, respectively (**Figure 12(d)**). Under the same D_{PS}, doping different CuO-NPs correspond to the same color hue but with different color purities. Three primary colors were obtained by using the CuO-NPs/PS hybrid CPhC color films, and iridescent colors were also achieved by modifying the D_{PS} of PS nanospheres (**Figure 12(d)**). Additionally, **Figure 12(d)** shows the larger color shift of sample A (□) and sample B (O) than that of sample C (Δ) because the reflection peak of sample C was considerably close to the cutoff wavelength (700 nm) of the CIE 1931 chromaticity diagram. The CuO-NPs/PS hybrid CPhC films demonstrated considerable potential to exhibit not only panchromatic colors but also holographic colors. Thus, CuO-NPs content and PS nanosphere size could influence the optical properties of CPhC color films.

4.2. Tunable structural color of the Ag-NPs/PS CPhC color films

Figure 13(a)–(c) shows the reflectance spectra of Ag-NPs/PS hybrid CPhC color films, in which the PS nanospheres with a D_{PS} were 170 (sample D), 215 (sample E), and 250 nm (sample F). Pure PS CPhC color films exhibited the reflectance peaks at 417 (sample D), 515 (sample E), and 601 nm (sample F), corresponding to blue, green, and red. After introducing 4.5 wt% of Ag-NPs into three kinds of PS CPhCs, the hybrid CPhC films showed the low reflection intensity and red-shift reflectance peaks at 421, 519, and 608 nm. The reflectance spectra of Ag-NPs/PS hybrid CPhC color films showed different reflection peak positions, consistent with the Bragg law. In addition, the PS CPhC films containing Ag-NPs exhibited weak reflectance because of the considerably high Ag-NP concentration (**Figure 13(a)–(c)**). The structural color was brilliant with the increase of Ag-NPs content, because the scattered light was strongly absorbed by the sediment on the bottom of Ag-NPs. Conversely, excessive Ag-NPs resulted in reduced reflection peak height and loss of brightness. Therefore, more photons could absorb in the PBG. The structural color of CPhC color films was effectively enhanced by introducing an appropriate amount of Ag-NPs. The effects of Ag-NPs content on the structural color of CPhC color films are summarized in **Figure 13(d)**.

The structural colors of Ag-NPs/PS hybrid CPhC films were calculated by using the corresponding chromaticity coordinates of the CIE 1931 standard colorimetric system. The visible light (380–700 nm) of the CIE 1931 chromaticity diagram visually calculated the changes in CPhC structural colors. Structural color of Ag-NPs/PS hybrid CPhC color films was measured by irradiating under a D65 light source. The reflection spectra of Ag-NPs/PS hybrid CPhC color films were used to determine CIE 1931 chromaticity coordinates. The structural color of hybrid CPhCs was visually presented in the CIE 1931 chromaticity diagram, in which the color coordinates indicated the corresponding structural color. At PS nanospheres with D_{PS} of 170 (sample D, ◇), 215 (sample E, ⬠), and 250 nm (sample F, ⬡), the structural colors were blue, green, and red, respectively (**Figure 13(d)**). Under the same D_{PS}, introducing different Ag-NPs corresponds to the same color hue but with different purities. The three primary colors were obtained by using the AgNPs/PS hybrid CPhC films, and iridescent colors were obtained by altering the D_{PS} (**Figure 13(d)**). Additionally, **Figure 13(d)** exhibits slight color shifts for samples D (◇), E (⬠), and F (⬡). The Ag-NPs/PS hybrid CPhC color films demonstrated a considerable potential to show not only panchromatic colors but also holographic colors. Thus, PS nanosphere size and Ag-NP content could influence the optical properties of CPhC color films.

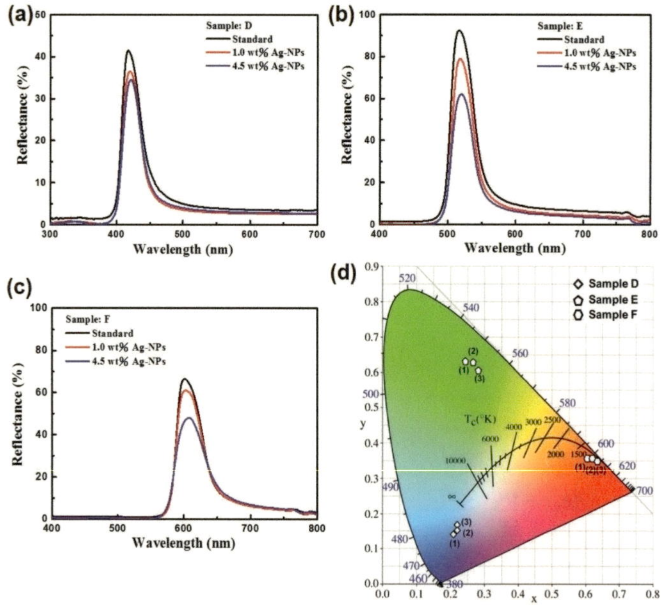

Figure 13. Reflection spectra of PS CPhC films with and without Ag-NPs: (a) sample D (◇), (b) sample E (⬠), and (c) sample F (⬡). (d) CIE chromaticity diagram of Ag-NPs/PS hybrid CPhC color films for different samples (D–F) and Ag-NP contents.

5. Conclusion

In this chapter, CuO-NPs and Ag-NPs doped into PS CPhC structures could absorb all wavelengths, remarkably increasing the color and producing brilliant tunable structural colors that are visible under natural lighting conditions. Considering that backscattered light was absorbed by the embedded metal NPs, the visual appearance of colloidal crystal coatings changed markedly from faint milky white to brilliant colors. The experimental results confirmed that introduction of low CuO-NP concentrations and high Ag-NP concentrations did not affect the structural quality of the PS CPhC films. In addition, the NPs/PS hybrid CPhC color films were measured and predicted by UV-Visible spectroscopy, FESEM, and EDS. The measured PBGs of the CPhCs were consistent with the theoretical calculation. The proposed novel method allows acquisition of tunable structural color for future applications in textile fabrics, bionic colors, catalysis, and paints.

Acknowledgements

This work is supported by the Ministry of Science and Technology (MOST) in Taiwan, under contract numbers MOST103-2221-E-035-029, and MOST105-2221-E-035-041. The authors appreciate the Precision Instrument Support Center of Feng Chia University for providing the fabrication and measurement facilities.

Author details

Chun-Feng Lai

Address all correspondence to: chunflai@fcu.edu.tw

Department of Photonics, Feng Chia University, Seatwen, Taichung, Taiwan

References

[1] M. Srinivasarao. Nano-optics in the biological world: beetles, butterflies, birds, and moths. Chem. Rev. 1999;99(7):1935–1961. DOI: 10.1021/cr970080y.

[2] P. Vukusic, and J. R. Sambles. Photonic structures in biology. Nature. 2003;424:852–855. DOI: 10.1038/nature01941

[3] R. A. Potyrailo, H. Ghiradella, A. Vertiatchikh, K. Dovidenko, J. R. Cournoyer, and E. Olson. Morpho butterfly wing scales demonstrate highly selective vapour response. Nat. Photon. 2007;1:123–128. DOI: 10.1038/nphoton.2007.2

[4] J. D. Forster, H. Noh, S. F. Liew, V. Saranathan, C. F. Schreck, L. Yang, J. G. Park, R. O. Prum, S. G. J. Mochrie, C. S. O'Hern, H. Cao, and E. R. Dufresne. Biomimetic isotropic nanostructures for structural coloration. Adv. Mater. 2010;22:2939–2944. DOI: 10.1002/adma.200903693

[5] Y. Takeoka. Angle-independent structural coloured amorphous arrays. J. Mater. Chem. 2012;22:23299–23309. DOI: 10.1039/C2JM33643J

[6] H. Noh, S. F. Liew, V. Saranathan, S. G. J. Mochrie, R. O. Prum, E. R. Dufresne, and H. Cao. How noniridescent colors are generated by quasi-ordered structures of bird feathers. Adv. Mater. 2010;22:2871–2880. DOI: 10.1002/adma.200903699

[7] C. F. Lai, Y. C. Wang, and H. C. Hsu. High transparency in the structural color resin films through quasi-amorphous arrays of colloidal silica nanospheres. J. Mater. Chem. C. 2016;4:398–406. DOI: 10.1039/C5TC03063C

[8] H. Cong, B. Yu, J. Tang, Z. Li, and X. Liu. Current status and future developments in preparation and application of colloidal crystals. Chem. Soc. Rev. 2013;42:7774–7800.

[9] A. Stein, B. E. Wilson, and S. G. Rudisill. Design and functionality of colloidal-crystal-templated materials-chemical applications of inverse opals. Chem. Soc. Rev. 2013;42:2763–2803.

[10] N. Vogel, M. Retsch, C. -A. Fustin, A. d. Campo, and U. Jonas. Advances in colloidal assembly: the design of structure and hierarchy in two and three dimensions. Chem. Rev. 2015;115(13):6265–6311. DOI: 10.1021/cr400081d

[11] C. I. Aguirre, E. Reguera, and A. Stein. Tunable colors in opals and inverse opal photonic crystals. Adv. Funct. Mater. 2010;20:2565–2578. DOI: 10.1002/adfm.201000143

[12] Y. Li, G. Duan, G. Liu, and W. Cai. Physical processes-aided periodic micro/nanostructured arrays by colloidal template technique: fabrication and applications. Chem. Soc. Rev. 2013;42:3614–3627.

[13] C. F. Lai, Y. C. Lee, and T. L. Tsai. Candlelight LEDs fabricated by using composite silica photonic crystals. Opt. Mater. Express. 2015;5(2):307–313. DOI: 10.1364/OME.5.000307

[14] C. F. Lai, Y. C. Lee, and C. T. Kuo. Saving phosphor by 150% and producing high color-rendering index candlelight LEDs containing composite photonic crystals. J. Light. Technol. 2014;32(10):1930–1935. DOI: 10.1109/JLT.2014.2309955

[15] S. M. Klein, V. N. Manoharan, D. J. Pine, and F. F. Lange. Preparation of monodisperse PMMA microspheres in nonpolar solvents by dispersion polymerization with a macromonomeric stabilizer. Colloid Polym. Sci. 2003;282(1):7–13. DOI: 10.1007/s00396-003-0915-0

[16] Y. Zhang, B. Dong, A. Chen, X. Liu, L. Shi, and J. Zi. Using cuttlefish ink as an additive to produce non-iridescent structural colors of high color visibility. Adv. Mater. 2015;27(32):4719–4724. DOI: 10.1002/adma.201501936

[17] B. Tang, Y. Xu, T. Lin, and S. Zhang. Polymer opal with brilliant structural color under natural light and white environment. J. Mater. Res. 2015;30(20):3134–3141. DOI: http://dx.doi.org/10.1557/jmr.2015.238

[18] F. Wang, Z. Gou, Y. Ge, and K. An. Preparation of tunable structural colour film by coating ps with titania on glass. IET Micro Nano Lett. 2016;11(1):50–53. DOI: 10.1049/mnl.2015.0264

[19] F. Wang, X. Zhang, L. Zhang, M. Cao, Y. Lin, and J. Zhu. Rapid fabrication of angle-independent structurally colored films with a superhydrophobic property. Dyes Pigment. 2016;130:202–208. DOI: 10.1016/j.dyepig.2016.03.022

[20] H. Cong, B. Yu, S. Wang, L. Qi, J. Wang, and Y. Ma. Preparation of iridescent colloidal crystal coatings with variable structural colors. Opt. Express. 2013;21(15):17831–17838. DOI: 10.1364/OE.21.017831

[21] C. I. Aguirre, E. Reguera, and A. Stein. Colloidal photonic crystal pigments with low angle dependence. ACS Appl. Mater. Interf. 2010;2(11):3257–3262. DOI: 10.1021/am100704f

[22] Z. Shen, L. Shi, B. You, L. Wu, and D. Zhao. Large-scale fabrication of three-dimensional ordered polymer films with strong structure colors and robust mechanical properties. J. Mater. Chem. 2012;22:8069–8075. DOI: 10.1039/C2JM30546A

[23] W. Wang, B. Tang, W. Ma, J. Zhang, B. Ju, and S. Zhang. Easy approach to assembling a biomimetic color film with tunable structural colors. J. Opt. Soc. Am. A. 2015;32(6): 1109–1117. DOI: 10.1364/JOSAA.32.001109

[24] O. L. J. Pursiainen, J. J. Baumberg, H. Winkler, B. Viel, P. Spahn, T. Ruhl. Nanoparticle-tuned structural color from polymer opals. Opt. Express. 2007;15(15):9553–9561. DOI: 10.1364/OE.15.009553

[25] Y. Yamada, M. Ishii, T. Nakamura, and K. Yano. Artificial black opal fabricated from nanoporous carbon spheres. Langmuir. 2010;26(12):10044–10049. DOI: 10.1021/la1001732

[26] F. Wang, X. Zhang, Y. Lin, L. Wang, and J. Zhu. Structural coloration pigments based on carbon modified ZnS@SiO_2 nanospheres with low-angle dependence, high color saturation, and enhanced stability. ACS Appl. Mater. Interf. 2016;8(7):5009–5016. DOI: 10.1021/acsami.5b11919

[27] F. Wang, X. Zhang, Y. Lin, L. Wang, Y. Qin, and J. Zhu. Fabrication and characterization of structurally colored pigments based on carbon-modified ZnS nanospheres. J. Mater. Chem. C. 2016;4:3321–3327. DOI: 10.1039/C5TC04192A

[28] M. O. A. Erola, A. Philip, T. Ahmed, S. Suvanto, and T. T. Pakkanen. Fabrication of Au- and Ag-SiO_2 inverse opals having both localized surface plasmon resonance and Bragg diffraction. J. Solid State Chem. 2015;230:209–217. DOI: 10.1016/j.jssc.2015.06.036

[29] B. Yu, F. Zhai, H. Cong, and D. Yang. Photosensitive polystyrene/silver bromide hybrid colloidal crystals as recoverable colorimetric naked eye probes for bromine gas sensing. J. Mater. Chem. C. 2016;4:1386–1391. DOI: 10.1039/C5TC02616D

[30] C. F. Lai, Y. C. Wang, C. L. Wu, J. Y. Zeng, and C. F. Lin. Preparation of a colloidal photonic crystal containing CuO nanoparticles with tunable structural colors. RSC Adv. 2015;5:105200–105205. DOI: 10.1039/C5RA21035F

[31] Z. Mu, X. Zhao, Y. Huang, M. Lu, and Z. Gu. Photonic crystal hydrogel enhanced plasmonic staining for multiplexed protein analysis. Small. 2015;11:6036–6043. DOI: 10.1002/smll.201501829

[32] X. Zhao, J. Xue, Z. Mu, Y. Huang, M. Lu, and Z. Gu. Gold nanoparticle incorporated inverse opal photonic crystal capillaries for optofluidic surface enhanced Raman spectroscopy. Biosens. Bioelectron. 2015;72:268–274. DOI: 10.1016/j.bios.2015.05.036

[33] C. F. Lai, and Y. C. Wang. Colloidal photonic crystals containing silver nanoparticles with tunable structural colors. Crystals. 2016;6(5):61. DOI: 10.3390/cryst6050061

[34] J. P. Ruparelia, A. K. Chatterjee, S. P. Duttagupta, and S. Mukherji. Strain specificity in antimicrobial activity of silver and copper nanoparticles. Acta Biomaterialia. 2008;4(3): 707–716. DOI: 10.1016/j.actbio.2007.11.006

[35] I. Sondi, and B. S. Sondi. Silver nanoparticles as antimicrobial agent: a case study on *E. coli* as a model for Gram-negative bacteria. J. Colloid Inter. Sci. 2004;275(1):177–182. DOI: 10.1016/j.jcis.2004.02.012

[36] A. J. Vizcaino, A. Carrero, and J. A. Calles. Hydrogen production by ethanol steam reforming over Cu–Ni supported catalysts. Int. J. Hydrogen Energy. 2007;32(10–11): 1450–1461. DOI: 10.1016/j.ijhydene.2006.10.024

[37] S. Ashokkumar, S. Ravi, V. Kathiravan, and S. Velmurugan. Synthesis, characterization and catalytic activity of silver nanoparticles using Tribulus terrestris leaf extract. Spectrochim. Acta Mol. Biomol. Spectrosc. 2014;121:88–93. DOI: 10.1016/j.saa. 2013.10.073

[38] G. H. Chan, J. Zhao, E. M. Hicks, G. C. Schatz, and R. P. V. Duyne. Plasmonic properties of copper nanoparticles fabricated by nanosphere lithography. Nano Lett. 2007;7(7): 1947–1952. DOI: 10.1021/nl070648a

[39] A. Haider, and I. K. Kang. Preparation of silver nanoparticles and their industrial and biomedical applications: a comprehensive review. Adv. Mater. Sci. Engine. 2015;2015:165257, http://dx.doi.org/10.1155/2015/165257.

[40] J. G. O. Mendoza, A. P. Vivanco, C. T. Quitl, P. Z. Moran, D. V. Hernandez, and F. Chavez. Optical fiber sensor based on localized surface plasmon resonance using silver nanoparticles photodeposited on the optical fiber end. Sensors. 2014;14:18701–18710. DOI: 10.3390/s141018701

[41] M. Balkanski, M. A. Nusimovici, and J. Reydellet. First order Raman spectrum of Cu_2O. Solid State Commun. 1969;7(11):815–818. DOI: 10.1016/0038-1098(69)90768-6

[42] W. T, Yao, S. H. Yu, Y. Zhou, J. Jiang, Q. S. Wu, L. Zhang, and J. Jiang. Formation of uniform CuO nanorods by spontaneous aggregation: selective synthesis of CuO, Cu_2O, and Cu nanoparticles by a solid-liquid phase arc discharge process. J. Phys. Chem. B. 2005;109(29):14011–14016. DOI: 10.1021/jp0517605

[43] A. Henglein. Colloidal silver nanoparticles: photochemical preparation and interaction with O_2, CCl_4, and some metal ions. Chem. Mater. 1998;10(1):444–450. DOI: 10.1021/cm970613j

[44] S. Pal, Y. K. Tak, and J. M. Song. Does the antibacterial activity of silver nanoparticles depend on the shape of the nanoparticle? A study of the gram-negative bacterium *Escherichia coli*. Appl. Environ. Microbiol. 2007;73(6):1712–1720. DOI: 10.1128/AEM.02218-06

Chapter 5

Adsorption of Ions at the Interface of Clay Minerals and Aqueous Solutions

Zengqiang Jia, Qian Wang, Chang Zhu and Gang Yang

Additional information is available at the end of the chapter

http://dx.doi.org/10.5772/65529

Abstract

Adsorption of ions at the interface of clay minerals and aqueous solutions plays a critical role in a wide spectrum of colloidal, chemical, physical, and geological processes. Owing to the particular complexity of related systems and the femtosecond scale of related processes, the direct experimental observations often become a challenging task. As a contrast, computer simulations have proven to be a competent and powerful approach therein and already realized fruitful and significant contributions. In this chapter, we attempt to draw a relatively comprehensive picture of the interfacial adsorption of ions mainly within the context of computer simulations. As elaborated, quantum mechanics (QM) and molecular dynamics (MD), two popular simulation techniques currently used, have respective advantages, and with their collaborative efforts, we are striding toward the in-depth and systematic understanding of adsorption configuration, distribution, stability, reaction thermodynamics and mechanism, dynamics, diffusivity as well as electric double layer and other fundamental issues that are closely associated with the adsorption of ions at the interface of clay minerals and aqueous solutions. In addition, we demonstrate that investigation of the interfacial adsorption of ions greatly helps to unravel the origin and mechanism of ion-specific effects, whose importance has been explicitly suggested to be no less than Gregor Mendel's work to genetics.

Keywords: interfacial adsorption, clay mineral, molecular dynamics, quantum mechanics, ion-specific effects

1. Introduction

Adsorption of ions at the interface of clay minerals and aqueous solutions is ubiquitous and plays a pivotal role in a wide spectrum of colloidal, chemical, physical, and geological processes, such as the transport and bioavailability of ions, nutrients and contaminants [1–3]. Owing to the large surface area, low permeability and high retention capability, clay minerals show outstanding adsorption performances for metal ions, organic matters and other substances that further control the transport and bioavailability of metal ions, nutrients, and contaminants; on the other hand, the surface properties of clay minerals are affected significantly by these adsorbents that may result in the aggregation, dissociation of clay minerals, and/or chemical reactions at the interface of clay minerals and aqueous solutions [4–7].

To date, it remains challenging for experimental techniques to observe the interfacial processes of clay minerals and aqueous solutions because of the complexity of related systems (involving aqueous solutions, ions, counterions, and clay minerals that may have the various surface structures and even mixtures) and the promptness of related processes (generally on the femtosecond scale). Computational simulations, which can provide the otherwise inaccessible details such as adsorption configuration, distribution, stability, reaction thermodynamics and mechanism, dynamics, diffusivity as well as other fundamental issues such as electric double layer began to emerge as a powerful research tool to complement the experimental results. In the recent few decades, we have witnessed the unprecedented advances of computer simulations within this context. Quantum mechanical (QM) methods are highly precise and it proves that the MP2 or MP2:DFT schemes can reproduce the reaction barriers with near chemical accuracy [8, 9]. Although QM methods are qualified for predicting reaction mechanism, thermodynamics, spectroscopic data, and other information, they are computationally costly and thereby restricted mainly for the treatment of relatively small systems. To overcome this disadvantage, Morokuma and collaborators [10] developed an ONIOM scheme that divides the systems into several layers. For the ONIOM scheme, each layer can freely choose the desired functional and basis set, and in many cases, molecular mechanics (MM) is employed for the low-level region, which is also referred to as the so-called QM/MM methodology. In this way, systems up to thousands of atoms become within the power of computer simulations. Classical molecular dynamics (MD) based on Newton's second law or the equation of motion ($F = m \times a$, where "F," "m," and "a" refer to the exerted force, mass, and acceleration, respectively) is ready to describe 10,000 atoms that approach the real systems; in the meanwhile, dynamic information on the femtosecond scale such as structural evolution, interaction, and diffusion can be monitored *in situ*. One of the representative achievements of MD simulations is toward the understanding of protein folding [11]. For complicated systems, computer simulations can consider the various influencing factors one by one (e.g., pH, substituent effect, temperature, pressure, water content, etc.), which often seem impossible to handle experimentally; in addition, QM and MD methods have their respective merits: QM is accurate, while MD provides dynamic information. *Ab initio* molecular dynamics (AIMD) has revolutionized the field of computer simulations by unifying molecular dynamics and electronic structure theories, allowing the colloidal, chemical, physical, or/and geological processes to be *in situ* monitored in an accurate and unbiased way. That is, AIMD integrates the merits of QM and MD methods by providing accurate and dynamic information within one simulation task.

Recently, these simulation techniques have been used in our group to tackle the adsorption behavior and mechanism of ions onto clay minerals, in the absence or the presence of aqueous solutions [12–16]. According to the QM-calculated results [12–14], metal ions such as K^+ and Ca^{2+} construct strong ionic bonds with the oxygen (O) atoms of hexagonal rings at the tetrahedral SiO_4 sheet of clay minerals, whereas OH^- deprotonates the clay minerals by depriving protons from the bridging O atoms and thus enhance the densities of negative charges of clay minerals. The presence of water molecules displaces metal ions from the center of the hexagonal rings [15]. Our MD simulations [15, 16] indicate that whether for metal ions or anions, adsorbed species can be classified into inner and outer-sphere modes: inner-sphere ions form direct interactions with the surfaces of clay minerals, while outer-sphere ones are separated by only one intermediate water molecule. In this chapter, we are aiming to draw a relatively comprehensive picture of the adsorption of ions at the interface of clay minerals and aqueous solutions, and hence, the discussions will not be limited to the computational results obtained from our group. Related results from other groups obtained thus far can also be included, and here, examples are given: Greathouse and Cygan [17] used MD simulations and found that $[UO_2(H2O)_5]^{2+}$ should be the major species for UO_2^{2+} adsorption onto the clay minerals of 2:1 layer type such as beidellite and montmorillonite and pyrophyllite. Yang et al. [18] continued the studies and demonstrated that the U^{6+} center and the O atoms from the hydration shell are the competitive binding sites during the interaction of $[UO_2(H_2O)_5]^{2+}$ with montmorillonite, and the adsorbed species using the U^{6+} center and coupled-water O atoms as the binding sites correspond, respectively, to the inner and outer-sphere modes.

In consequence, this chapter focuses on the studies of adsorption of metal ions and anions at the interface of clay minerals and salt solutions, with inclusion of both QM and MD results. To broaden the horizon, the underlying ion-specific effects and the removal of heavy metal ions and anions are discussed as well. As aforementioned, the adsorption of metal ions onto clay minerals can exert a positive influence for a wide range of colloidal, chemical, physical, and geological processes [1–3]; on the other hand, the adsorption of heavy metal ions and certain anions is known to cause contamination to soils that further results in the serious environmental pollutions and threatens the health of human beings. Biochar, produced by the pyrolysis of biomass, has proven to be effective to improve soil properties, remediate soils with heave metal and organic pollutants and increase crop biomass [19]. With use of density functional calculations, we have demonstrated that biochar is potential to remediate soils with anionic pollutants [20], in addition to heavy metal ions that are known to us all.

The scientific status of ion-specific effects has been well established, and Kunz et al. [21] explicitly suggested that the importance of ion-specific effects should be no less than Gregor Mendel's work to genetics. As a matter of fact, ion-specific effects occur ubiquitously and play a vital role during a wide range of interfacial processes of clay minerals and salt solutions, and a number of these processes can be associated closely with the adsorption processes. We have clearly demonstrated the presence of Hofmeister series for both metal ions [12, 22–24] and anions [14, 16] during the adsorption of ions onto clay minerals and the resulting aggregation of clay minerals; in addition, mechanism of ion-specific effects was explored and for metal ions, it can be ascribed to the difference of polarization effects responding to the electric field of colloidal particles [12].

2. QM calculations

Isomorphous substitutions occur in almost all types of clay minerals, especially 2:1 layer type such as mica, montmorillonite, and pyrophyllite. A plethora of negative charges are thus created that determine the adsorption properties for ions. First principles density functional theory (DFT) calculations of pyrophyllite and montmorillonite have been carried out by Vamsee et al. [25], with consideration of octahedral Mg^{2+}/Al^{3+} substitutions. The octahedral Mg^{2+}/Al^{3+} substitutions lead to the enrichment of negatively charges at the interlayer, which then strongly bind counterbalancing ions such as Li^+, Na^+, K^+, Mg^{2+}, and Ca^{2+}. For montmorillonite, it shows that the interlayer distance varies in a nearly direct proportion to the radius of intercalated counterions (**Figure 1**). Compaction of the interlayer space triggers the Li^+ ions to migrate into the tetrahedral sheet and be closely coordinated with the basal O atoms of the tetrahedral sheet. The electron densities on these basal O atoms are increased mainly as a result of isomorphous substitutions at the octahedral sheet instead of associated counterions. The authors also point out that inclusion of dispersion corrections is necessary in order to achieve satisfying agreements with the experimental results with respect to lattice parameters and bulk modulus.

Shi et al. [26] study the adsorption of Na^+, NH_4^+ and protonated formamide (FAH) ions onto the dry and hydrated montmorillonite surfaces at various interlayer distances. Complex interactions that vary with the identity of counterions, volume, and interlayer space are detected among the counterions, lower and upper montmorillonite surfaces, such as ionic bonding, H-bonding, electrostatic attraction and repulsion as well as vdW forces. Na^+ ions form direct ionic bonds with the surfaces of montmorillonite by situating at approximately the center of six O atoms of hexagonal rings with the shortest Na-O distances of ca. 2.46 Å, whereas NH_4^+ and FAH ions construct strong H bonds with the surfaces of montmorillonite, with the shortest H-bond distance being equal to 2.31 and 2.13 Å, respectively. Different from Na^+ ions that have been trapped by one of montmorillonite surfaces, NH_4^+ and FAH ions construct strong interactions with both montmorillonite surfaces through forceful H bonds. The expansion energy potentials near the equilibrium interlayer distances are described by the nearly quadratic functions, and at larger interlayer distances, the attraction interactions among the counterions, lower and upper surfaces turn to be predominant and the energy potentials approach the constant values.

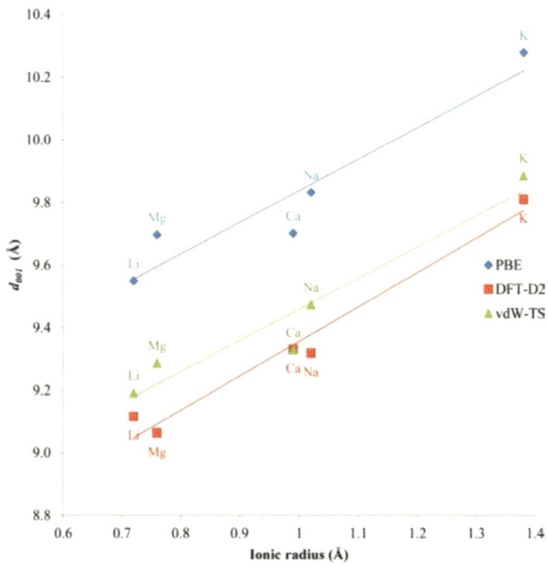

Figure 1. The interlayer distance (d_{001}) versus the ionic radius of counterions for montmorillonite with octahedral Mg^{2+}/Al^{3+} substitutions.

Isomorphous substitutions and resulting effects on the adsorption of ions have been studied by several different groups [27–32]. When montmorillonite is acid treated, the adsorption of protons (H^+) takes place at the interlayer, and these protons act as counterions and can be exchangeable. Owing to the acidic treatment, the lattice Al atoms of montmorillonite undergo the rapid autotransformation and migrate to the ion-exchanged sites at the surfaces of clay minerals, at a comparable rate for lattice Al atoms at the octahedral and tetrahedral sheets [27]. At the same time, the cation-exchange capacity of montmorillonite declines significantly. The effects of isomorphous substitutions on the Brønsted acid sites are demonstrated by Claudia and Esther [28]. It shows that isomorphous substitutions of the octahedral Mg^{2+}/Al^{3+} sites result in stronger acidities than those of tetrahedral Al^{3+}/Si^{4+} sites, which further lead to the more effective protonation of probe molecules such as methylamine, ammonia, and pyridine and acetonitrile in the former case. The interaction strengths of probe molecules abide by the same sequence for both octahedral Mg^{2+}/Al^{3+} and tetrahedral Al^{3+}/Si^{4+} substitutions, as methylamine > ammonia > pyridine > acetonitrile, which seems to be not directly related to the proton affinities of probe molecules. The Al^{3+}/Si^{4+} substitutions lead to substantial alterations to the atomic arrangements along the c-axis direct and charge distribution at montmorillonite surfaces, and counterions interact with the internal surfaces by forming a specific configuration associated with the six O atoms of hexagonal rings at the tetrahedral SiO_4 surface [29]. Counterions with smaller radii or/and higher charges are likely to lead to strong binding with the internal surfaces of clay minerals. This is apparent because metal ions with smaller size are more apt to enter into the hexagonal rings and/or metal ions with higher charges are beneficial for stronger electrostatic interactions. The Mg^{2+}/Al^{3+} substitutions cause the vicinities to be the most negative charged and thus can qualitatively predict the most possible sites for the adsorption of counterions [30]. The adsorption energetics of metal ions on the layer surfaces are found to be linearly correlated with the inverse cation-substitution distances (**Figure 2**), further substantializing the negative point-like character of Mg^{2+}/Al^{3+} substitutions. The change of different adsorption sites causes remarkable differences in the adsorption energetics, up to 126 kJ/mol. Adsorption of Na^+ ions on montmorillonite surfaces is somewhat preferential when Na^+ ions are located in proximity to the Mg^{2+}/Al^{3+} rather than Al^{3+}/Si^{4+} substitution sites, and the energy difference reaches approximately 15 kJ/mol [31]. Three different types of isomorphous substitutions (tetrahedral, octahedral and both) have been calculated by Mignon et al. [32], and it clearly indicates the distinct adsorption behaviors for the various metal ions: K^+ ions remain closely bound to the surfaces of each type of isomorphous substitution (i.e., inner-sphere) and this is consistent with the role as a good swelling inhibitor; Li^+ ions are preferential for hydration and coordinates to four water molecules in all types of substitutions (i.e., outer-sphere), except the emergence of inner-sphere species in the case of isomorphous substitutions at both tetrahedral and octahedral sites; Na^+ ions display an intermediate adsorption behavior, bound to the surface in the case of tetrahedral substitution (i.e., inner-sphere) while hydrated in the case of octahedral substitution (i.e., outer-sphere). It is due to that the binding energies of alkali ions with water molecules decrease in the order $Li^+ > Na^+ > K^+$ while with the surfaces of clay minerals change in the reverse order ($K^+ > Na^+ > Li^+$).

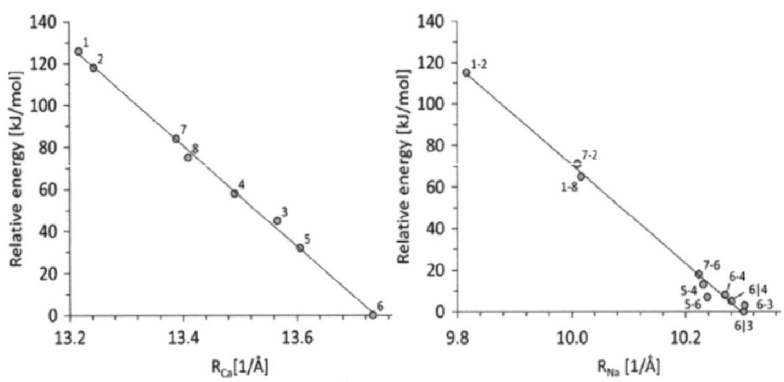

Figure 2. Correlation of the relative adsorption energetics for Ca^{2+} (up) and Na^+ (bottom) ions with the sum of inverse cation-substitution distances (R_M, M = Ca, Na). Details for the R_M calculations can be found in the supplementary information of Ref. [30].

With isomorphous substitutions, counterions are strongly adsorbed onto the surface of clay minerals, and the hydration of counterions occurs facilely in humid environments, carrying the water molecules into the interlayer space and causing the swelling or collapse of the lamina structures [33–42]. It is apparently seen that the interactions among counterions, water molecules, and surfaces in the adsorption processes of metal ions are of great scientific importance. Periodic DFT calculations have been performed to investigate the hydration of Ca^{2+} ions in the interlayer space of montmorillonite [43]. The hydration of metal ions is significantly affected by the interior surfaces of clay minerals that possess a plethora of negative charges. The hydration structures of Ca^{2+} ions in the interlayer space are distinct from those in gas phase, especially at higher water loadings. Nonetheless, at all water contents, Ca^{2+} ions are closely bound to one of the interior surfaces of montmorillonite and the hydrated Ca^{2+} ions in the interlayer space of montmorillonite have a tendency to construct a H-bond (HB) network that links the upper and lower surfaces, which is achieved as long as the number of water molecules (N) reaches three (N = 3). In gas phase, Ca^{2+} ions tend to form direct bonds with as many as water molecules until the first shell has been saturated (N ≥ 6).

The distribution and migration of actinide compounds in the environmental circumstances are within the context of risk assessment for long-time radioactive waste repositories. Alena et al. [44] perform the periodic DFT calculations on the adsorption of uranyl (UO_2^{2+}) on solvated (110) and (010) edge surfaces of pyrophyllite. The bidentate adsorption configurations on the various partially deprotonated sites of pyrophyllite (i.e., octahedral Al(O,OH), tetrahedral Si(O,OH), and mixed AlO–SiO). It shows that the preferred sites for adsorption of UO_2^{2+} are different for (110) and (010) edge surfaces, respectively, as the octahedral Al(O,OH) and mixed AlO–SiO sites. The authors of the same group [45] further show that structural parameters of uranyl adsorption complexes are essentially determined by the surface chemical groups that construct the local adsorption sites rather than type of clay minerals, and relatively slight effects on structural parameters are exerted even if substituted cations occur at the adsorption sites. Adsorption of uranyl can induce the hydrolysis of -AlOH and result in the formation of uranyl mono-hydroxide. The frayed edge sites formed in micaceous clays are indicated to have a crucial role with respect to the long-term stability of radioisotopes of Cs on the topsoil surfaces [46].

It is known to us that clay minerals have outstanding adsorption performances for metal ions, including heavy metal ions that may rigorously pollute the environmental circumstances and further the heath of human beings. On the other hand, anions such as Cr(VI)-oxo and phosphate are also widely spread in soil systems as metal ions that can result in environmental pollutions [47–49]. There are obviously fewer computational reports on the adsorption of anions on clay minerals in contrast to the extensive studies on metal ions. DFT calculations are performed by Zhu et al. [50] to investigate the adsorption of As(V)- and As(III)-oxo anions and related reactions on Mn(II) and Mn(IV) sites of birnessite. It shows that whether on Mn(II) or Mn(IV) site, the adsorption of As(V)- rather than As(III)-oxo complexes is more thermodynamically favorable and thereby the oxidation of As(III) is expected to be inhibited due to the blocking of adsorption sites by As(V)-oxo complexes. For kaolinite surface, the adsorption of $SiO(OH)_3^-$ takes place primarily on the Al-terminated (001) surfaces, similar to the situation of $Si(OH)_4$ adsorption. The various adsorption configurations of $Si(OH)_4$ and $SiO(OH)_3^-$ on the Si- and Al-terminated. Owing the more reactive nature of unbounded O atom, $SiO(OH)_3^-$ forms stronger electrostatic interactions than $Si(OH)_4$ and result in more stable adsorption configurations [51]. As a result of $SiO(OH)_3^-$ adsorption, the surfaces of kaolinite become more hydrophilic and carry more negative charges, which may significantly facilitate the dispersion of kaolinite particles in aqueous solutions.

Nowadays, biochar has been listed as one of the most concerned materials due to the high capacity for ion adsorption and facile recyclability, showing excellent performances to remove heavy metal ions from clay minerals and remediate soil ecosystems [52–55]. It is essential to acquire an in-depth mechanistic understanding of ion adsorption on biochar. However, biochar has no specific structures and it is suggested to use other carbon-based materials for prototypes. Graphene is a two-dimensional honeycomb structure composed by sp^2-hybridized carbon atoms; that is, it is similar to other carbon-based materials such as biochar in that all of them are polycyclic aromatic hydrocarbons, mainly built upon the structural unit of benzene ring. Accordingly, grapheme can be an idea model for comprehension of ion adsorption on carbon-based materials. Colherinhas et al. [56] conduct DFT calculations and indicate that the adsorption strength of metal ions on grapheme is closely associated with the radius and charge of metal ions. The adsorption configurations with the highest stability correspond to the hollow site (*H*-site) where metal ions are situated at the center of an aromatic ring, as a result of maximizing the cation–π interactions [57]. As indicated by Addition of water molecules may significantly weaken the binding between metal ions on grapheme [58]. Different from metal ions, halide ions are electron-rich and can be adsorbed on graphene that are generally used as electron donors? Zhu et al. [20] and Shi et al. [59] clearly show that halide ions are potentially adsorbed on graphene, where halide ions are electron donors while graphene turn to be electron acceptors. Zhu et al. [20] demonstrate the binding strengths of metal ions on pristine graphene are stronger than the halide ion when they are adsorbed; however, the edge-fluorination alters the adsorption priority of metal ions versus halide ions and the adsorption strengths of metal ions/anions change in direct/reverse proportion with increase in functionalization degrees. In consequence, the preferential adsorption and selective removal of certain metal ions or anions can be facilely realized by choice of an appropriate functionalization degree. It is suggested by us that the binding strengths of hydrated ions should be calculated as follows:

$$E_b = E_{GN-ion-nW} - (E_{GN} + E_{ion-nW}) \tag{1}$$

where E_{GN}, $E_{ion\text{-}nW}$, and $E_{GN\text{-}ion\text{-}nW}$ are the energies of graphene, hydrated ions, and hydrated ions interacting with graphene, respectively, and n refers to the number of water molecules.

That is, the hydrated ions are treated as a whole instead of separately. Because of the different adsorption configurations of hydrated halide ions and metal ions that water molecules around form H-bonds with halide ions and at the same time interact with graphene through their O atoms, while metal ions construct direct ionic bonds with the O atoms of water molecules, and hence, only the H atoms of water molecules can be available to interact with graphene. It shows that in the case of edge fluorination, the binding energies of halide ions remain substantial while those of metal ions decline apparently, and the change trends are opposite in the case of pristine graphene.

3. MD simulations

As claimed by MD simulations, the adsorption of ions (whether metal ions or anions) onto the basal surfaces of clay minerals can proceed via two potential modes, respectively as the inner-sphere species that form direct interactions with the surfaces of clay minerals and the outer-sphere species that are separated from the surfaces by only one intermediate water molecule [15, 16, 60–72]. With regard to the inner-sphere metal ions, direct bonds are constructed with the surface of clay minerals, while forceful and complicated H-bond networks are detected in the case of anions such as halide ions (X^-) and hydroxide ion (OH^-) [16]. The adsorption behaviors and performances of metal ions are largely associated with the identity of clay minerals because their surface properties may vary significantly. With regard to regular clay minerals (with no isomorphous substitutions or defect sites), a majority of metal ions are found to adsorb primarily as the outer-sphere surface complexes, and to the best of our knowledge, only the weakly solvated alkali metals (K^+, Rb^+, and Cs^+) are inclined to form mainly the inner-sphere surface complexes [73, 74].

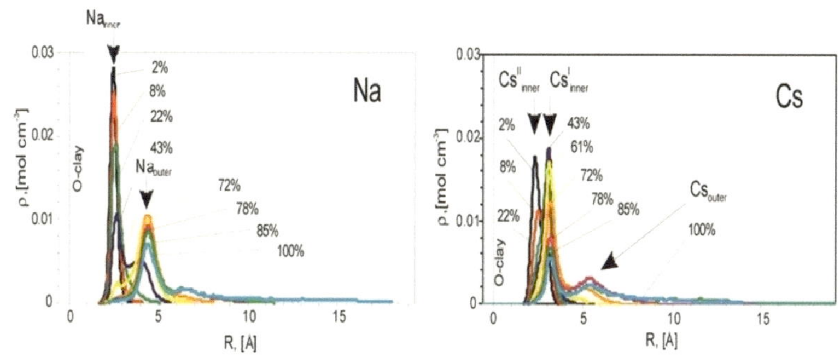

Figure 3. Na^+ and Cs^+ density distributions at the external surface of montmorillonite as the function of relative humidity (RH). Black arrows show position of inner-sphere and outer-sphere ion complexes [69].

The mechanism of metal ions adsorbed onto the montmorillonite surfaces has been extensively investigated by MD simulations. Churakov [69] studies the Na^+ and Cs^+ migration on the external surfaces of montmorillonite under partially saturated and saturated conditions and demonstrates that the surface complexation of Na^+ ions is driven substantially by the availability of water molecules. The density distributions of Na^+ and Cs^+ normal to the external surfaces are displayed in **Figure 3**. Under saturated conditions, Na^+ ions are adsorbed principally as the outer-sphere surface complexes, and when the relative humidity (RH) declines and falls within the range of 43–61%, the thickness of water film reduces from two to one layer, which causes Na^+ ions to approach the surfaces of montmorillonite and construct the so-called inner-sphere surface complexes; meanwhile, the coordination number of water molecules with Na^+ ions shows an obvious decrease and as a compensation, the surface O atoms become part of the Na^+ coordination shell. Rather than, Cs^+ ions display a distinct adsorption behavior: They form both outer and inner-sphere surface complexes under saturated conditions and are presented exclusively as the inner-sphere surface complexes when the RH decreases below 60%. The adsorption differences between Cs^+ and Na^+ ions are assumed to be due to their different ionic sizes that can result in the obviously different strengths of ion–water and ion–surface interactions. Marry et al. [64] compare the dynamic properties of Na^+ and Cs^+ ions in the monohydrated montmorillonite interlayers. The distribution of Li^+, Na^+, and K^+ ions on montmorillonite surfaces is also investigated by the combination of MD and Monte Carlo (MC) simulations [65–68], which are in agreement with the results of spectroscopic observations [70–80].

Hydration occurs readily for metal ions at the interfaces of clay minerals and aqueous solutions. Water molecules can become an essential portion of coordination for metal ions. Greathouse and Cygan [17] perform a systematic MD study and show that $[UO_2(H2O)_5]^{2+}$ should be the major species for uranyl adsorption onto the surface of clay minerals of 2:1 layer type. There is no adsorption near the surface of pyrophyllite, while the adsorption behaviors are similar on the surfaces of beidellite and montmorillonite. Ologomeric uranyl complexes are observed at dense solutions, particularly near the surfaces of pyrophyllite and montmorillonite instead of beidellite. Subsequently, Yang et al. [18] demonstrate that the U^{6+} center and the O atoms used to form the hydration shell are the two competitive binding sites during the interaction of $[UO_2(H_2O)_5]^{2+}$ with montmorillonite, and the adsorbed species using the U^{6+} center and coupled-water O atoms as the binding sites correspond, respectively to the inner and outer-sphere modes. In addition, the work of adhesion between the uranyl species and montmorillonite surfaces is evaluated and can reach 2.348 J/m^2, larger than the value between two montmorillonite sheets. Vasconcelos et al. [81] study the interactions between aqueous metal ions (Cs^+, Na^+, Pb^{2+}, and Cd^{2+}) and the basal surfaces of kaolinite, showing that Cs^+ ions can form strong inner-sphere complexes at the ditrigonal cavities of the tetrahedral SiO_4 surface (**Figure 4**). The outer-sphere Cs^+ ions are also present while highly mobile indicating the occurrence of weak adsorption. For Na^+ ions, only a small portion is adsorbed directly on the tetrahedral SiO_4 surface as inner-sphere complexes that are not clearly defined sites and very weak outer-sphere complexes predominate instead. The inner-sphere complexes of Pb^{2+} and Cd^{2+} do not occur on the tetrahedral SiO_4 surface, and only the outer-sphere mode is resulted from the adsorption of these two heavy metal ions; however, the corresponding outer-sphere surface complexes are apparently stronger than the outer-sphere Na^+ ions due to the apparently enhanced interactions of Pb^{2+} and Cd^{2+} versus Na^+ with surrounding water molecules. Sakuma et al. [82] investigate the distribution and dynamics of monovalent cations (Li^+, Na^+, K^+, Cs^+) on muscovite surfaces and show that hydrated Li^+ ions close to muscovite surfaces form two types of inner-sphere complexes and one type of outer-sphere complex. The two types of inner-sphere Li^+ complexes are respectively situated at the cavity and rim of the ditrigonal rings of muscovite. Na^+, K^+, Cs^+, and H_3O^+ form stable inner-sphere complexes on muscovite surfaces. Adsorbed Li^+, Na^+, K^+, and H_3O^+ species have resembling coordination conditions with water molecules, while Cs^+ ions have an obviously lower binding number with water molecules. Steele and collaborators [83] investigate the interactions of the (001) surface of muscovite with toxic divalent metal ions (Cu^{2+}, Zn^{2+}, and Cd^{2+}) and find that none of them

http://dx.doi.org/10.5772/65529

can form strong bonds with the regular surface of muscovite, which is similar to the scenarios of Pb^{2+} and Na^+ ions discussed earlier.

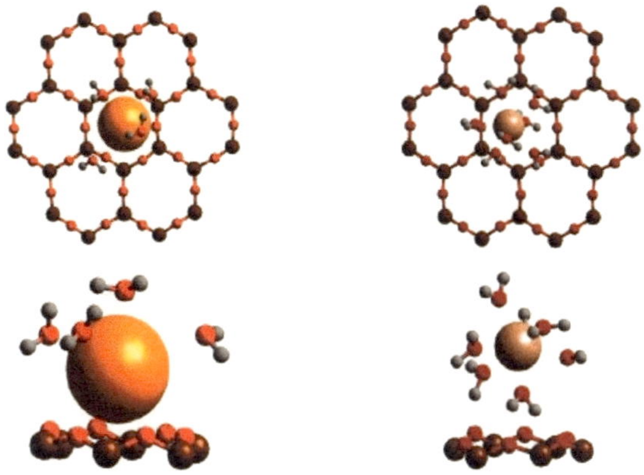

Figure 4. Representative Cs^+ and Cd^{2+} adsorption complexes [81].

MD simulations can also provide the dynamic and diffusion information for up to real systems that seems challenging to acquire by the structural optimization tasks using QM-based methods. As indicated by the time-evolution trajectories, the inner-sphere Cs^+ ions have little mobility and are kept rather stable at the hexagonal cavities of the tetrahedral SiO_4 surface of regular kaolinite, whereas the outer-sphere Cs^+ ions are liable to migrate on the tetrahedral SiO_4 surface. The trajectories of outer-sphere Na^+ ions are quite diffusive and resemble those of outer-sphere Cs^+ ions. Although more focused at the adsorption sites than the outer-sphere species, the inner-sphere Na^+ ions that are asymmetrical with respect to the hexagonal rings are not stable enough to maintain at the adsorption sites during the MD simulations, which is quite different from the case of inner-sphere Cs^+ ions. Such discrepancy should be caused by the radius difference between these two alkali ions that result in the remarkable variations of ion-surface and ion-water interactions. As aforementioned, Cd^{2+} and Pb^{2+} ions that fall above the hexagonal cavities of the tetrahedral SiO_4 surface are exclusively outer-sphere adsorbed while their stabilities are apparently higher than those of other outer-sphere ions.

The diffusional dynamics of interlayer species such as Cs+ ions and water molecules increases within the content of interlayer water molecules and with the distance from clay surfaces [84]. The mobility of ions in the water films on the external surfaces of clay minerals is similar to that at the surface of a water-saturated pore, as long as the thickness of water film is more than two layers. At the lower water pressure (i.e., thinner water films), the mobility of metal ions shows a dramatic decrease, and this is associated with the change in the surface complexation behaviors [85–89]. Zhang et al. [85] investigate the mobility of Na^+ and Ca^{2+} ions in the interlayer of montmorillonite, showing that the self-diffusion coefficient of

Ca^{2+} is lower than that of Na^+ due to the much more stable and larger hydration shell for Ca^{2+} versus Na^+ ions. In Na montmorillonite with high water contents, the interlayer Na^+ ions can displace in a relatively free way, while when Na^+ and Ca^{2+} ions coexist in the interlayer of montmorillonite, the moving regions of Na^+ ions have apparently been restricted by the hydration shell of Ca^{2+} ions as well as the interlayer spatial confinement, and as a result, the mobility of Na^+ ions in (Na, Ca)-montmorillonite is obviously inferior to that of Na-montmorillonite.

A number of factors can exert potential influences on the adsorption of metal ions onto clay minerals, such as isomorphous substitutions, temperature, water pressure, and counterions. It seems almost impractical for experimental operations to adjust one factor and at the same time keep all the others constant, while molecular simulations can easily realize such tasks. Below, we will analyze these factors one by one.

(1) Isomorphous substitutions, which are a general phenomenon for clay minerals of 2:1 layer type such as mica and montmorillonite [81, 84, 86–90]. Ngouana et al. [84] investigate the effects of substitutional disorder on the mobility of aqueous solvated species within the montmorillonite interlayers using the differently distributed Al^{3+}/Si^{4+} and Mg^{2+}/Al^{3+} substitutions, concluding that the specific localization of isomorphous substitutions has only a minor effect on the thermodynamic, structural, and transport properties. Liu et al. [91] use three montmorillonite models with different layer charges in the octahedral and tetrahedral sheets to study the thermodynamics and structural properties of Cs smectites, with use of the advanced CLAYFF force field [92]. All these smectites are found to swell in a similar way, with production of consistent swelling plateaus and energetic profiles. This indicates that the layer-charge distribution affects very slightly on the swelling behaviors of Cs smectites. Both ionic species and water molecules within the interlayer exhibit pronounced confining effects and correspond to obviously slower diffusivities than the corresponding bulk species, and smectite with the highest octahedral charges results in the lowest mobility of Cs^+ ions confined in the interlayer pores, whether at high or at low water contents. The layer charges can significantly affect the distribution of Cs^+ ions in the interlayer pores: inner-sphere Cs^+ ions in the interlayer pores of smectite can be adsorbed on the hexagonal (H-site) and triangular (T-site) sites, with the H-site being energetically preferential, (**Figure 5**). The relative stabilities of different adsorption sites may show variations, and the preference sequence is as tetrahedrally substituted H-site > nonsubstituted H-site > tetrahedrally substituted T-site > nonsubstituted T-site, while no conspicuous adsorption preference is detected for the various octahedral substitutions. MC is combined with MD demonstrating that the layer-charge location has a pronounced effect on the interlayer structure [93]. The full hydration of Ca^{2+} and Mg^{2+} ions is accompanied with the interlayer expansion into the 2W state. All layer charges in beidellite are created in the tetrahedral sheet, and metal ions remain adsorbed to the surfaces as the inner-sphere complexes at much higher water contents than in the case of montmorillonite, which may be caused by strong ion-surface interactions. For montmorillonite, the layer charges are exclusively assigned to the octahedral sheet, and all metal ions have been completely hydrated as the outer-sphere complexes even at low water content (2W state).

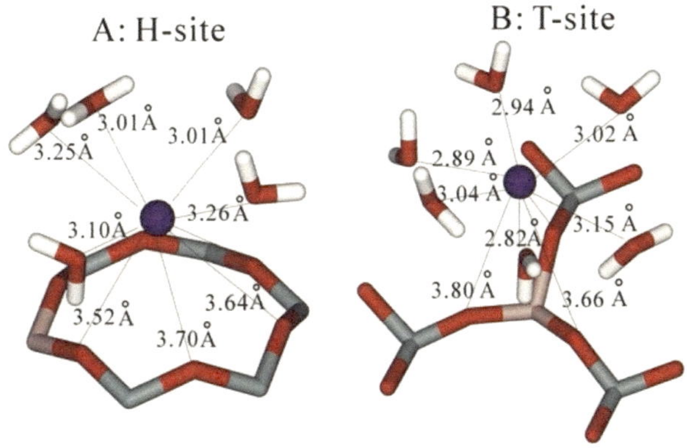

Figure 5. Two predominant coordination environments for the inner-sphere Cs^+ ions confined in the interlayer pores of smectite: (A) hexagonal site (H-site) and (B) triangular site (T-site) [91].

(2) Temperature: Malikova et al. [94] study the effect of temperature fluctuations (0~150°C) on the adsorption of metal ions in the interlayer of montmorillonite. The diffusion coefficients for Na^+ and Cs^+ ions increase by an order of magnitude over the covered temperature range, and the clear differences at these temperatures indicate that Na^+ and Cs^+ ions have different modes of diffusion in the monohydrated montmorillonite: at both low and high temperatures, Cs^+ ions exhibit a site-to-site diffusion mode that allows the coordination to six surface O atoms of the tetrahedral SiO_4 surface of clay minerals, while no definitely preferential sites are present for the adsorption of Na^+ ions, even at low temperatures. Water phase behaves similarly in Na- and Cs-montmorillonites, and a rapid transition from the surface to bulk dynamics has been detected in the case of Na-montmorillonite. Zheng and Zaoui [95] further state that the size and mass of hydrated metal ions play a crucial role on their diffusion behaviors. The hydration number of metal ions shows a gradual decrease with the increase of temperatures, and at relatively high temperatures, more water molecules from the hydrates of metal ions are released that elevates the amount of free water molecules into aqueous solutions and enhances the mobility of metal ions to a certain degree.

(3) Defect sites. Defect sites are ubiquitous in all types of clay minerals and play a vital role during the adsorption and reaction processes [15, 83, 96–98]. Li et al. [15] perform a systematic investigation on the adsorption behavior, stability, and mechanism of metal ions on the defective clay minerals, showing that the adsorption quantities and stabilities of metal ions on kaolinite can be fundamentally promoted by the occurrence of defect sites. In the case of regular kaolinite, Na^+ ions are adsorbed mainly as the outer-sphere species and only at dense solutions, the inner-sphere Na^+ ions can be detected sporadically. The presence of defect sites reverses the adsorption preference of outer versus inner-sphere Na^+ ions and causes the inner-sphere mode to be predominant for defective models with the various sizes. Pb^{2+} ions on regular kaolinite exist exclusively as the outer-sphere species, and the presence of defect sites causes the emergence of a small quantity of the inner-sphere species and the outer-sphere Pb^{2+} ions govern the adsorption processes consistently. As indicated in **Figure 6**, the adsorption configurations

of metal ions are significantly altered by the presence of defect sites, while the change of defect sizes or contents causes very limited influences. In the case of $\mathbf{Si_4}$ (a large defect site with 10 lattice Si atoms having been leached), the inner-sphere Pb^{2+} ions are nonexistent because no direct interactions with kaolinite surfaces can be detected, while interestingly, a very small number of Pb^{2+} ions have distances from kaolinite surfaces that can be ascribed to the inner-sphere species. After structural analyses, we show that these Pb^{2+} ions are associated with the water molecules that enter into the cavities of $\mathbf{Si_4}$ and accordingly are nominated as quasi inner-sphere species. Defect sites improve the numbers and stabilities of both inner and outer-sphere adsorption, and the inner-sphere Pb^{2+} ions occur only at defect sites that reinforce the ion-surface interactions, which also evidences the vital role of defect sites during the adsorption processes. The MD simulated results satisfactorily negotiate the experimental "conflicts" that some consider only the weak adsorption of metal ions on kaolinite surfaces, while others insist the presence of strong adsorption (i.e., inner-sphere species)—actually, structural differences such as the presence of absence of defect sites can cause the inconsistency in different experimental observations. Steele et al. [83] study the interactions of Cu^{2+}, Zn^{2+}, and Cd^{2+} with the regular surface of muscovite and the surface containing defect sites. The results indicate that no strong bonds have formed between regular surface and any of the three heavy metal ions. However, Cu^{2+} and Zn^{2+} ions can be strongly attached on the surface with defect sites, while Cd^{2+} ions will not bind directly to either regular surface or defective surface and instead, persist to exist exclusively as the outer-sphere complexes in defective surface as in the case regular surface.

(4) Water content. The dynamics of the interlayer species in clays is critically dependent on the water content in the system [84]. Zheng and Zaoui [99] compare the diffusion behaviors of counterions in montmorillonite at the one-, two-, and three-layer hydrated states, and the results show that the diffusion coefficients of counterions increase in a reverse proportion to the quantity of water content, corroborating that water content within the interlayer of clay minerals plays a very important role on the diffusion behavior of counterions. As discussed earlier, Churakov [69] demonstrate that the distribution of Na^+ ions on montmorillonite surfaces is strongly dependent on the relative humidity (RH): At relatively high RH, Na^+ ions form exclusively the outer-sphere complexes, and inner-sphere Na^+ surface complexes will be constructed. With the addition of water molecules, the inner-sphere Na^+ surface complexes that have formed at low RH will be reversibly transformed to the outer-sphere species, while K^+ ions show a distinct response to the increase of RH: K^+ ions at low RH form stable inner-sphere surface complexes and with increase of RH, remain tightly attached to the surface as inner-sphere species [100]. Zhang et al. [85] investigate the swelling properties, hydration behaviors, and mobility of interlayer ions (Na^+ and Cs^+) in montmorillonite at different water contents. It indicates that both Na^+ and Ca^{2+} ions are apt to construct the inner-sphere surface complexes at low water contents and the outer-sphere complexes at high water contents. The coordination numbers of water molecules in the first hydration shell of Na^+ and Ca^{2+} ions show a gradual increase as the water content increases, while the contribution of surface O atoms presents an opposite tendency. The

transitions from the inner-sphere surface complexes to outer-sphere complexes occur approximately at the water content of 170 mg_{water}/g_{clay} that correspond to the swelling processes of (Na, Ca)-montmorillonite from the monolayer hydrated to bilayer hydrated states.

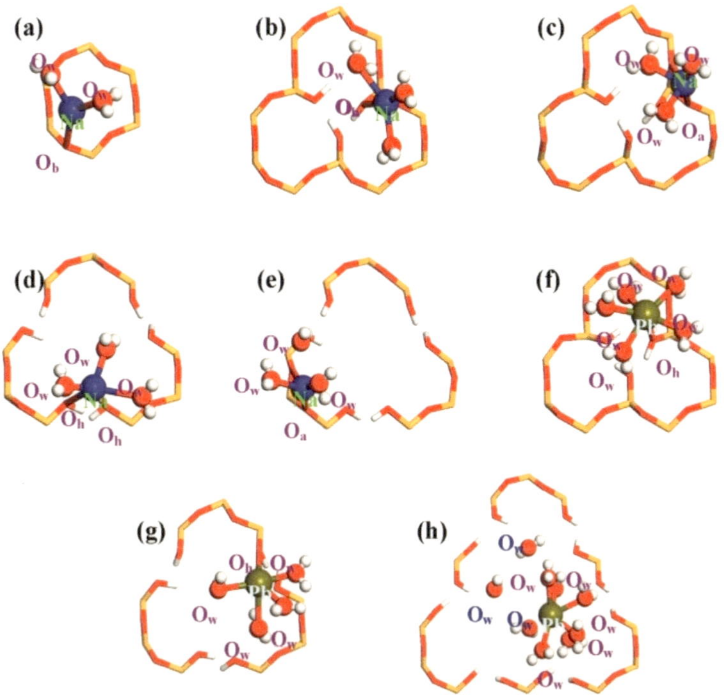

Figure 6. Local structures of inner-sphere Na^+ and $Pb^{2}+$ ions adsorbed on the tetrahedral SiO_4 surface of regular (**Si_0**) and defective kaolinite (**Si_1**, **Si_2** and **Si_4**): (a) for **Si_0**(Na^+), (a–c) for **Si_1**(Na^+), (a, d, e) for **Si_2**(Na^+) as well as (f) for **Si_1**($Pb^{2}+$), (g) for **Si_2**($Pb^{2}+$) and (h) for **Si_4**($Pb^{2}+$), respectively. Note that **Si_4**($Pb^{2}+$) is for quasi inner-sphere $Pb^{2}+$ ions adsorbed on the water molecules that enter into the silanol cavities of **Si_4** [15].

(5) Counterion. Metal ions and anions in aqueous solutions are referred to each other as "counterion." It indicates that the distribution and stability of metal ions adsorbed onto the surface of clay minerals are affected substantially by different counterions such as the various halide ions (A^- = F^-, Cl^-, I^-) [16]. However, the adsorption modes of metal ions seem not affected by the choice of counterions; for example, inner-sphere (mainly) and outer-sphere adsorbed for Na^+ ions and exclusively outer-sphere adsorbed for Pb^{2+} ions for all halide ions as counterions. Na^+ ions in dilute NaOH solutions are adsorbed only at tetrahedral SiO_4 surface, in line with the scenarios of halide ion solutions, while when the NaOH concentration increases to around 0.48 mol/L, a portion of Na^+ ions begin to emerge at the octahedral AlO_6 surface. For all halide ions as counterions, the local structures of adsorbed metal ions are similar, and when changing to the OH^- ion, several other structures will be generated for metal ions that are coordinated with OH^- ions. The OH^- ions can even drag metal ions toward the octahedral AlO_6 surface. Different from halide ions, OH^- can act not only as H-bond acceptor but also H-

bond donor, in a similar way as water molecules. In consequence, OH^- is capable of constructing forceful H-bond networks together with water molecules in aqueous solutions, which results in the distinct counterion effects as compared to halide ions. The stabilities of adsorbed Na^+ ions are affected by counterions and decrease in the order $OH^- > F^- > Cl^- > I^-$. The adsorption quantities and affinities of Cu^{2+} ions on clay soils are experimentally observed to be dependent on the choice of anions [101], which are consistent with the simulated results of counterion effects.

The diffusion coefficients of metal ions (D_M) in dilute solutions seem insensitive to the change of counterions; for example, the inner-sphere D_{Na+} values are very close to each other when the counterion is I^-, Cl^-, F^-, or OH^-. However, the D_M values are significantly affected by counterions in dense solutions; for example, for the adsorption of 0.96 mol/L NaA solutions ($A^- = F^-, Cl^-, I^-, OH^-$) on the tetrahedral SiO_4 surface of kaolinite, the inner-sphere diffusion coefficients of Na^+ ions vary dramatically and decrease in the order 0.74 ± 0.06 (I^-) > 0.67 ± 0.04 (Cl^-) > 0.42 ± 0.06 (F^-) > 0.28 ± 0.03 (OH^-), with units being 10^{-9} $m^2 \cdot s^{-1}$. Both inner and outer-sphere diffusion coefficients of metal ions show a gradual increase with salt concentrations suggesting the enhancement of mobility of adsorbed metal ions. The outer-sphere Pb^{2+} rather than Na^+ complexes have different coefficients and thus correspond to higher stability, which is due to the stronger interactions of Pb^{2+} ions with surrounding water molecules.

Although obviously less attention has been given, anions are clearly shown to be capably adsorbed onto the surface of clay minerals as in the case of metal ions [16]. Anions construct only H-bonds with the surface clay minerals while form stable inner and outer-sphere complexes, even if when the metal ions of constituent electrolytes correspond solely to the outer-sphere adsorption mode; for example, F^- in PbF_2 where Pb^{2+} ions are presented exclusively as outer-sphere complexes. Albeit forming only H-bonds with the surface of clay minerals, adsorbed anions have surprisingly high stabilities that are comparable to those of metal ions. In the case of F^- and OH^- ions, an appreciable amount show even superior stabilities suggesting the formation of especially stable anion-mineral complexes. All anions can form mono-dispersive inner and outer-sphere complexes at the octahedral AlO_6 surface, and OH^- ions can form several new adsorption structures that are absence for halide ions, forming OH^- clusters at the octahedral AlO_6 surface as well as Na^+ and OH^- pairs at both tetrahedral SiO_4 and octahedral AlO_6 surfaces. Adsorbed anions have comparable diffusion coefficients with metal ions and depend significantly on the choice of metal ions, with those from PbA_2 instead of NaA solutions responding more sluggishly to the change of salt concentrations.

4. Concluding remarks

With the advent of computer simulations, probing the complicated and prompt adsorption processes of ions at the interface of clay minerals and aqueous solutions becomes viable that complement the experimental observations. In the recent three decades, computer performances have improved unprecedentedly and sophisticated computational techniques (including methodologies and software) have emerged consecutively, which make computer simulations grow into an independent discipline. Nowadays, two types of computational simulations are popular in the studies of the adsorption processes of ions at the interface of clay minerals and aqueous solutions: QM methods accurately predict the adsorption behavior, reaction thermodynamics and mechanism, whereas MD methods provide the dynamic information for up to real systems on the femtosecond scale. With these computational techniques, a plethora of important details regarding to the adsorption configuration, distribution, stability, reaction thermodynamics and mechanism, dynamics, diffusivity as well as electric double layer and other fundamental issues have been brought to us, as has been elaborated in the chapter.

Ion-specific effects, also known as Hofmeister effects, are detected over 130 years ago when Hofmeister [102] finds a series of salt ions have consistent effects on the solubility and stability of proteins. Recently, ion-specific effects have experienced a renaissance and as suggested by Kunz et al. [21], the importance of ion-specific effects should be no less than Gregor Mendel's work to genetics. As a matter of fact, ion-specific effects control a wide spectrum of chemical, physical, biological, and colloidal processes, such as the interfacial tension of electrolyte solutions [103], chemical reaction rate [104, 105], protein conformation and stability [106–108], colloidal interaction [107] and even mudslides [109]. Ion-specific effects continue to defy all-encompassing theories [110]. Ionic size [111], hydration [111, 112], dispersion force [113–115] and surface charge [116, 117] are successively proposed to interpret the mechanism of ion-specific effects. Recently, our group [23, 118] reveals that the polarization effect could be the driving force of ion-specific effects for the adsorption of different metal ions onto the surface of clay minerals. With use of DFT calculations, we [12] have clearly shown that the polarization effect increases in a direct proportion with the negative charge of clay minerals and should be the driving force for the aggregation of colloidal soil particles. The aggregation mechanisms of colloidal soil particles with one- and bi-components are given and can be easily extended to more complicated colloidal soil particles. For the bi-component colloidal soil particles, the colloidal soil particles with more negative charges shows the larger polarization effect to metal ions and hence binds metal ions firstly and tightly, which then attract the colloidal soil particles with fewer negative charges. The aggregation processes are cycled and finished this way, and in consequence, the polarization effect is assumed to be responsible for the underlying ion-specific effects.

Although the aggregation of colloidal soil particles (usually carrying abundant negative charges) is dominated by metal ions [22], anions are found to also play a significant role and have clear Hofmeister series, as $H_2PO_4^- < Cl^- < NO_3^- < SO_4^{2-} < HPO_4^{2-} < PO_4^{3-}$ [14]. The DFT calculations indicate that interaction energies of anions with metal ions and proton affinities of anions are two influencing factors for such anion specificities. OH^- can facilely deprive protons from the bridging O atoms and increase the negative charges of colloidal soil particles, which further result in the significant retardation of the aggregation processes. The MD simulations demonstrate that clear anion-specific effects are manifested during the adsorption of anions at the interface of clay minerals and aqueous solutions that abide by the sequence of $OH^- > F^- > Cl^- > I^-$ [16]. Accordingly, computer simulations have contributed remarkably to the mechanistic understanding of ion-specific effects.

As commented by Nostro and Ninham [119], "if we had to sum up in a line or two what we have learned over the past decade, we could say this: previously biologists and physical chemists lived in parallel universes, almost disjunct." Ion-specific effects may probably be the bridge that connects soil science with other disciplines such as biology, chemistry, and physics, and the progresses made therein are likely to bring about revolutionary breakthroughs in colloidal and soil sciences. It is just in time that we are lucky to encounter the renaissance of ion-specific effects [120], and computer simulations are destined to play a critical role within the context.

Acknowledgements

This work was supported by the National Natural Science Foundation of China (41371249) and the Fourth Excellent Talents Program of Higher Education in Chongqing (2014-03).

Author details

Zengqiang Jia, Qian Wang, Chang Zhu and Gang Yang*

*Address all correspondence to: theobiochem@gmail.com

College of Resources and Environment and Chongqing Key Laboratory of Soil Multi-scale Interfacial Process, Southwest University, Chongqing, China

References

[1] Jungwirth P, Tobias DJ. Specific ion effects at the air/water interface. Chem. Rev. 2006; 106: 1259-1281.

[2] Geckeis HJ, Lützenkirchen R., Polly T., Rabung T, Schmidt, M. Mineral-water interface reactions of actinides. Chem. Rev. 2013; 113: 1016-1062.

[3] Gupta, SS, Bhattacharyya KG. Adsorption of metal ions by clays and inorganic solids. RSC. Adv. 2014; 4: 28537-28586.

[4] Sposito G, Skipper NT. Sutton R, Park SH, Soper AK, Greathouse JA. Surface geochemistry of the clay minerals. Proc. Natl. Acad. Sci. USA 1999; 96: 3358-3364.

[5] Rotenberg B, Marry V, Malikova N, Turq P. Molecular simulation of aqueous solutions at clay surfaces. J. Phys. Condens. Matter. 2010; 22: 284114.

[6] Bhattacharyya KG, Gupta SS. Adsorption of a few heavy metals on natural and modified kaolinite and montmorillonite: A review. Adv. Colloid. Interface. Sci. 2008; 140: 114-131.

[7] Bergaya F, Theng BGK, Lagaly G. Handbook of Clay Science; Elsevier: Amsterdam, 2006.

[8] Svelle S, Tuma C, Rozanska X, Kerber T, Sauer J, Quantum chemical modeling of zeolite-catalyzed methylation reactions: Toward chemical accuracy for barriers. J. Am. Chem. Soc. 2009; 131: 816-825.

[9] Yang G, Zhou LJ, Liu XC, Han XW, Bao XH. Adsorption, reduction and storage of hydrogen within ZSM-5 zeolite exchanged with various ions: A comparative theoretical study. Micropor. Mesopor. Mater. 2012; 161: 168-178.

[10] Lundberg M, Kawatsu T, Vreven T, Frisch MJ, Morokuma K. Transition states in a protein environment—ONIOM QM:MM modeling of isopenicillin N synthesis. J. Chem. Theory Comput. 2009; 5: 222-234.

[11] Dill1 KA, MacCallum JL. The protein-folding problem, 50 years on. Science. 2012; 338: 1042-1046.

[12] Tian R, Yang G, Tang Y, Liu XM, Li R, Zhu HL, Li H. Origin of Hofmeister effects for complex systems. PLoS One. 2015; 10: 0128602.

[13] Gao XD, Yang G, Tian R, Ding WQ, Hu FN, Liu XM, Li H. Formation of sandwich structure through ion adsorption at the mineral and humic interfaces: A combined experimental computational study. J. Mol. Struct. 2015; 1093: 96-100.

[14] Tian R, Yang G, Zhu C, Liu XM, Li, H. Specific anion effects for aggregation of colloidal minerals: A joint experimental and theoretical study. J. Phys. Chem. C 2015; 119: 4856-4864.

[15] Li X, Li H, Yang G. Promoting the adsorption of metal ions on kaolinite by defect sites: A molecular dynamics study. Sci. Rep. 2015; 5: 14377.

[16] Li X, Li H, Yang G. Configuration, anion-specific effects, diffusion, and impact on counterions for adsorption of salt anions at the interfaces of clay minerals. J. Phys. Chem. C 2016; 120: 14621-14630.

[17] Greathouse JA, Cygan RT. Water structure and aqueous uranyl (VI) adsorption equilibria onto external surfaces of beidellite, montmorillonite, and pyrophyllite: Results from molecular simulations. Environ. Sci. Technol. 2006; 40: 3865-3871.

[18] Yang W, Zaoui A. Behind adhesion of uranyl onto montmorillonite surface: A molecular dynamics study. J. Hazard. Mater. 2013; 261: 224-234.

[19] Tang JC, Zhu WY, Kookana R, Katayama A. Characteristics of biochar and its application in remediation of contaminated soil. J. Biosci. Bioeng. 2013; 116: 653-659.

[20] Zhu C, Yang G. Insights from the adsorption of halide ions on graphene materials. Chem. Phys. Chem. 2016; 17: 2482-2488.

[21] Kunz W, Lo Nostro P, Ninham BW. The present state of affairs with Hofmeister effects. Curr. Opin. Colloid. Interface. Sci. 2004; 9: 1-18.

[22] Tian R, Yang G, Li H, Gao XD, Liu XM, Zhu HL, Tang Y. Activation energies of colloidal particle aggregation: Towards a quantitative characterization of specific ion effects. Phys. Chem. Chem. Phys. 2014; 16: 8828-8836.

[23] Liu X, Yang G, Li H, Tian R, Li R, Jiang XJ, Ni JP, Xie DT. Observation of significant steric, valence and polarization effects and their interplay: A modified theory for electric double layers. RSC. Adv. 2014; 4: 1189-1192.

[24] Tian R, Yang G, Liu XM, Huang CZ, Gao XD, Li H. Observation of the unusual aggregation kinetics of colloidal minerals in acidic solutions. J. Chem. Sci. 2015; 127: 1083-1089.

[25] Vamsee KV, Al-Saidi WA, Kenneth DJ. Density functional theory study of pyrophyllite and M-montmorillonites (M = Li, Na, K, Mg, and Ca): Role of dispersion interactions. J. Phys. Chem. A 2011; 115: 9695-9703.

[26] Shi J, Lou ZY, Yang ML, Zhang Y, Liu HB, Meng YF. An interlayer expansion model for counterion-intercalated montmorillonite from first-principles calculations. Comput. Mater. Sci. 2015; 96: 134-139.

[27] Rhodes CN, Brown DR. Autotransformation and ageing of acid-treated montmorillonite catalysts: A solid-state 27Al NMR study. J. Chem. Soc. Faraday Trans. 1995; 91: 1031-1035.

[28] Claudia BJ, Esther AV. Brønsted sites on acid-treated montmorillonite: A theoretical study with probe molecules. J. Phys. Chem. A 2009; 113: 8994-9001.

[29] Shi J, Liu HB, Lou ZY, Zhang Y, Meng YF, Zeng Q, Yang ML. Effect of interlayer counterions on the structures of dry montmorillonites with Al3+/Si4+ substitution. Comput. Mater. Sci. 2013; 69: 95-99.

[30] Lasse PL, Jukka TT, Timothy S, Seppo K, Tapani AP. Montmorillonite interlayer surface chemistry: Effect of magnesium ion substitution on cation adsorption. Theor. Chem. Acc. 2015; 134: 51.

[31] Pascal C, Wanda A, Alessandro C, Eric H, Christopher JGP. Adsorption of low-molecular-weight molecules on a dry clay surface: An ab initio study. J. Phys. Chem. C 2009; 113: 12293-12300.

[32] Mignon P, Ugliengo P, Sodupe M, Hernandez ER. Ab initio molecular dynamics study of the hydration of Li+, Na+ and K+ in a montmorillonite model. Influence of isomorphic substitution. Phys. Chem. Chem. Phys. 2010; 12: 688-697.

[33] Karaborni S, Smit B, Heidug W, Urai J, van Oort E. The swelling of clays: Molecular simulations of the hydration of montmorillonite, Science. 1996; 271: 1102-1104.

[34] Boek ES, Coveney PV, Skipper NT. Monte Carlo molecular modeling studies of hydrated Li-, Na-, and K-smectites: Understanding the role of potassium as a clay swelling inhibitor. J. Am. Chem. Soc. 1995; 117: 12608-12617.

[35] Odriozola G, Aguilar JF. Stability of K-montmorillonite hydrates: Hybrid MC Simulations. J. Chem. Theory Comput. 2005; 1: 1211-1220.

[36] Tambach TJ, Hensen EJM, Smit B. Molecular simulations of swelling clay minerals. J. Phys. Chem. B 2004; 108: 7586-7596.

[37] Segad M, Jonsson B, Akesson T, Cabane B. Ca/Na montmorillonite: Structure, forces and swelling properties. Langmuir. 2010; 26: 5782-5790.

[38] Suter JL, Sprik M, Boek ES. Free energies of absorption of alkali ions onto beidellite and montmorillonite surfaces from constrained molecular dynamics simulations. Geochim. Cosmochim. Acta. 2012; 91: 109-119.

[39] Whitley HD, Smith DE. Free energy, energy, and entropy of swelling in Cs-, Na-, and Sr-montmorillonite clays. J. Chem. Phys. 2004; 120: 5387-5395.

[40] Chavez-Paez M, dePablo L, dePablo JJ. Monte Carlo simulations of Camontmorillonite hydrates. J. Chem. Phys. 2001; 114: 10948-10953.

[41] Chatterjee A, Ebina T, Onodera Y, Mizukami F. Effect of exchangeable cation on the swelling property of 2:1 dioctahedral smectite—A periodic first principle study. J. Chem. Phys. 2004; 120: 3414-3424.

[42] Liu XD, Lu XC. A thermodynamic understanding of clay-swelling inhibition by potassium ions. Angew. Chem. Int. Ed. 2006; 45: 6300-6303.

[43] Lou ZY, Liu HB, Zhang Y, Meng YF, Zeng Q, Shi J, Yang ML. A density functional theory study of the hydration of calcium ions confined in the interlayer space of montmorillonites. J. Theor. Comput. Chem. 2014; 13: 1450028.

[44] Alena K, Benjamí M, Sven K, Notker R. Uranyl adsorption on solvated edge surfaces of pyrophyllite: A DFT model study. Phys. Chem. Chem. Phys. 2012; 14: 5815-5823.

[45] Alena K, Sven K, Notker R. Uranyl adsorption at solvated edge surfaces of 2:1 smectites. A density functional study. Phys. Chem. Chem. Phys. 2015; 17: 13757-13768.

[46] Masahiko O, Hiroki N, Masahiko M. Mechanism of strong affinity of clay minerals to radioactive cesium: First-principles calculation study for adsorption of cesium at frayed edge sites in muscovite. J. Phys. Soc. Jpn. 2013; 82: 033802.

[47] Zhang SQ, Hou WG. Adsorption behavior of Pb(II) on montmorillonite. Colloids Surf. A: Physicochem. Eng. Aspects. 2008; 320: 92-97.

[48] Jiang MQ, Jin XY, Lu XQ, Chen ZL. Adsorption of Pb(II), Cd(II), Ni(II) and Cu(II) onto natural kaolinite clay. Desalination. 2010; 252: 33-39.

[49] Öztürk N, Bektaş TE. Nitrate removal from aqueous solution by adsorption onto various materials. J. Hazard. Mater. 2004; 112: 155-162.

[50] Zhu MQ, Paul KW, Kubicki JD, Sparks DL. Quantum chemical study of Arsenic (III, V) adsorption on Mn-oxides: Implications for arsenic (III) oxidation. Environ. Sci. Technol. 2009; 43: 6655-6661.

[51] Han YH, Liu WL, Chen JH. DFT simulation of the adsorption of sodium silicate species on kaolinite surfaces. Appl. Surf. Sci. 2016; 370: 403-409.

[52] Sun J, Lian F, Liu Z, Zhu L, Song Z. Biochars derived from various crop straws: Characterization and Cd(II) removal potential. Ecotoxicol. Environ. Saf. 2014; 106: 226-231.

[53] Meng J, Feng X, Dai Z, Liu X, Wu J, Xu J. Adsorption characteristics of Cu(II) from aqueous solution onto biochar derived from swine manure. Environ. Sci. Pollut. R. 2014; 21: 7035-7046.

[54] Madadrang C J, Kim H Y, Gao G, et al. Adsorption behavior of EDTA-graphene oxide for Pb (II) removal. ACS Appl. Mater. Inter. 2012; 4: 1186-1193.

[55] Li L, Fan L, Sun M, et al. Adsorbent for chromium removal based on graphene oxide functionalized with magnetic cyclodextrin–chitosan. Colloid. Surf. B 2013, 107: 76-83.

[56] Colherinhas G, Fileti EE, Chaban VV. The band gap of graphene is efficiently tuned by monovalent ions. J. Phys. Chem. Lett. 2015; 6: 302-307.

[57] Peles-Lemli B, Kánnár D, Nie J C, et al. Some unexpected behavior of the adsorption of alkali metal ions onto the graphene surface under the effect of external electric field. J. Phys. Chem. C 2013; 117: 21509-21515.

[58] Shi GS, Wang ZG, Zhao JJ, Hu J, Fang HP. Adsorption of sodium ions and hydrated sodium ions on a hydrophobic graphite surface via cation-π interactions. Chinese Phys. B 2011; 20: 068101.

[59] Shi G, Ding Y, Fang H. Unexpectedly strong anion–π interactions on the graphene flakes. J. Comput. Chem. 2012; 33: 1328-1337.

[60] Bourg IC, Sposito G. Molecular dynamics simulations of the electrical double layer on smectite surfaces contacting concentrated mixed electrolyte ($NaCl$-$CaCl_2$) solutions. J. Colloid Interface. Sci. 2011; 360: 701-715.

[61] Kozaki T, Sato H, Sato S, Ohashi, H. Diffusion mechanism of cesium ions in compacted montmorillonite. Eng. Geol. 1999; 54: 223-230.

[62] Nakashima Y. Diffusivity measurement of heavy ions in Wyoming montmorillonite gels by X-ray computed tomography. J. Contam. Hydrol. 2003; 61: 147-156.

[63] Sato H. Effects of the orientation of smectite particles and ionic strength on diffusion and activation enthalpies of I^- and Cs^+ ions in compacted smectite. Appl. Clay. Sci. 2005; 29: 267-281.

[64] Appelo CAJ, van Loon LR, Wersin P. Multicomponent diffusion of a suite of tracers (HTO, Cl, Br, I, Na, Sr, Cs) in a single sample of Opalinus Clay. Geochim. Cosmochim. Acta. 2010; 74: 1201-1219.

[65] Jakob A, Pfingsten W, Van Loon L. Effects of sorption competition on caesium diffusion through compacted argillaceous rock. Geochim. Cosmochim. Acta. 2009; 73: 2441-2456.

[66] van Loon LR, Glaus MA. Mechanical compaction of smectite clays increases ion exchange selectivity for cesium. Environ. Sci. Technol. 2008; 42: 1600-1604.

[67] Wersin P, van Loon LR, Soler JM, Yllera A, Eikenberg J, Gimmi Th, Hernan P, Boisson JY. Long-term diffusion experiment at Mont Terri: First results from field and laboratory data. Appl. Clay Sci. 2004; 26: 123-135.

[68] Bourg IC, Bourg ACM, Sposito G. Modeling diffusion and adsorption in compacted bentonite: A critical review. J. Contam. Hydrol. 2003; 61: 293-302.

[69] Churakov SV. Mobility of Na and Cs on montmorillonite surface under partially saturated conditions. Environ. Sci. Technol. 2013; 47: 9816-9823.

[70] Sposito G. Molecular models of iron adsorption on mineral surfaces. Rev. Mineral. Geochem. 1990; 23: 261-279.

[71] Spark KM, Wells JD, Johnson BB. Characterizing trace metal adsorption on kaolinite. Eur. J. Soil. Sci. 1995; 46: 633-640.

[72] Ma YM, Zhang H, Zhang BJ. Structure of sodium sulphate aqueous solution/quartz interface: A molecular dynamics simulation. Mol. Simul. 2014; 40: 634-639.

[73] Bourg IC, Sposito, G. In Handbook of Soil Science, Ion Exchange Phenomena, 2nd ed.; Sumner, ME., Ed.; CRC Press: Boca Raton, FL, 2011.

[74] Marry V, Turq P, Cartailler T, Levesque D. Microscopic simulation of structure and dynamics of water and counterions in a monohydrated montmorillonite. J. Chem. Phys. 2002, 117: 3454-3463.

[75] Park S H, Sposito G. Monte Carlo simulation of total radial distribution functions for interlayer water in Li-, Na-, and K-montmorillonite hydrates. J. Phys. Chem. B. 2000; 104: 4642-4648.

[76] Hensen EJM, Tambach TJ, Bliek A, Smit B. Adsorption isotherms of water in Li-, Na-, and K-montmorillonite by molecular simulation. J. Chem. Phys. 2001; 115: 3322-3329.

[77] Tournassat C, Chapron Y. Leroy P, Bizi M, Boulahya F. Comparison of molecular dynamics simulations with triple layer and modified Gouy-Chapman models in a 0.1 M NaCl-montmorillonite system. J. Colloid. Interface. Sci. 2009; 339: 533-541.

[78] Tachi Y, Yotsuji K. Diffusion and sorption of Cs^+, Na^+, I^- and HTO in compacted sodium montmorillonite as a function of pore water salinity: Integrated sorption and diffusion model. Geochim. Cosmochim. Acta 2014, 132: 75-93.

[79] Kim Y, Cygan RT, Kirkpatrick RJ. ^{133}Cs NMR and XPS investigation of cesium adsorbed on clay minerals and related phases. Geochim. Cosmochim. Acta. 1996; 60: 1041-1052.

[80] Nakano M, Kawamura K, Ichikawa Y. Local structural information of Cs in smectite hydrates by means of an EXAFS study and molecular dynamics simulations. Appl. Clay. Sci. 2003; 23: 15-23.

[81] Vasconcelos IF, Bunker BA, Cygan RT. Molecular dynamics modeling of ion adsorption to the basal surfaces of kaolinite. J. Phys. Chem. C. 2007; 111: 6753-6762.

[82] Sakuma H, Kawamura K. Structure and dynamics of water on Li^+-, Na^+-, K^+-, Cs^+-, H_3O^+-exchanged muscovite surfaces: a molecular dynamics study. Geochim. Cosmochim. Acta. 2011; 75: 63-81.

[83] Steele HM, Wright K, Nygren MA, Hillier IH. Interactions of the (001) surface of muscovite with Cu (II), Zn (II), and Cd (II): A computer simulation study. Geochim. Cosmochim. Acta. 2000; 64: 257-262.

[84] Ngouana WBF, Kalinichev AG. Structural arrangements of isomorphic substitutions in smectites: Molecular simulation of the swelling properties, interlayer structure, and dynamics of hydrated Cs-montmorillonite revisited with new clay models. J. Phys. Chem. C 2014; 118: 12758-12773.

[85] Zhang L, Lu X, Liu XD, Zhou JH, Zhou HQ. Hydration and mobility of interlayer ions of (Na_X, Ca_Y)-montmorillonite: A molecular dynamics study. J. Phys. Chem. C. 2014; 118: 29811-29821.

[86] Smith DE. Molecular computer simulations of the swelling properties and interlayer structure of cesium montmorillonite. Langmuir 1998; 14: 5959-5967.

[87] Young DA, Smith DE. Simulations of clay mineral swelling and hydration: Dependence upon interlayer ion size and charge. J. Phys. Chem. B. 2000; 104: 9163-9170.

[88] Sutton R, Sposito G. Molecular simulation of interlayer structure and dynamics in 12.4 Å Cs-smectite hydrates. J. Colloid Interface Sci. 2001; 237: 174-184.

[89] Sutton R, Sposito G. Animated molecular dynamics simulations of hydrated caesium-smectite interlayers. Geochem. Trans. 2002; 3: 73-80.

[90] Whitley HD, Smith DE. Free energy, energy, and entropy of swelling in Cs-, Na-, and Sr-montmorillonite clays. J. Chem. Phys. 2004; 120: 5387-5395.

[91] Liu XD, Lu XC, Wang RC, Zhou HQ. Effects of layer-charge distribution on the thermodynamic and microscopic properties of Cs-smectite. Geochim. Cosmochim. Acta 2008, 72: 1837-1847.

[92] Cygan RT, Liang JJ, Kalinichev AG. Molecular models of hydroxide, oxyhydroxide, and clay phases and the development of a general force field. J. Phys. Chem. B 2004; 108: 1255-1266.

[93] Teich-McGoldrick SL, Greathouse JA, Jové-Colón CF, Cygan RT. Swelling properties of montmorillonite and beidellite clay minerals from molecular simulation: Comparison of temperature, interlayer cation, and charge location effects. J. Phys. Chem. C 2015; 119: 20880-20891.

[94] Malikova N, Marry V, Dufrêche JF, Turq P. Na/Cs montmorillonite: Temperature activation of diffusion by simulation. Curr. Opin. Colloid Interface Sci. 2004; 9: 124-127.

[95] Zheng Y, Zaoui A. Temperature effects on the diffusion of water and monovalent counterions in the hydrated montmorillonite. Physica A: Stat. Mech. Appl. 2013; 392: 5994-6001.

[96] Gupta SS, Bhattacharyya KG. Adsorption of heavy metals on kaolinite and montmorillonite: A review. Phys. Chem. Chem. Phys. 2012; 14: 6698-6723.

[97] Bhattacharyya KG, Gupta SS. Adsorption of a few heavy metals on natural and modified kaolinite and montmorillonite: A review. Adv. Colloid Interface Sci. 2008; 140: 114-131.

[98] Croteau T, Bertram AK, Patey GN. Water adsorption on kaolinite surfaces containing trenches. J. Phys. Chem. A 2010; 114: 2171-2178.

[99] Zheng Y, Zaoui A. How water and counterions diffuse into the hydrated montmorillonite. Solid State Ion. 2011; 203: 80-85.

[100] Hensen EJM, Smit B. Why clays swell. J. Phys. Chem. B. 2002; 106: 12664-12667.

[101] Yu S, He ZL, Huang CY, Chen GC, Calvert DV. Effects of anions on the capacity and affinity of copper adsorption in two variable charge soils. Biogeochem. 2005; 75: 1-18.

[102] Hofmeister F. Zur lehre von der wirkung der salze. Archiv. Für. Experimentelle. Pathologie. und. Pharmakologie. 1888; 25: 1-30.

[103] Para G, Jarek E, Warszynski P. The Hofmeister series effect in adsorption of cationic surfactants—theoretical description and experimental results. Adv. Colloid. Interface. Sci. 2006; 122: 39-55.

[104] Kropman M, Bakker H. Dynamics of water molecules in aqueous solvation shells. Science. 2001; 291: 2118-2120.

[105] Maroncelli M, Macinnis J, Fleming GR. Polar solvent dynamics and electron-transfer reactions. Science. 1989; 243, 1674-1681.

[106] Broering JM, Bommarius AS. Evaluation of Hofmeister effects on the kinetic stability of proteins. J. Phys. Chem. B. 2005; 109: 20612-20619.

[107] Cacace M, Landau E, Ramsden J, The Hofmeister series: Salt and solvent effects on interfacial phenomena. Q. Rev. Biophys. 1997; 30: 241-277.

[108] Yang Z. Hofmeister effects: An explanation for the impact of ionic liquids on biocatalysis. J. Biotechnol. 2009; 144: 12-22.

[109] Chávez-Páez M, van Workum K, de Pablo L, de Pablo JJ. Monte Carlo simulations of Wyoming sodium montmorillonite hydrates. J. Chem. Phys. 2001; 114, 1405-1413.

[110] Tobias DJ, Hemminger JG. Getting specific about specific ion effects. Science. 2008; 319: 1197-1198.

[111] Parsons DF, Boström M, Nostro PL, Ninham BW. Hofmeister effects: Interplay of hydration, nonelectrostatic potentials, and ion size. Phys. Chem. Chem. Phys. 2011; 13: 12352-12367.

[112] Collins KD. Charge density-dependent strength of hydration and biological structure. Biophys. J. 1997; 72: 65-76.

[113] Duignan TT, Parsons DF, Ninham BW. A continuum solvent model of the multipolar dispersion solvation energy. J. Phys. Chem. B. 2014; 117: 9412-9420.

[114] Ninham BW, Yaminsky V. Ion binding and ion specificity: The Hofmeister effect and Onsager and Lifshitz theories. Langmuir. 1997; 13: 2097-2108.

[115] Parsons DF, Deniz V, Ninham BW. Nonelectrostatic interactions between ions with anisotropic ab initio dynamic polarisabilities. Colloid Surf. A: Physicochem. Eng. Asp. 2009; 343: 57-63.

[116] dos Santos AP, Levin Y. Ion specificity and the theory of stability of colloidal suspensions. Phys. Rev. Lett. 2011; 106: 167801-167804.

[117] Levin Y, dos Santos AP, Diehl A. Ions at the air-water interface: An end to a hundred-year-old mystery? Phys. Rev. Lett. 2009; 103: 257802-257805.

[118] Liu XM, Li H, Li R, Xie DT, Ni JP, Wu LS. Strong non-classical induction forces in ion-surface interactions: General origin of Hofmeister effects. Sci. Rep. 2014; 4: 5047.

[119] Nostro PL, Ninham BW. Hofmeister phenomena: An update on ion specificity in biology. Chem. Rev. 2012; 112: 2286-2322.

[120] Jungwirth P, Cremer PS. Beyond Hofmeister. Nature. Chem. 2014; 6: 261-263.

Chapter 6

Manganese Sulfide (MnS) Nanocrystals: Synthesis, Properties, and Applications

Anna M. Ferretti, Sara Mondini and
Alessandro Ponti

Additional information is available at the end of the chapter

http://dx.doi.org/10.5772/65092

Abstract

Manganese(II) sulfide (MnS) is an interesting material for both fundamental and applicative research, especially when its bulk properties are modulated by reducing the size into the nanometric region (< 100 nm). Due to its polymorphism, MnS is an attractive material to develop synthetic strategies for polymorphism control. We have reviewed the literature concerning MnS nanosystems having at least one dimension smaller than 100 nm. Successful synthetic techniques for the preparation of zero- and one-dimensional MnS nanosystems (either homogeneous and heterogeneous) with size, shape, and polymorphism control are presented with emphasis on solvothermal techniques and on studies devoted to understanding the growth mechanism and the polymorphism. Properties and applications are collected in three broad areas corresponding to nanosize MnS used as an optical, electric, and magnetic material. MnS has attracting properties such as its large bandgap, which makes it promising for emission in the ultraviolet region. The magnetic properties have also arisen attention since MnS is antiferromagnetic at low temperature and (super)paramagnetic at room temperature. Finally, the layered structure of the hexagonal polymorph is responsible for the good performance of nanosize MnS as a lithium-ion battery electrode or supercapacitor material since the insertion/exchange of small ions is easy.

Keywords: manganese sulfide, MnS, nanocrystal, one-dimensional nanostructure, polymorphism control, solvothermal synthesis

1. Introduction

Manganese(II) sulfide (MnS) is a p-type semiconductor with a wide bandgap (E_g = ~3 eV) [1] that undergoes a transition to an antiferromagnetically (AFM) ordered phase below room temperature [2]. These features attracted interest for the potential applications of nanosized MnS in the field of short wavelength optoelectronics [3–5], or as a photoluminescent component [6], photoreduction catalysts [7], and contrast agent for magnetic resonance imaging (MRI) [8, 9]. The optical properties of MnS have also raised considerable interest in protobiology since it is deemed to have played a key role (along with ZnS in mixed haloes formed around primeval sub-aerial hot springs) in the prebiotic photosynthesis development, thanks to its ability to photoreduce CO_2 [7]. More recently, interesting energy-related applications have been demonstrated, where MnS was employed as an electrode material in Li-ion batteries [10–13] and as a supercapacitor material [14, 15]. These interesting properties and promising applications have raised considerable attention and research on small MnS particles can be considered to begin with the submission of three papers within the middle 6 months of 2001. In March, the revised version of a paper by Qian and coworkers [16] describing the synthesis of submicrometric particles of all three polymorphs of MnS was submitted and appeared in print later that year. Truly nanometric (<100 nm) MnS nanocrystals (NCs) were independently reported by Banin and Cheon research groups in 2001. It is noteworthy that the paper by Cheon and coworkers [17] was received by the publisher just 2 weeks before the paper by Banin and coworkers [18] arrived at a different publisher. However, the latter article appeared in print in 2001 while the former in 2002.

Three polymorphs (α, β, and γ) of MnS are known, and their crystal structure is represented in **Figure 1**. In the cubic rock-salt α-MnS structure, sulfide anions an expanded fcc lattice, and the manganese cations occupy all octahedral sites. The metastable form β-MnS has the cubic zincblende structure. Similarly to α-MnS, β-MnS comprises an expanded fcc lattice of S^{2-}anions, but in this polymorph, the Mn^{2+} cations reside on half of the tetrahedral sites. Finally, γ-MnS has the hexagonal wurtzite structure, based on a slightly compressed (c/a = 1.618) hcp lattice of sulfide anions with manganese cations occupying half of the tetrahedral sites. The polymorphs have similar but unequal physical properties; for instance, all the three forms of MnS undergo transition to an AFM ordered structure [19], but the transition temperature varies from T_N ~80 K for γ-MnS to ~100 K for β-MnS and 154 K for α-MnS [2]. The magnetic structures below T_N were determined by neutron diffraction and could be explained by the interplay between the crystal structure and the superexchange interaction between the magnetic Mn^{2+} cations [19]. α-MnS displays the well-known Type-II structure where the magnetic moments are ferromagnetically (FM) coupled within the (111) planes and successive (111) planes form an AFM structure. Both β- and γ-MnS display Type-III ordering since in these structures the Mn^{2+} ions are coordinated tetrahedrally. For the details of these more complicated magnetic structures refer to Reference [19].

Naturally occurring MnS minerals with α, β, and γ structures are known, but they have curiously been discovered in a large time interval. Natural α-MnS has been known as alabandite since the beginning of the nineteenth century [20] and is a widespread manganese ore.

http://dx.doi.org/10.5772/65092

MnS minerals with β and γ structures have been much more recently reported upon: rambergite (γ-MnS) has been discovered in 1996 in Sweden [21] whereas browneite (β-MnS) was found in a meteorite collected in Poland in 2012 [22].

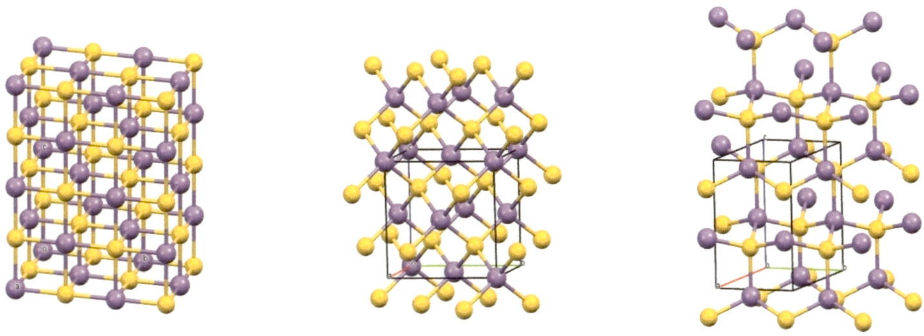

Figure 1. Crystal structure of the three polymorphs of MnS. Left: cubic α-MnS (rock salt); middle: cubic β-MnS (zincblende); right: hexagonal γ-MnS (wurtzite). The primitive cell is also shown. Color code: Mn, violet; S, yellow.

We limit our review to homogeneous or heterogeneous nanostructures that comprise pure MnS with at least one dimension smaller than 100 nm (some MnS sub-microcrystals are included when of particular interest). Both zero- (0D) and one-dimensional (1D) MnS nanosystems have been prepared using several bottom-up methods ranging from solvothermal and hydrothermal "wet" techniques to chemical vapor deposition (CVD). Though we will review MnS nanostructures prepared by any technique, we will give some detail of solvothermal methods, which found the most widespread application in the synthesis of 0D nanostructures and show a nice diversity in the choice of precursors and experimental conditions. Conversely, 1D nanosystems (rods, wires, saws) were often obtained prepared by the CVD. Some heterostructures, both 0D and 1D, comprising MnS have also been prepared by solvothermal or CVD techniques and displayed peculiar properties. The issue of polymorphism control in 0D NCs has been the subject of several reports showing that both physical and chemical parameters can be used to obtain the desired crystal structure.

2. Synthesis and characterization

In this section, we will review both the synthetic methods for the preparation of zero-dimensional (0D) and one-dimensional (1D) MnS nanosystems and their morphological [usually performed resorted to scanning electron microscopy (SEM) and transmission electron microscopy (TEM)], structural [X-ray (XRD) and electron (ED) diffraction], and compositional [usually by energy dispersive X-ray spectroscopy (EDX, a.k.a. EDS), more rarely by electron energy loss spectroscopy (EELS) and its imaging counterpart electron spectroscopy imaging (ESI)] characterization. The optical, electric, and magnetic properties of the MnS nanosystems will be discussed in the following sections. The present section will separately deal with 0D and 1D MnS nanostructures, and the reports specifically addressing the control of the MnS polymorphism are collected at the end of the section. Synthetic studies involving systematic variation in the reaction conditions are deemed particularly interesting and will be given particular attention.

2.1. Zero-dimensional (0D) nanosystems

In this subsection, we focus on nanostructures comprising MnS NCs with aspect ratio close to 1 and start with chemically homogeneous nanosystems, i.e., pure MnS NCs. The subsection is concluded by reviewing heterogeneous nanosystems comprising a 0D MnS NC.

2.1.1. Homogeneous nanocrystals

Of course, most examples of 0D nanosystems are homogeneous MnS NCs or a mixture of two or three polymorphic MnS NCs. We separately review NCs obtained by solvothermal techniques, that is, where one or more precursors are decomposed at high temperature in non-aqueous solvent in an inert atmosphere, possibly in the presence of other compounds acting as a growth regulator, protective coating, shape- or polymorph-inducer, etc.

2.1.1.1. Solvothermal methods

To the best of our knowledge, the synthesis of 0D MnS NCs with at least one dimension below 100 nm was independently reported by two research groups in 2001. It is noteworthy that one paper was received by the publisher just 2 weeks before the other one arrived at a different publisher. However, the latter article appeared in print in 2001 while the former in 2002. In the presented by Banin group, α-MnS NCs were obtained using a solvothermal technique employing reagents rarely found in subsequent work [18]. The procedure involved the injection of a trioctylphosphine/trioctylphosphine oxide (TOP/TOPO) solution of *both* precursors $MnCl_2$ and $(Me_3Si)_2S$ (Mn:S = 1:0.67 to 1:0.77, defect of S) in hot TOPO (360°C). Size control was achieved by varying the concentration of Mn from 0.066 to 0.13 and 0.2 M, so that 80, 40, and 20 nm nanosystems were obtained. XRD showed that the product was pure α-MnS NCs, but the crystallite size turned out to be 12 nm. This result and the grainy appearance of the TEM images suggest that the nanosystems are different-sized agglomerates of smaller MnS crystallites. The first magnetic characterization of MnS NCs was also reported. The paper presented by the Cheon group [17] describes 0D α-MnS NCs and "composite" β-MnS/γ-MnS nanowires and multipodes obtained by a very different solvothermal protocol. It involved a single precursor, manganese bis(*N,N*-diethylcarbamate) $Mn(Et_2N\text{-}CS_2)_2$ (Mn:S = 1:4), which was dissolved in hexadecylamine, then injected into hot hexadecylamine and aged for 20 min. By varying the reaction temperature, the MnS NCs shape and crystal structure could be controlled, but not independently. At 120°C, γ-MnS nanowires and γ-MnS multipodes with the β-MnS core were obtained whereas ~30 nm α-MnS nanocubes were produced at 250°C. In subsequent research, it has often been confirmed that low/high reaction temperature favors the (β, γ)/α crystal structure, as expected on the basis of the relative thermodynamic stability. The first optical characterization of MnS NCs was also reported.

Using a single molecule as a precursor for both Mn and S could offer a few advantages, e.g., Mn–S bonds are already present and likely to survive the precursor decomposition, and the Mn:S molar ratio is exactly fixed. Moreover, precursor synthesis is usually easy. Thus, several papers reported the use of a single precursor, often chemically similar to $Mn(Et_2N\text{-}CS_2)_2$, for the preparation of MnS NCs. In a paper regarding the synthesis of NCs of many divalent metal sulfides [23], manganese *O*-hexadecylxanthate $Mn(C_{16}H_{33}O\text{-}CS_2)_2$ (Mn:S = 1:4) was used to obtain α-MnS NCs. The morphology is irregular and no size information was given. The authors noted that the precursor is easily attacked by hexadecylamine at room temperature, so that trioctylamine had to be employed as a reaction solvent at 150°C. In the last example [24] involving a single precursor, $Mn(Ph\text{-}COS)_2TMEDA$ (Mn:S = 1:2, Ph-COS = thiobenzoate, TMEDA = *N,N,N′,N′*-tetramethylethylenediamine), the roles of solvent and ligand are separated when non-coordinating 1-octadecene was used as a solvent while oleylamine was used as a ligand (TMEDA could also act as a Mn ligand). Octahedral NCs (size ca. 24 nm) were obtained by heating the precursor solution at 300–350°C for 2 h. XRD data showed pure α-MnS, as expected on the basis of the high reaction temperature, and EDX data confirmed the MnS stoichiometry. Single molecule precursors were also used to synthesize MnS NCs within mesoporous silica [25] (see Section 2.1.1.2), to study polymorphism control [26] (see Section 2.3), and to prepare β-MnS nanowires [27] (see Section 2.2.2).

The fixed Mn:S ratio inherent to the use of single-molecule precursors might also be regarded as a limitation while exploring synthetic conditions. This consideration, and the effort saved when precursors are commercially available, prompted many groups to employ a two-precursor strategy to synthesize MnS NCs. Of course, in this case, more care should be taken since the chemistry is more complex: the Mn–S bond has yet to be established and the reactive Mn and S species must be simultaneously present in the reaction environment. A convenient sulfide precursor is elemental sulfur (yellow powder, S_8), which is readily soluble in apolar solvents. The octameric molecule must be converted to sulfide S^{2-} anions during the reaction. Both 1-octadecene [28] and long-chain primary amines [29] have been shown to be able to produce the reactive sulfur species leading to metal sulfide NCs. MnS NCs were synthesized by heating an oleylamine solution of sulfur powder and $MnCl_2$ [30]. NC size and shape were varied by changing the Mn:S molar ratio. Nanorods with low aspect ratio (20 × 37 nm) were obtained using Mn:S = 1:1 at 240°C while long (17 × 44 nm) and short (23 × 37 nm) bullet-shaped NCs were synthesized using Mn:S = 2:1 and 3:1, respectively, at 280°C. Unfortunately, the NC composition and crystal structure seem to be not provided.

Sulfur can be inactivated when a long-chain carboxylic acid RCOOH is present in the reaction mixture, presumably due to the formation of RCOSH species [31]. It was shown that, in order to produce pure MnS NCs, the molar ratio Mn:S ≥ (Mn:RCOOH + 1) is required, e.g., if 1 equivalent of RCOOH is present, at least 2 equivalents of S are required (see **Figure 2**).

The synthesis of octahedral MnS NCs using a solvothermal protocol involving heating a solution of manganese carboxylate and sulfur powder in 1-octadecene has been studied in detail in Ref. [32]. Three manganese precursors were synthesized: Mn(II) distearate ($MnSt_2$), dioleate ($MnOl_2$), and hydroxyoleate [Mn(OH)(Ol)]. Whenever Mn:S ≥ 2, XRD showed that the produced NCs had α-MnS structure. Rietveld analysis of XRD profiles yielded NC size in agreement with TEM data, thus showing that the NCs are single crystals. The NC size could be finely controlled by changing the precursor and the reaction temperature and time. Larger size is achieved by using $Mn(Ol)_2$ at higher temperature with longer reaction time . In this manner, α-MnS NCs with a size from 9.3 to 29.5 nm (and 10 other intermediate sizes) could be obtained (see **Figure 3**). The size dispersion was also investigated and it was interestingly found that $MnOl_2$ and Mn(OH)(Ol) behave oppositely. Size dispersion is minimum and maximum at 300°C for $MnOl_2$ and Mn(OH)(Ol), respectively. Furthermore, the change in size dispersion when the reaction time is varied is decreasing for $MnOl_2$ and volcano-shaped for Mn(OH)(Ol).

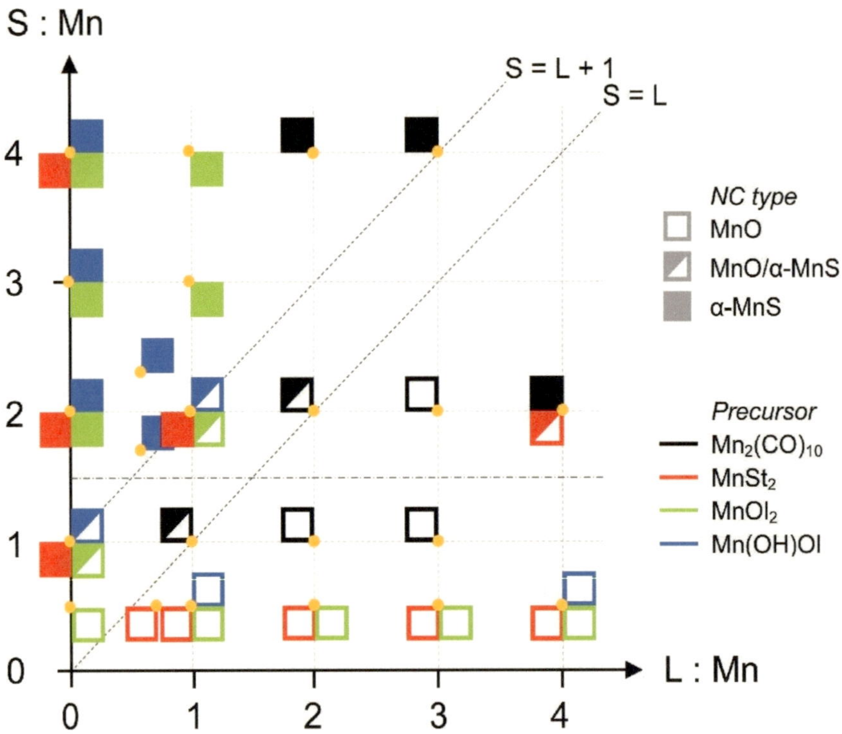

Figure 2. Synthetic outcome from solvothermal synthesis using different Mn precursors as a function of the L:Mn (L = surfactant) and S:Mn molar ratios. The type of NC is encoded as follows: open squares: MnO NCs; half-filled squares: mixture of MnO and α-MnS NCs; solid squares: α-MnS NCs. The precursors are color coded as follows. Black: $Mn_2(CO)_{10}$, red: $MnSt_2$, green: $MnOl_2$, blue: Mn(OH)Ol. The free surfactant, L, was stearic acid for $Mn_2(CO)_{10}$ and Mn distearate, and oleic acid for Mn dioleate and Mn hydroxyl oleate, respectively. Reprinted from Ref. [31] under the terms of the Creative Commons Attribution License.

$MnOl_2$ and sulfur powder were also used to prepare α-MnS NCs with a peculiar stellated-octahedral shape [33]. When a solution of $MnOl_2$ and S (1:2) in 1-octadecene was heated at 300°C, 15 nm spherical α-MnS NCs were obtained, but when a 1:1 v/v mixture of oleic acid and oleylamine was present in the reaction mixture, 45 nm α-MnS NCs with stellated-octahedral shape were produced. High-resolution TEM (HRTEM) and ED experiments showed that the latter are single crystals with elongated branches pointing along the six <001>

http://dx.doi.org/10.5772/65092

equivalent directions. A further solvothermal procedure involving sulfur powder and Mn(II) nitrate [34] can be found in Section 2.3; elemental S (and Mn) was also used in a hydrothermal protocol for α-MnS nanorods [35].

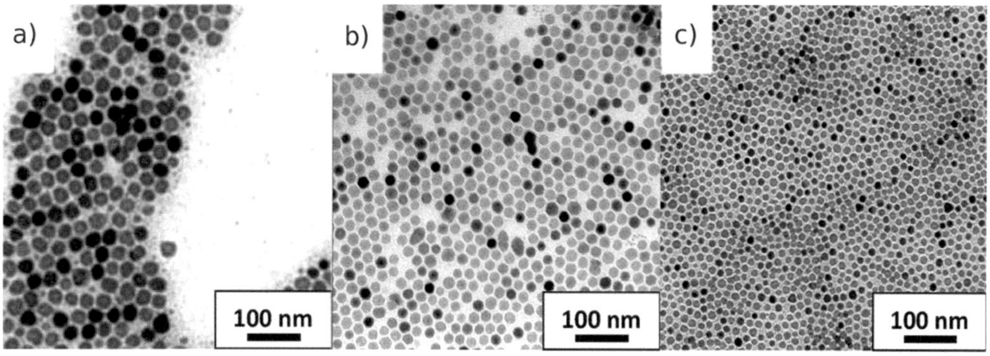

Figure 3. α-MnS NCs by a solvothermal method using sulfur powder in 1-octadecene. (a) 30 nm α-MnS NCs obtained using $Mn(Ol)_2$ as a metal precursor at 300°C. (b) 21 nm α-MnS NCs obtained using $Mn(Ol)_2$ at 320°C. (c) 14 nm α-MnS NCs obtained using Mn(OH)(Ol) at 300°C. Adapted from Puglisi A, Mondini S, Cenedese S, Ferretti AM, Santo N, Ponti A. Chem Mater. 2010;**22**:2804–2813. Ref. [32] Copyright (2010) American Chemical Society.

Since yellow sulfur powder (S_8) must be converted to sulfide anion S^{2-} during the formation of MnS NCs, using an inorganic sulfide as a precursor seemed appealing. In this manner, the synthesis of MnS NCs would not rely on the complex redox chemistry to produce sulfide anions from elemental sulfur. For instance, hydrated sodium sulfide has been employed in the synthesis of ZnS@MnS@ZnS heterostructures by co-precipitation with metal acetates in a hydroalcoholic medium (see Section 2.1.2) [6]. However, inorganic sulfides are poorly soluble in apolar solvents, which limits their application in solvothermal synthesis. This problem can be resolved by using ammonium sulfide $(NH_4)_2S$, which can be dissolved into dry oleylamine with a concentration of up to 0.5 M [36]. By injecting such solution into hot oleylamine solutions of metal precursors, the authors succeeded in the synthesis of NCs of several metal sulfides, including MnS. In this case, the sulfide (0.92 mmol) solution was injected into a solution of $MnCl_2$ (0.44 mmol, Mn:S = 1:2.1) in a 1:6 v/v (oleic acid)/oleylamine mixture at 250°C and aged for 5 min. Trapezoid-shaped MnS NCs with an average size of ca. 20 × 12 nm were obtained. The crystal structure was claimed to be γ; no further characterization data were provided. In a later report [13], a closely related reaction gave ca. 14 nm MnS NCs with trapezoid shape. In this case, the crystal structure was claimed to be β. This is surprising since the second reaction is just scaled up by a factor 2, the other synthetic conditions being unchanged. Re-examination of the XRD plots, especially the peak intensities, led us to the conclusion that both reactions produced a mixture of β- and γ-MnS NCs. In the first reaction, γ-MnS NCs were predominant and the reverse was true for the second reaction. The predominantly β-MnS NCs were then investigated as an electrode material for batteries (see below).

Another class of sulfide precursors, which can be considered as an intermediate between the extremes of elemental sulfur and inorganic sulfides, comprises small, typically organic, sulfur-containing molecules that can easily release sulfide anions upon hydrolysis. Sodium thiocyanate was used for the synthesis of α-MnS nanobelts [37] (see below) and thiourea $(NH_2)_2C{=}S$ proved to be useful in hydrothermal protocols for the preparation of 1D MnS nanosystems [38, 39] (see below). In the context of the solvothermal synthesis of 0D MnS NCs, thioacetamide $CH_3C(S)\text{-}NH_2$ gave significant results. In a heat-up approach, involving heating at 250°C a solution of $MnCl_2$ (0.50 mmol) and thioacetamide (0.50 mmol, Mn:S = 1:1) in a 1:5 v/v (oleic acid)/oleylamine mixture, monodisperse α-MnS nanocubes were synthesized [52]. Size control was achieved by varying the reaction time. A cube edge size of 14 ± 1, 26 ± 2, and 40 ± 3 nm was obtained when the reaction mixture was heated for 30, 60, and 120 min, respectively; note that the size dispersion (mean/std.dev.) is nearly constant. EDX showed that Mn:S = 1:0.9 and the two elements are homogeneously distributed. Based on the reaction carried out at different temperatures, the authors proposed that nanocubes formed via the oriented aggregation of very small spherical NCs followed by intraparticle ripening. It is interesting to note that the final nanocubes also spontaneously form meta-crystals with the simple cubic packing driven by strong interactions between the {001} faces of the nanocubes. Yang et al. [40] used thiourea in their investigation aimed at understanding the effect of reaction temperature and pressure on the crystal structure of MnS nanosystems (see Section 2.3). NC aggregation was also observed in another synthetic example where thiourea was used as a precursor [9]. In this case, $MnCl_2$ (0.20 mmol) and thioacetamide (0.60 mmol, Mn:S = 1:3) were dissolved in a complex mixture of 1-octanol/octylamine/(oleic acid)/acetone 3:3:3:1 v/v and autoclaved at 150°C for 1 h. The product comprised large (ca. 90 nm) spherical aggregates of small (ca. 5 nm) α-MnS NCs. The authors claim that NC aggregation is induced by water generated by the decomposition of thioacetamide, but it should be noted that the hydrolysis of thioacetamide to acetamide and H_2S consumes (not produces) water. Interestingly, when the hexane dispersion of these α-MnS NC aggregates was shaken with an aqueous sodium citrate solution at 85°C, the aggregates could be separated and the individual α-MnS NCs became water dispersible. These NCs were next investigated as a possible MRI contrast agent (see below).

To conclude this subsection, we mention an interesting report describing how γ-MnS NCs can be produced by cation exchange from $Cu_{2\text{-}x}S$ NCs thanks to the similarity of the roxbyite and wurtzite crystal structures [41]. Disk-shaped (ca. 6 × 16 nm) $Cu_{2\text{-}x}S$ NCs suspended in TOP were injected into a solution of Mn chloride tetrahydrate in an oleylamine/toluene mixture at 100°C and reacted for 10 min before cooling. The NCs retained their size and shape. Despite EDX showed about 10% residual Cu, both XRD and ED displayed a pure γ-MnS pattern. The interesting point is that, starting from the roxbyite structure (distorted hcp lattice of sulfide anions with copper cations in tetrahedral and trigonal sites), γ-MnS NCs with wurtzite structure (hcp lattice of S^{2-} anions with Mn^{2+} cations in tetrahedral sites) were only obtained. Since both α- and β-MnS have a ccp lattice of S^{2-} anions, the authors could conclude that cation exchange preserved the compact sulfide lattice, as expected. However, the comparison of these experiments with cation exchange assays using cobalt showed that also the cation site occupation (tetrahedral *vs.* octahedral) was preserved.

2.1.1.2. *Other methods*

A few examples reporting non-solvothermal synthesis of 0D MnS NCs could be found. Sub-microcystals of all three MnS polymorphs were prepared by a hydrothermal method [42]. Since this report describes in detail how to control the crystal structure of the product, it is reviewed in Section 2.3. Hydrothermal methods were used to synthesize 1D MnS NCs, as reviewed in the next section.

Another method used to synthesize 0D MnS NCs involves the growth of the NCs within the pores of mesoporous silica, thanks to heat or chemical treatment following impregnation of the porous materials. Spherical γ-MnS NCs were synthesized inside the pores of MCM-41 or SBA-15 silica host material [43]. It should be noted that the authors used the old nomenclature (e.g., see Ref. [19]) and denominated γ-MnS as "wurtzite β-MnS". Unfortunately, for synthetic and morphological details, the reader is referred to a previous publication [44], which describes the synthesis of Mn-doped ZnS NCs. We then provide synthetic and morphological data with this *caveat* in mind. Mesoporous silica was impregnated with Mn(II) acetate and, after drying, treated with H_2S gas at 100°C for 24 h. The impregnation/reaction cycle was carried out two or three times. The porous structure of the host material was not affected by these treatments. Using such moderate temperature for the chemical treatment, reaction with the silica wall and formation of other MnS polymorphs could be avoided. γ-MnS NCs were obtained with diameter 3, 6, 8, and 11 nm employing silica with different pore sizes.

A similar method was used to synthesize MnS NCs within mesoporous silica [25]. The main differences are the use of $Mn(EtO\text{-}CS_2)_2TMEDA$ as a xanthate single precursor with Mn:S = 1:4 and the heat (instead of chemical) treatment of the impregnated silica. As we will soon see, this causes large difference in the outcome. The P123 mesoporous silica (pore diameter 7 nm) impregnated with $Mn(EtO\text{-}CS_2)_2TMEDA$ was dried and heated in a N_2 atmosphere at 450°C for 12 h. In this case also, the ordered porous morphology of the host material was unchanged after the heat treatment. EDX showed that the products had Mn:S = 1:1 stoichiometry, but in contrast to the previous case, the XRD pattern contained peaks from all three MnS polymorphs. XRD data estimated that the MnS NC size ranged between 3.5 and 5.1 nm, smaller than the pore diameter. In this case, the influence of the mesoporous environment on the reaction mechanism was such that NCs with α, β, and γ structures were obtained, as supported by the fact that, in the absence of P123 silica, the heat treatment of $Mn(EtO\text{-}CS_2)_2TMEDA$ gave 200 nm MnS NCs with purely zinc blend β structure.

2.1.2. *Heterostructures*

A few examples of 0D heterostructured nanosystems including a MnS component have been reported, mostly related to optical applications. Nanosystems including both MnS and ZnS have gathered some attention thanks to the isostructurality of ZnS (zincblende, $F\bar{4}3m$, a = 0.54 nm) and β-MnS (a = 0.56 nm) and the similar cell size. MnS@ZnS nanosystems were prepared by a two-step solvothermal method [45]. The MnS core NCs were synthesized by injecting a sulfur solution in 1-octadecene (1.2 mmol of S) into a hot (270°C) solution of Mn(II) stearate (0.048 mmol, Mn:S = 1:25) and rapidly cooling the mixture. MnS NCs isolated after this first step are very small (diameter 2.3 nm). The ZnS shell was prepared by re-heating the MnS cores at 250°C and there injecting varying amounts of Zn(II) stearate. These nanosystems were subjected to a detailed investigation of their optical properties, but other characterization data are not abundant. The MnS@ZnS heterostructures have a diameter ranging from 4.2 to 6.0 nm, so that the ZnS shell thickness can be estimated to be in the 0.9–1.8 nm interval. The HRTEM and ED data of 5.4 nm MnS@ZnS nanosystems suggest that the MnS core has the β structure. This was confirmed in a later article by the same group [46], at least for the naked MnS NCs, by means of XRD. Recall that isostructural ZnS and β-MnS have cell size differing by about 0.7%, so that the XRD patterns of nanosized ZnS and β-MnS cannot be discerned. The detailed optical investigation of these systems is reviewed below.

Aiming at improving the optical properties of these MnS@ZnS nanosystems, ZnS@MnS@ZnS heterostructures were synthesized by a three-step hydroalcoholic co-precipitation protocol [6]. To a solution of Zn(II) acetate dihydrate (25 mmol) in an ethanol/water 1:1 mixture, a hydro-alcoholic solution of Na_2S (25 mmol) was added dropwise and a white precipitate formed. Next, a hydroalcoholic solution of Mn(II) acetate tetrahydrate (5–20 mmol) was added to form the intermediate MnS layer. Finally, the ZnS shell was prepared by a similar procedure involving the addition of 12.5 mmol of Zn followed by 25 mmol of S. Again, a detailed optical investigation (see below) came with limited morphostructural characterization. The XRD pattern of the ZnS@MnS@ZnS heterostructures is very close to that of the ZnS cores, suggesting that the MnS layer could be amorphous or β-MnS, which is isostructural with ZnS. The nanosystem size calculated from the XRD data is in agreement with that obtained from TEM images, i.e., the ZnS core has a diameter of 6.5 nm and the ZnS@MnS@ZnS nanosystems range from 5.5 to 6.3 nm, showing that leaching occurs in the second/third step. Note that no sulfur is added in the second synthetic step, thus suggesting that Mn diffuses into the lattice of the ZnS core, as also supported by optical measurements.

Finally, $Cu_{1.94}S$/γ-MnS heterostructures have been solvothermally synthesized and subjected to a detailed structural investigation [50]. The solvothermal protocol comprises two steps: first, $Cu_{1.94}S$ NCs are formed by heating a solution of anhydrous Cu(II) acetate (0.31 mmol) in a 1:2 v/v mixture of oleylamine and dodecanethiol at 220°C for 15 min, the latter one also acts as a sulfur source. Then, anhydrous Mn(II) acetate (0.70 mmol) dissolved in oleylamine is injected into the hot mixture. After aging for further 10–20 min, the mixture was cooled and the nanosystems isolated. When no Mn is added, one recovers $Cu_{1.94}S$ NCs with monoclinic structure and disk shape (diameter ~15 nm, height ~7 nm). The flat disk surfaces correspond to the (800) planes of the monoclinic structure. When the second synthetic step is carried out, it turns out that γ-MnS nanocylinders heteroepitaxially grew on *one* of the $Cu_{1.94}S$ nanodisk flat surfaces. The γ-MnS nanocylinder has the same diameter as the $Cu_{1.94}S$ nanodisk and is 14, 16, and 21 nm long for 10, 15, 20 min aging time, respectively. It is interesting that the thickness of the $Cu_{1.94}S$ nanodisk reduced to ~4 nm. HRTEM images show that both components are single crystals and that the interface, where $Cu_{1.94}S$ (800) and γ-MnS (001) planes face each other, is very abrupt. HRTEM images clearly show that the (101) planes of γ-MnS smoothly connect with the (442) planes of $Cu_{1.94}S$ across the interface, the interplanar mismatch being very small (0.3%). The authors proposed that the heterostructure initially forms by cation exchange, favored by the presence of a hcp lattice of sulfide anions in both crystals and, perhaps, by the fact that Cu^{+} and Mn^{2+} ions occupy different sites in the S^{2-} lattice. However, the question why γ-MnS grows on *one* nanodisk side only remains open.

2.2. One-dimensional (1D) nanosystems

Most 1D MnS nanowires (high aspect ratio) have been synthesized by CVD, a technique involving the decomposition of volatile thermally-labile precursors on a substrate thus producing very pure, thin-film, possibly nanostructured coatings. Because of the spatially anisotropic synthetic environment, CVD is apt for inducing the growth of 1D nanosystems. However, examples of 1D MnS nanosystems prepared by solvothermal techniques could also be found, usually nanorods with a lower aspect ratio than nanowires. Regarding the product crystal structure, 1D nanosystems crystallized in the γ-MnS structure, which is prone to anisotropic growth having hexagonal (uniaxial) structure, and in the cubic α-MnS structure. We could not find examples of β-MnS 1D nanosystems.

2.2.1. *Nanobelts, nanorods, and nanosaws*

We start with the notable example of α-MnS nanobelts synthesized by a solvothermal method. Indeed, it is noteworthy that 1D nanosystems were obtained notwithstanding that both the crystal structure and the synthetic environment are isotropic. The preparation of α-MnS nanobelts [37] involved dissolving Mn(II) acetate tetrahydrate (1 mmol) and ammonium thiocyanate (2 mmol, Mn:S = 1:2) in molten dodecylamine (40 ml) and autoclaving this solution for 72 h at 220°C. XRD showed that the product is pure α-MnS. As observed by SEM, the product comprises belt-like nanostructures, several μm long and 50–150 nm wide with an estimated thickness of 25 nm. The nanobelts have constant width along their entire length. TEM confirmed the nanobelt morphology and on the basis of ED data, the authors claimed that the nanobelt growth direction is [100] but the presented data are also consistent with the [110] growth (actually with growth along any [*hk*0] direction). The authors interestingly explored how the outcome depends on the synthetic conditions. Upon changing reaction time or temperature, the quality of the nanobelts was poor. When Mn(II) chloride or sulfate was used, the obtained product was bulk α-MnS. Replacing ammonium thiocyanate with sulfur powder or thioacetamide yielded α-MnS nanocubes and the latter were also obtained when dodecylamine was substituted with diluted ammonia. Use of sodium sulfide gave α-MnS microcrystals. It is clear that formation of the nanobelts critically depends on both physical and chemical synthetic conditions. The authors proposed that at least three conditions must be met for the formation of α-MnS nanobelts: temperature higher than 200°C to form the α-MnS phase, slow nucleation and growth (here obtained thanks to the slow release of sulfur from thiocyanate and the coordination of Mn(II) by dodecylamine), and selective surfactant adsorption on the different crystal faces of MnS. We also note that adsorption of acetate anions on the NC surface seems to be important as Mn(II) chloride or sulfate gave bulk α-MnS. The question whether α-MnS nanobelts originate from γ-MnS 1D nanostructures by a phase transformation remains open though one should note that the peculiar nanobelt morphology has not been observed in γ-MnS 1D nanostructures.

Starting from elemental Mn and S, α-MnS nanorods (40-60 nm × 0.5-1.2 μm) were prepared by a hydrothermal route [35]. Mn metal (10 mmol) and S powder (10 mmol, Mn:S = 1:1) were suspended in water and then heated at 240–260°C for 14–20 h. XPS showed that the product composition is Mn:S = 1.1:1 and XRD gave a clean α-MnS pattern. The product observed in the SEM and TEM images had nanorod morphology (diameter 40–60 nm and length 0.5–1.2 μm) with a constant width and a high aspect ratio (13–20). The authors investigated the effect of various synthetic conditions. At temperature lower than 240°C, $Mn(OH)_2$ or impure products formed and a low yield was observed when the reaction time was shorter than 10 h. When no sulfur was used, hexagonal $Mn(OH)_2$ nanorods were obtained. The authors then proposed that α-MnS nanorods formed in a two-step mechanism where $Mn(OH)_2$ nanorods first formed which then converted to α-MnS nanorods upon the action of sulfur, a process driven by the low aqueous solubility of MnS. MnS nanorods [39] were synthesized by autoclaving for 20 h at 150°C in an aqueous solution of $MnCl_2$ and thiourea (both 1 M, Mn:S 1:1) in the presence of a through-hole anodized aluminum oxide (AAO) template previously impregnated with $MnCl_2$. The produced nanorods have homogeneous morphology (diameter 60–80 nm, length 700–800 nm) and are densely packed perpendicularly to the AAO template. It is surprising that such homogeneous morphology does not correspond to a pure crystal phase. Indeed, XRD showed the presence of both α- and γ-MnS in comparable amounts. The authors also showed that the morphology can be controlled by varying the concentration of the precursors. When the latter is increased to 2 M, arrays of MnS nanowires (diameter 60–80 nm, length 30–40 μm) are obtained while MnS nanosheets (thickness 150–190 nm, width 10–15 μm) are produced when the concentration is lowered than 0.5 M. A similar hydrothermal technique was used to prepare γ-MnS nanorods-tetrapods [14] by autoclaving an aqueous solution of Mn(II) chloride tetrahydrate (21.4 mM), Na_2S (42.8 mM, Mn:S 1:2), and ammonia (~1 M). The tetrapods have reasonably good morphology: each branch is a nanorod with a diameter of ~50 nm and a length of ~250 nm. XRD showed that they are pure γ-MnS and the growth direction along the [001] direction was determined by HRTEM. Additional experiments showed that the tetrapods formed by the self-assembly of independent nanorods. These nanosystems are promising materials for supercapacitor applications (see below).

CVD has been employed to synthesize γ-MnS 1D nanosystems with a peculiar comb-shaped morphology [10]. $MnCl_2$ and sulfur powder (Mn:S = 1:3) were used as precursors and $AuCl_3$ treated (001) silicon was used as a substrate. In this way, nanoribbons with one smooth edge and one saw-toothed edge were obtained and referred to as nanosaws. The product is mainly γ-MnS but a small amount of β-MnS was also observed by XRD. The nanosaws are over 25 μm long and 100–350 nm wide (the width is uniform all along the ribbon). The saw-teeth protrudes 10–50 nm from the (imaginary) flat baseline. Combining ED and HRTEM data, it was found out that the nanosaw long axis is parallel to the $[01\bar{1}0]$ direction and that the nanosaws have (0001) side faces and $(2\bar{1}\bar{1}0)$ top and bottom faces; the teeth direction is along [0001]. The chemical composition of the nanosaws was investigated at nanometer resolution using EELS and EDX and it turned out that the nanosaws have homogeneous composition with Mn:S close to 1:1. Further accurate HRTEM investigations showed that the thickness of the nanosaws decreases from 30-50 nm at the smooth edge to a few nanometers at the teeth and provided enough information for the authors to propose the following two-step growth mechanism. First, nanoribbons are formed along the [01$\bar{1}$0] direction with (0001) side faces. Subsequently, the teeth grow in the [0001] direction starting from a Mn-terminated polar (0001) side face due to the polarization and termination of these crystal surfaces. The photoluminescence and potentiality as Li-ion battery electrode materials of γ-MnS nanosaws have also been investigated (see below).

2.2.2. Nanowires

A brief description of α-MnS nanowires obtained by atmospheric pressure CVD using $MnCl_2$ and sulfur powder as precursors and $AuCl_3$ treated (001) silicon as a substrate was reported in Refs. [47, 48] but unfortunately only SEM images and XRD pattern were given. A more detailed study regarding Cd-doped α-MnS nanowires synthesized by CVD was later published [49]. In this case, $MnCl_2$ and CdS were used as precursors and the Cd content was controlled by varying the temperature of the Mn source between 600 and 700°C. We here focus on the almost pure MnS nanowires that gave a clean α-MnS pattern in the XRD plot. The nanowires are about 20 μm long and have an average diameter of 70 nm. ED and HRTEM data showed that the growth direction is along [110]. The chemical composition was studied by EDX: a line scan perpendicular to the nanowires long axis gave a Mn:S ratio close to 1 and a Cd content lower than 1%. These conclusions were confirmed by XPS. The photoluminescence and magnetic properties of these α-MnS nanowires have also been investigated (see below).

To improve the performance of a supercapacitor material comprising γ-MnS nanorods-tetrapods, the same group synthesized another 1D γ-MnS nanomaterial by a hydrothermal method involving autoclaving for 2 h at 120°C an aqueous solution of Mn(II) chloride tetrahydrate (8.4 mM), thiourea (44 mM, Mn:S 1:5), and KOH (3.3 mM) [14]. They obtained a mixture of γ-MnS nanowires (diameter 10–20 nm, length 0.5–3 μm, growth direction [102]) and mixed-phase MnS NCs (10–20 nm). The authors attributed the presence of the NCs to fracture of the nanowires induced by lattice expansion due to the formation of manganese hydroxysulfide during the synthesis. This mixed product was used to fabricate an asymmetric supercapacitor (see below).

Purer samples of γ-MnS nanowires have been prepared by autoclaving for 18 h at 120°C a solution of Mn(II) acetate (1.44 mmol), thiourea (7.88 mmol, Mn:S ≅ 1:5.5), and ethylenediamine in water (pH ~10) [56]. XRD showed that the product is predominantly γ-MnS but some α-MnS is present. The product has nanowire morphology and interestingly comprises two types of nanowires. The primary nanowires occur in bundles and have a hexagonal cross section and [0001] growth direction. The second-type nanowires do not form bundles and have a rectangular cross section and [1$\bar{1}$00] growth direction. Unfortunately, quantitative size data are not provided and we can only roughly estimate the width of nanowires at 20 and 70 nm for [0001] and [1$\bar{1}$00] nanowires, respectively. When the nanowires deposited on a TEM copper grid were placed in a NaCl aqueous solution, copper from the grid substituted for manganese in the wurtzite γ-MnS lattice, but very differently for the two nanowires types. Primary γ-MnS nanowires pseudomorphically transformed into Cu_2S nanowires (no size or shape change) thanks to the similar structure of the two crystals (recall the above described $Cu_{1.94}S$/MnS dimeric heterostructure [50]) and to the nanowire growth direction. **Figure 1** shows that the γ-MnS structure has a layered structure comprising rather compact MnS planes perpendicular to the [0001] direction which are separated by about 0.35 nm. This interlayer space is easily accessible from the exterior, thanks to the large lateral surface of the nanowires. The substitution of Mn^{2+} with two Cu^{+} ions is thus favored by the [0001] elongated morphology of the nanowires. Conversely, in the case of [1$\bar{1}$00] nanowires, the interlayer space is less accessible and ion substitution is not favored. Indeed, as nicely demonstrated by EDX and EELS imaging (ESI), the cation exchange occurred in well-defined bands several tens of nanometer wide along the nanowires axis. Heterostructured nanowires with γ-MnS/Cu_2S nanojunctions were thus synthesized.

Another 1D nanoheterostructure involving γ-MnS nanowires has been synthesized to improve the performance of 1D γ-MnS nanosystems as electrode materials for Li-ion batteries [11]. The preparation method is similar to that used by the same authors to prepare γ-MnS nanosaws [10], but the used Mn:S ratio of 1:3 and methane gas(partial flow = 5 sccm) added to the argon flow in the furnace. The products are nanowires comprising an elongated γ-MnS nanocore conformally coated with graphitic carbon and with hexagonal cross section (see **Figure 4**). The nanowires have diameter in the range of 60–100 nm and are more than 100 μm long but they easily break into shorter nanorods upon manipulation. The heterogeneous composition of the nanowires is already visible in TEM images and HRTEM confirmed the presence of both γ-MnS, as indicated by XRD data, and graphitic carbon, as indicated by Raman spectra. HRTEM also showed that the γ-MnS nanocore growth direction is [0001] and that the γ-MnS/C interface is narrower than 0.5 nm. The nanometer-resolved chemical composition of the nanowires was deduced from EDX and EELS data, which showed that the γ-MnS nanocore has Mn:S ratio close to 1:1 and diameter in the range of 20–50 nm. Extensive mechanical and electrochemical investigations of these heterostructured nanowires are reviewed below.

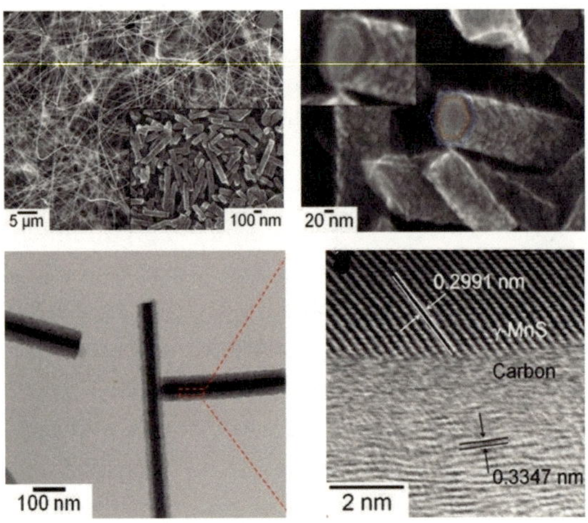

Figure 4. Electron microscopy images of γ-MnS/C nanowires. Top left:Low magnification FE-SEM image; the inset shows the nanowires after transfer onto a TEM grid. Top right: High magnification FE-SEM image showing the hexagonal cross section of the nanowires; the inset shows a close-up of an individual nanowire. TEM image (bottom left) of representative γ-MnS/C nanowires and HRTEM image (bottom right) of the γ-MnS/C interface. Adapted with permission from [11]. Copyright (2014) American Chemical Society.

http://dx.doi.org/10.5772/65092

2.3. Polymorphism control

In this section, we review reports where the issue of polymorphism control was purposely investigated. As early as 2001, Lu et al. [16] found that the crystal structure of the MnS sub-microcrystals, produced by autoclaving at 190–200°C a solution/suspension of $MnCl_2$ tetrahydrate and thiourea (Mn:S = 1:4), could be regulated by the solvent. In particular, γ-MnS nanorods and tetrapods were obtained using benzene as a solvent, β-MnS was prepared using THF, and α-MnS was obtained with aqueous solvents (water, aqueous ammonia, aqueous ethylenediamine). Unfortunately, the morphology of the α- and β-MnS crystallites was not given. These results were later confirmed by Ref. [38].

The formation of MnS NCs with a given crystal structure and their polymorphic interconversion could be controlled by changing thermal parameters, as expected on the basis of the different thermodynamical stability of the three polymorphs. In 2002, it was reported [17] that when MnS nanosystems were solvothermally prepared by decomposing Mn(II) di(ethyldithiocarbamate) (Mn:S = 1:4) in hexadecylamine, cubic α-MnS NCs was formed at 250°C, whereas γ-MnS nanowires and multipods with γ-MnS branches grown from a β-MnS seed were obtained at temperature lower than 150°C. The heating rate was found to affect the crystal structure of the synthetic outcome in a similar reaction where the same single precursor was dissolved in a 1:1:2 mixture of oleic acid, oleylamine and 1-octadecene and heated at 320°C [26]. By heating a reaction mixture at a rate of 15°C/min, α-MnS NCs were obtained, but larger pencil-shaped γ-MnS NCs were produced at a heating rate of 25 and 35°C/min.

The tendency toward α-MnS when the approach to equilibrium is favored was also reported in an investigation of the crystal structure of MnS sub-microcrystals obtained by injecting an aqueous solution of $MnCl_2$ in a pre-heated alkaline solution of NaHS (Mn:S ≅ 4:1) in a pressurized reactor (note the absence of organic reagents) [42]. The initial MnS polymorph formed upon mixing is dependent on the reaction temperature. α-MnS sub-microcrystals formed within 5 min after injection when the reaction temperature was higher than 325°C, whereas a pure γ-MnS product was obtained when the reaction temperature was 235°C. At room temperature, the initial product is a mixture of γ- and β-MnS NCs, which underwent crystal phase transformation upon heating. Pure γ-MnS crystallites were obtained by treating the reaction mixture at 150°C for 3 days while complete transformation into the thermodynamically stable α-MnS phase required temperature higher than 200°C for 3 days. In view of these results, the authors noted that it is unclear whether α-MnS sub-microcrystals prepared by injection at 325°C directly formed as such or precipitated as γ-MnS crystallites that quickly transformed into α-MnS within the 5-minute reaction time.

Pure samples of all three MnS polymorphs were synthesized again by changing the reaction temperature in a solvothermal protocol [40]. In this investigation, pressure-induced crystal phase transformation of MnS nanosystems was also reported. Anhydrous $MnCl_2$ (0.5 mmol) and thioacetamide (0.5 mmol, Mn:S = 1:1) were dissolved in oleylamine (5 ml) and autoclaved for 30 min at various temperatures. At 200°C, 15 nm β-MnS NCs were obtained while γ-MnS bipods with β-MnS cores were produced at 230 and 250°C. Pure γ-MnS rods (340 × 50 nm) were obtained at 270°C and large (ca. 500 nm) spherical α-MnS NCs formed at 280°C. These results show that small temperature differences strongly affect the product crystal structure.

However, the effect of pressure is the most interesting part of this report. Samples of β-MnS NCs and γ-MnS bipods (with β-MnS core) were compressed in a diamond anvil cell and their transformation into denser α-MnS was followed by XRD (see **Figure 5**). β-MnS NCs were more resistant to pressure (transformation in the range of 5-3–8.3 GPa) compared to γ-MnS bipods (2.9-4.7 GPa). The authors also showed that these transformations are first-order phase transitions with volume decrease of 18% (β-MnS NCs) and 25% (γ-MnS bipods).

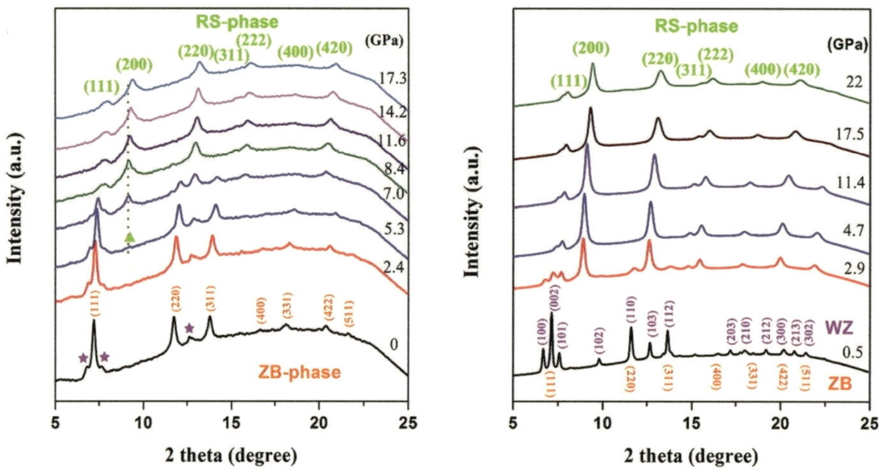

Figure 5. High-pressure XRD patterns of β-MnS NCs (left) and γ-MnS bipods with β-MnS core (right). Legend: RS = α-MnS, ZB = β-MnS, WZ = γ-MnS. The stars and triangle in the left panel represent diffraction peaks from γ-MnS and the (200) peak of α-MnS, respectively. Adapted with permission from [40]. Copyright (2012) American Chemical Society.

In addition to the variation of physical parameters, control of the polymorphism of MnS nanosystems can also be achieved by chemical means. In Ref. [34], it was briefly mentioned that using a higher (lower) sulfur content in a solvothermal protocol involving heating at 200°C a solution of Mn(II) nitrate and sulfur powder in octadecylamine produces large α-MnS NCs (micrometer-size γ-MnS nanorods) but no detail was reported. The importance of both physical and chemical parameters in polymorphism control was ascertained for a solvothermal reaction involving autoclaving a solution of $MnCl_2$ (3.2 mmol), sulfur powder (3.0 mmol, Mn:S = 1:0.94), and KBH_4 (5.0 mmol) as a reducing agent in ethylene glycol. At low temperature (≤180°C), hollow 300–500 nm spherical aggregates composed of 30–40 nm γ-MnS NCs formed, as determined by SEM, TEM, and XRD, with composition Mn:S = 1:1.03 from ICP data. These are the optimum conditions for the preparation of γ-MnS, any change in reaction conditions induced the partial or complete formation of α-MnS. This of course includes an increase in temperature but also the solvent as ethanolamine and ethylenediamine yielded α-MnS. The role of the reducing agent is also interesting. KBH_4 is necessary to prepare γ-MnS but the stronger reductant hydrazine produced α-MnS. In general, it seems that the presence of amino groups, either in the solvent molecule or in the reductant, favors the formation of α-MnS. Similar results were obtained for the hydrothermal reaction [51] of $MnCl_2$ tetrahydrate (1 mmol) with Na_2S nonahydrate (1.5 mmol, Mn:S = 1:3) at 180°C for 9 h yielding γ-MnS submicrorods (diameter 200–300 nm, length 1–1.5 μm) with a hexagonal cross section. However, when 34 mmol of hydrazine was added to the reaction mixture, α-MnS octahedral sub-microcrystals (edge length 170–200 nm) were formed. It was ascertained that neither pH nor temperature (120°C–180°C) affects the synthetic outcome. However, carrying out experiments with different reaction times and observing that treatment of the γ-MnS nanorods with hydrazine gave α-MnS led the authors to propose that α-MnS crystallites originated from redissolution of γ-MnS nanorods mediated by the formation of a hydrazine-manganese complex.

Finally, an extensive study [31] on the effect of the surfactant on the crystal phase of the MnS NCs resulting from a solvothermal protocol was carried out. The amount of sulfur needed to obtain MnS NCs by a solvothermal synthesis depends on the amount of surfactant present. A particular reaction was investigated where a mixture of (Mn precursor):sulfur:surfactant 2:4:1 dissolved in 1-octadecene was heated to 320°C (heating rate = 10°C/min). As a Mn precursor, Mn distearate and $Mn_2(CO)_{10}$ were used and a wide range of surfactants were investigated, that is, stearic acid, oleic acid, hexadecylamine, dodecylamine, octadecylamine, oleylamine, oleylalcohol, and dodecanethiol. It was concluded that to prepare γ-MnS NCs from a solution of elemental sulfur and a manganese precursor in 1-octadecene, it is necessary that the reaction mixture comprises both amine and carboxylic acid surfactants, otherwise α-MnS NCs are formed. Furthermore, whether the carboxylic acid is present as a free surfactant or as a carboxylate ligand within the Mn precursor is irrelevant to the crystal structure of the resulting NCs.

3. Optical properties

Because of the growing interest in optical and optoelectronic applications, several studies reported on the optical absorption and photoluminescence (PL) spectra of MnS nanosystems, which can be useful as light emitting and optoelectronic devices in the ultraviolet spectral region [52], thanks to the wide bandgap of MnS and blue-shifting quantum confinement effects.

The optical absorption spectra of 0D α-MnS NCs were described in several reports. Hydrothermally synthesized 30-nm spherical α-MnS NCs displayed a well-defined absorption band peaking at 261 nm (4.75 eV) [38]. Larger NCs showed broader absorption peaks. Star-shaped ~100-nm hexapods exhibited a broad peak at ca. 360 nm (3.4 eV) and ~200-nm hexagonal NCs had a similar peak but red-shifted to ca. 370 nm (3.3 eV) whereas the absorption spectrum of intermediate-size octahedral NCs is almost featureless [26]. Size-dependence of the absorption spectrum was observed for ordered aggregates of α-MnS nanocubes with edges 14, 26, and 40 nm [52]. The spectra are almost featureless but peaks in the near ultraviolet (337, 346, and 355 nm for 14, 26, and 40 nm nanocubes, respectively) were discerned and attributed to the transition to the excitonic state. The blue-shift from the 388 nm transition of the bulk material is sizeable and modulated via the nanocubes edge length.

Photoluminescence of 0D α-MnS NCs was only reported in the case of size-controlled nanocubes, [52], which when excited at 300 nm emit in the near ultraviolet range at room temperature. The PL maximum is size dependent and ranges from 356 to 373 nm (3.48–3.23 eV) when the cube edge grows from 14 to 40 nm, showing the effects of quantum confinement. This phenomenon and the NC size control afforded by wet nanocrystal chemistry techniques make size-controlled α-MnS NCs promising materials for emitting devices in the ultraviolet region. PL was also investigated in 1D α-MnS NCs. Hydrothermally synthesized α-MnS nanorods (40–60 nm × 0.5–1.2 μm) excited at 273 nm at room temperature displayed a strong, narrow PL peak at 400 nm (3.1 eV), corresponding to the good size dispersion and crystal quality of the nanorods [35]. A more in-depth investigation of the PL of α-MnS nanowires (pure and Cd^{2+}-doped) was later reported in Ref. [49]. We here focus on the pure α-MnS nanowires that are about 70 nm in diameter and grown along the [110] direction up to about 20 μm in length. Their PL was observed after excitation at 325 nm between 7 and 300 K and comprised two bands. The broad and weak band at ca. 2.9 eV (428 nm) was attributed to band-edge emission involving the excited states of the Mn^{2+} ions. The other band centered about 1.6 eV (775 nm) has strongly temperature-dependent intensity and vanishes at temperature above ~150 K (see **Figure 6**). This band was attributed to the decay of impurity-perturbed Mn^{2+} excited states. Time-resolved PL of the 1.6 eV band after excitation at 266 nm showed that its average decay time at 7 K is 40 μs.

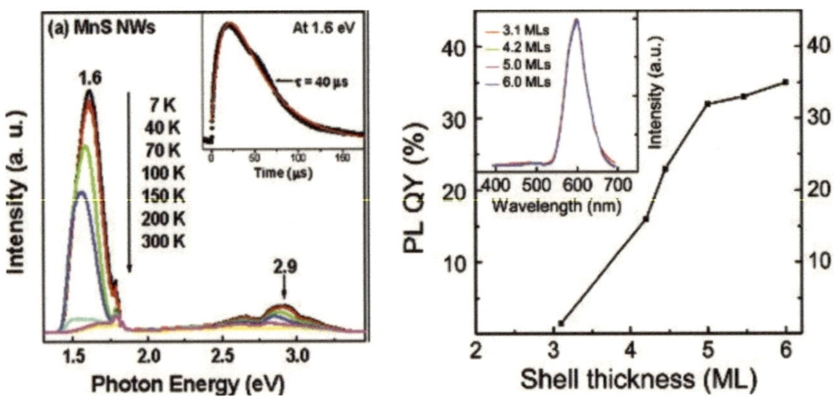

Figure 6. Photoluminescence of MnS nanosystems. Left: Temperature dependence of the photoluminescence spectrum of α-MnS nanowires measured between 7 and 300 K after excitation at 325 nm; the time resolved photoluminescence of the 1.6 eV band after excitation at 266 nm (4.66 eV) is shown in the inset. Adapted with permission from [49]. Copyright (2006) American Chemical Society. Right: Photoluminescence quantum yield of β-MnS@ZnS core–shell nanoparticles as a function of the ZnS shell thickness (1 ML = 0.31 nm); the normalized photoluminescence spectra at room temperature are shown in the inset. Reprinted with permission from [45]. Copyright (2009) American Chemical Society.

The PL of β-MnS nanosystems received some attention. The photoluminescence excitation (PLE) spectra, obtained by monitoring the intensity of the PL while varying the excitation wavelength, showed several well-resolved peaks attributed to the transition from the ground $^6A_1$ state to the excited states of Mn^{2+} ions. Interestingly, the $^6A_1 \rightarrow {}^4T_2(4D)$ transition undergoes a shift at temperatures below the Néel temperature of the bulk material revealing the strong correlation between the energy of the Mn-internal optical transitions and the magnetic transition from the paramagnetic state to the antiferromagnetic state. The optical properties of β-MnS@ZnS core-shell nanoparticles were studied in detail in Ref. [45]. By a two-step one-pot solvothermal procedure, β-MnS@ZnS core-shell heterostructures were synthesized with equally-sized β-MnS NC cores (diameter 2.3 nm) and the ZnS shell of varying thickness (0.9–1.8 nm). The PL spectrum (excitation at 325 nm) comprises the "orange" peak at ca. 600 nm (2.1 eV), the position of which is almost independent of temperature and shell thickness. The PL quantum yield instead grows rapidly with ZnS thickness, from 1.5% for the 0.9 nm thick shell to 35% for the thickest shell (see **Figure 6**). The authors showed that a large increase in PL quantum yield is mainly due to the enhancement of the energy transfer from the ZnS shell to the core Mn^{2+} ions, though a small contribution from a slight decrease in non-radiative relaxation from Mn^{2+} ions to surface states/traps brought about by the thicker ZnS shell cannot be excluded. Interestingly, while the PL quantum yield has a strong dependence on the ZnS shell thickness, the PL lifetimes at room temperature are rather insensitive to the shell thickness. In a second report [46], the same group studied the relationship between PL and the diffusion of Mn^{2+} ions into the ZnS shell occurring when a final annealing step was added to the synthetic protocol. Annealing had a detrimental effect on the PL quantum yield, which was attributed to the diffusion of Mn^{2+} ions into the ZnS shell that causes an increase in non-radiative relaxation, as also evidenced by the significant decrease of the PL lifetime upon annealing. It was shown that Mn^{2+} ions can diffuse into the shell at temperatures as low as 220°C.

Based on this concept, another group prepared small (5.5–6.5 nm) 0D ZnS@MnS@ZnS heterostructures by a hydroalcoholic co-precipitation protocol [6]. These triple-layer nanoparticles display a narrow intense "orange" PL band in addition to the short wavelength peaks attributed to defects in the ZnS lattice. The wavelength of the "orange" peak gradually shifts from 742 to 750 nm upon increasing the amount of Mn precursors in the synthesis, which presumably corresponds to increasing thickness of the MnS layer. The authors propose several explanations for the systematic red shift upon increasing the Mn content. This effect could be due to size-dependent electron-phonon coupling and crystal-field effects but they can also be related to the incorporation of Mn^{2+} ions into the ZnS lattice. Finally, long β-MnS nanowires (diameter ca. 25 nm) have a clearly defined absorption band at 371 nm (3.34 eV), described as an excitonic transition, and a Gaussian-shaped PL peak at 488 nm (2.54 eV), the shape of which was interpreted by the authors as evidence of the good monodispersion and surface passivation of these solvothermally prepared nanowires [27].

We are aware of a single report on the optical properties of 0D γ-MnS NCs, whereas 1D γ-MnS nanosystems have attracted some attention. Small (3–11 nm) γ-MnS NCs grown inside mesoporous silica exhibited the so-called "orange" band at 2.1 eV (590 nm) when excited at 325 nm [43]. Thin γ-MnS nanowires (diameter 2.2 nm, aspect ratio ~80) displayed an absorption band edge at 3.64 eV (341 nm) and a well-defined PL peak centered at 3.34 eV (371 nm), which therefore seem to be promising as near ultraviolet emitters [17]. A morphologically heterogeneous sample of γ-MnS rod-like and branched nanosystems (with an aspect ratio of 6–10) displayed a very broad absorption band at 278 nm (4.46 eV) and very weak PL with peaks at 368 and 438 nm (3.37 and 2.83 eV), both attributed to band edge emission (no trap state emission was detected) [38]. The optical properties of CVD-synthesized γ-MnS nanosaws (length ~25 μm in the [01$\bar{1}$0] direction, width = 100-350 nm) were analyzed in detail in Ref. [10]. The PL spectrum of the nanosaws was recorded at temperature between 10 and 300 K after excitation at 325 nm. Three PL bands were described. Two relatively weak bands were ascribed to (blue-shifted) exciton-related band edge emission (371 nm, 3.34 eV) and surface defect emission (430 nm, 2.88 eV). The predominant band is the "orange" band at 582 nm (2.13 eV), which is attributed to the radiative recombination of electrons in surface state shallow traps with photogenerated holes caused by lattice stacking faults. The "orange" band is 10^4 times more intense than the exciton-related band, which is in agreement with the high density of stacking faults evidenced by HRTEM. The PL intensity of the "orange" band decreased quickly when the temperature was below 100 K, to be compared with the Néel temperature of bulk γ-MnS (~80 K). Indeed, the antiferromagnetic ordering of the Mn^{2+} ions contributes to the carrier transfer of electrons from the conduction band to donor states. By fitting the thermally activated carrier transfer model to the temperature dependence of the "orange" band intensity, it was determined that the donor states are ~40 meV below the conduction band. This result supports the claim that the donor states are associated with surface states. The lifetime of the "orange" band at 300 K turned out to be 44 μs, shorter that the bulk value (140 μs) further supporting the view that the high surface/volume ratio of the nanosaws facilitates energy transfer from Mn^{2+} excited states to surface states and thus accelerates non-radiative relaxation.

In summary, all α-MnS NCs show band-edge emission but only 14–40 nm nanocubes are able to emit in the ultraviolet range. The "orange" band was only observed in the spectrum of the thicker and longer α-MnS nanorods. Small (2–11 nm) 0D β-MnS NCs, both uncoated and ZnS-coated, display the "orange" peak at 2.1 eV in the PL spectrum, while β-MnS nanowires (diameter 25 nm) have a significantly blue-shifted PL peak at 2.54 eV; ZnS-coated β-MnS NCs have an attractive PL quantum yield of 35%. Very thin γ-MnS nanowires emit in the near ultraviolet, whereas large nanosaws are good "orange" emitters. In conclusion, the optical properties of MnS nanostructures are very sensitive to morphology and crystal defects so that tight synthetic control is required to obtain good performance. Interesting results, especially in the near ultraviolet region, have been demonstrated, and MnS remains a promising candidate as an emitter in the range 340–380 nm.

4. Electrical properties and applications

Recently, electrical properties and applications of MnS nanosystems have attracted much interest. Indeed, several studies showed that MnS NCs are promising materials as electrodes for lithium ion batteries (LIBs) [10–13, 53] and as supercapacitors [14, 15]. However, the first electrical characterization in 2008 was the measurement of the current-voltage (*I-V*) curve of α-MnS nanobelts [37] obtained by an autoclave solvothermal technique. The nanobelts grow along the [100] direction and are longer than 1 μm, 50–150 nm wide and 25 nm thick. Several measurements of single nanobelts gave a linear *I-V* response between −1 and +1 V (ohmic behavior) with a specific conductance of 3.4×10^5 S/m, comparable to that of 1D ZnO or In_2O_3 nanosystems grown by physical methods.

The use of crystalline nanomaterials to build electrodes for LIBs has gathered much attention since nanomaterials have a high surface/volume ratio, reduce Li diffusion length, are able to accommodate crystal strain, can be processed from solution, and can be produced by the rich toolbox of colloidal chemistry, which enables one to control the nanomaterial composition, size, and shape. It is noteworthy that all three MnS polymorphs have been investigated as nanosized LIB electrode materials. To put the capacity values reported below in the appropriate framework, we recall that the theoretical capacity of MnS is 616 mAh g^{-1}. Sub-micrometric α-MnS crystals, hydrothermally-grown at different temperatures, were mixed with carbon black and a polymeric binder (8:1:1 w/w) and spread on copper foil. These electrodes showed good performance [53]. All samples exhibited a lithiation plateau at about 0.7 V versus Li/Li^+, which corresponds to the Li^+ insertion in MnS and evidences that the α-MnS sub-microcrystals are suitable anode materials for LIBs. The α-MnS sample prepared at the lowest temperature (120°C) displayed the highest initial lithiation capacity (1327 mAh g^{-1}), thanks to the small particle size (150–600 nm). In all cases, however, the lithiation capacity is sharply reduced in the second cycle and then it decreased slowly upon cycling. The large irreversible decrease between the first and second cycles was attributed to the formation of a solid electrolyte interface film on the surface of the electrode. Considering the overall behavior, the best electrode is prepared using particles synthesized at 160°C (about 700 nm in size) which maintained a capacity of 578 mAh g^{-1}, corresponding to ca. 81% of the second lithiation capacity, after 20 cycles. The authors attributed this performance to the better crystallinity of the 160°C α-MnS sub-microcrystals.

Recently, β-MnS NCs were used to prepare promising electrode materials for LIBs without employing any binder or conductive filler (e.g., carbon black) [13]. The solvothermally prepared ~15 nm β-MnS NCs were first electrophoretically deposited on copper plates and then stripped of their organic ligand coating by either heating at 300°C or dipping in a methanolic solution of ammonium sulfide. The charge-discharge curves and the capacity retention on cycling of both heat-treated and dipped β-MnS NC film electrodes are shown in **Figure 7**. These electrodes showed a lithiation plateau at 1.7 V, stable for more than 50 cycles, which makes them suitable as cathodes in LIBs. The electrodes showed excellent stability with a final capacity of 420 (dipped) and 470 mAh g^{-1} (heated) after more than 50 charge-discharge cycles. The good performance of these electrodes was attributed to the improved interparticle coupling and electrical properties (e.g., electrical conductivity) obtained by the combination of electrophoretic deposition and ligand stripping which allowed the authors to prepare electrodes free of additives and organic ligands.

Finally, also 1D γ-MnS nanosystems were used as electrodes due to their potential application for LIBs. CVD synthesized γ-MnS nanosaws (length ~25 μm in the $[01\bar{1}0]$ direction, width = 100–350 nm) were mixed with carbon black and a polymeric binder (7:2:1 w/w) and spread on copper foil [10]. This electrode showed good performance in the initial cycle with a capacity of 1150 mAh g^{-1} and an extended plateau at ~0.5 V. The performance, however, degraded rapidly with cycling (second cycle capacity = 780 mAh g^{-1}, after 40 cycles capacity ~200 mAh g^{-1}) and the voltage plateau disappeared. To improve the electrochemical performance, the authors succeeded in synthesizing γ-MnS nanowires (aspect ratio ≈1000) conformally coated with graphitic carbon by a single-step CVD technique. The nanowires have a hexagonal section [60–100 nm, inner (γ-MnS) diameter is 20–50 nm] and are over 100 μm long in the [0001] direction. The electrodes were prepared by mixing the carbon-coated nanosaws with a polymeric binder (92:8 w/w) and spreading the mixture on a copper foil. The charge-discharge curves displayed no voltage plateau. The initial capacity of 1036 mAh g^{-1}decreased to 747 mAh g^{-1} at the second cycle and reached a good 503 mAh g^{-1} after 25 cycles. The improved results were ascribed to the presence of a highly conducting carbon layer, which is thin enough to allow Li^+ to reach the inner γ-MnS core, prevents the decomposition of polysulfides, and provides more active regions for the electrochemical reaction.

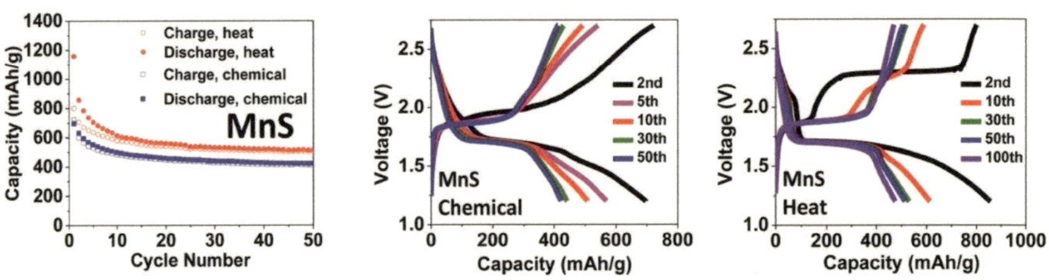

Figure 7. Electrochemical performance of β-MnS NC film electrodes without a conductive filler and a polymeric binder. Left: Capacity retention on cycling. Charge-discharge curves of chemical- (middle) and heat-treated (right) β-MnS NC film electrodes; the plateau at 1.7 V can be clearly seen. Adapted with permission from [13]. Copyright (2012) American Chemical Society.

Nanosized γ-MnS has very recently turned out to have exciting properties for supercapacitor applications. A supercapacitor is a high-capacity electric capacitor showing performance between conventional electrolytic capacitors and rechargeable batteries. They typically store 10–20% of the energy that one can put into a battery but supercapacitors store and release energy much faster than a battery and can withstand many more cycles because they do not rely on electrochemical reactions to charge-discharge. Supercapacitors have advantages in applications where a large amount of power is needed for a relatively short time and when a very high number of charge-discharge cycles or a longer lifetime is required. A typical application is power supply stabilization in consumer electronics: supercapacitor-powered devices (e.g. portable speakers, screwdrivers) have already reached the market. Supercapacitors are widely used to reduce energy consumption in transportation, e.g., recovery of braking energy since they can quickly store and release energy over long times with a high cycle rate.

Hydrothermally synthesized γ-MnS tetrapods (with 50 × 250 nm branches) were mixed with acetylene black and a fluorinated polymeric binder (70:15:15 w/w) and used to coat a nickel foam current collector [14]. Cyclic voltammetry showed that these electrodes have high specific capacitance even at a high rate, e.g., of 705 F g^{-1} at a scan rate of 1 mV/s and 323 F g^{-1} at 100 mV/s. This excellent performance was attributed to the intercalation of

http://dx.doi.org/10.5772/65092

hydroxyl ions between the MnS layers parallel to the basal crystal plane, which are easily accessible from the exterior since they are perpendicular to the nanorod axis along the [001] direction. The charge-discharge curves maintain their quasi-triangular shape at a current density as high as 10 A g^{-1}. The γ-MnS tetrapod capacitors had very good cycling performance with a capacity retention of 80% after 1000 cycles and 63% after 5000 cycles, with high and constant coulombic efficiency. The XRD investigation of cycled electrodes indicated that the performance degradation can be attributed to the transformation of the NCs from the γ- to the stable α-crystal structure. Asymmetric capacitors prepared by coupling a γ-MnS tetrapod positive electrode with a negative activated carbon electrode were shown to be purely capacitive and maintain the high rate capacity of the symmetric capacitor. The specific capacity is of course lower (59.8 F g^{-1} at 1 mV/s scan rate and 37.6 F g^{-1} at 100 mV/s) but the achieved energy density (11.7 Wh kg^{-1} at the power density of 4.45 kW kg^{-1}) is higher than that of carbon-based symmetric capacitors in aqueous electrolytes. In order to achieve better performance by improving the carbon-based-electrode, the same group later reported on an asymmetric supercapacitor comprising inhomogeneous MnS nanomaterials (NCs and nanowires, all three polymorphs present) as a positive electrode and highly porous amorphous carbon derived from eggplants [15]. These capacitors (with almost purely capacitive behavior) exhibited high specific capacitance (around 100 mAh g^{-1}) even at a rate of 100 mV s^{-1} and at current as high as 50 mA and retained 90% of the capacitance after 5000 cycles. The authors attributed this stability to the ionic accessibility and stability of the layered crystal structure of γ-MnS and showed that after being fully charged, two capacitors in series can light up a red LED indicator for 15 minutes.

5. Magnetic properties and applications

As anticipated, all three MnS polymorphs have antiferromagnetic (AFM) structure at low temperature, since the high-spins d^5 Mn^{2+} (S = 5/2) ions are antiferromagnetically coupled by superexchange interaction. The ordering (Neél) temperature T_N of bulk samples is highest for α-MnS (154 K), followed by β-MnS (~100 K) and γ-MnS (~80 K) [2]. Above T_N, all three forms of MnS are paramagnetic (PM) and follow the Curie-Weiss law $\chi = N(p_{eff}\mu_B)^2/[3k(T-\Theta)]$ with Θ = −465, −982, and −932 K for the α, β, and γ structures, respectively [19]. The effective magnetic moment corresponds to five unpaired electrons, e.g., p_{eff}= 4.54 for α-MnS [54]. The magnetic properties of MnS nanosystems have been investigated by recording the isothermal magnetization $M(H)$, useful to observe the low-temperature hysteresis loop and to measure the Curie-Weiss temperature Θ in the paramagnetic phase, and the thermal dependence of the zero-field-cooled (ZFC) and field-cooled (FC) magnetization $M(T)$, which provide information about the AFM/FM transition.

Several reports focused on α-MnS 0D NCs prepared by solvothermal techniques. Puglisi et al. [32] studied octahedral α-MnS NCs with size 14, 20, and 29 nm. The general behavior of the NCs corresponds to that observed for AFM materials. The high-temperature magnetization obeys the Curie-Weiss law with size-dependent Θ = −149 K (14 nm), −227 K (20 nm), −272 K (29 nm), much higher than the bulk value. The effective magnetic moment is close to the bulk value, ranging from 4.6 to 4.9 μ_B. These results clearly indicate that the strength of the exchange interaction supporting the AFM order is strongly size-dependent whereas the local electronic structure of the Mn^{2+} ions is similar to that in the bulk. The low-temperature magnetization does not show cusps related to the AFM/PM phase transition but suggests the presence of some ferromagnetic (FM) materials, in particular for the 29 nm NCs which display a maximum in the ZFC magnetization at 25 K. This was confirmed by the isothermal magnetization at 5 K displaying an open loop with strongly size-dependent coercivity $H_c = 9$ Oe (14 nm), 81 Oe (20 nm), 180 Oe (29 nm). To ascertain whether this is due to an FM impurity or to an FM region in the NC, field-cooled hysteresis loops were recorded (see **Figure 8**). These were shifted in the direction opposite to the field by a size-dependent amount $H_{shift} = 0$ Oe (14 nm), −17 Oe (20 nm), −44 Oe (29 nm). This exchange bias effect is the hallmark of an exchange interaction across a FM/AFM interface and is evidence of the presence of a FM region in the NCs. In accordance with the interpretation of a similar behavior observed in several magnetic NC types, it was proposed that α-MnS displays a magnetic core-shell structure, where the outer shell of the NC is (at least partially) FM because of the under-coordination of the Mn^{2+} ions and lattice distortion near the NC surface. The FM shell is exchange coupled to the AFM core and gives rise to the open and shifted hysteresis loops observed.

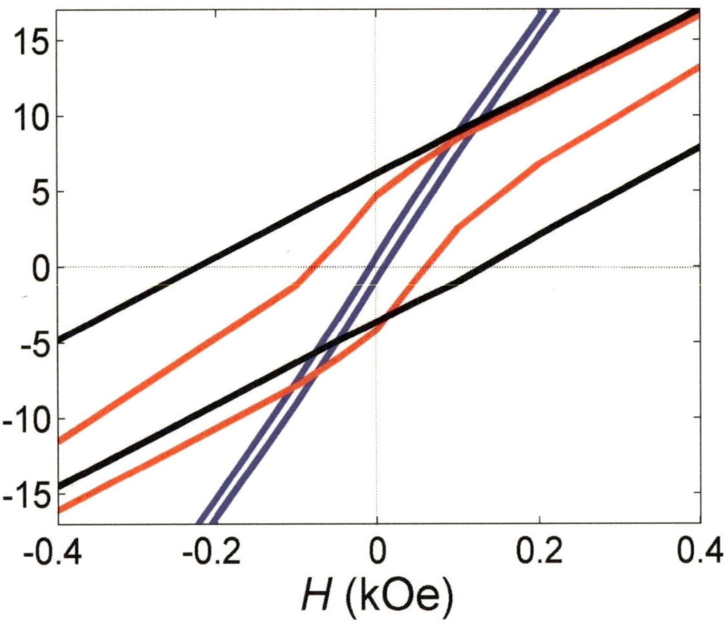

Figure 8. Low field part of the field-cooled isothermal magnetization (hysteresis loop) of 14 nm (blue), 20 nm (red), and 29 nm (black) α-MnS NC. Both the open loop and the loop shift are clearly seen.

Size-dependent magnetic properties were also measured for α-MnS cubic NCs with an edge size of 14, 26, and 40 nm [52] and further supported the core-shell model. The ZFC/FC magnetization is similar to that observed for octahedral NCs, showing at 32 K a maximum of the ZFC magnetization of 26 and 40 nm NCs. The largest NCs also reveal an increase of the FC magnetization below 28 K related to the freezing of the surface spins. The low-temperature hysteresis loops confirmed the presence of an FM region since they display open loops with size-dependent coercivity ranging from 10 Oe (14 nm) to 1265 Oe (40 nm), which is in good agreement with the coercivity of octahedral NCs. The study of the magnetic properties of 45 nm star-shaped α-MnS NCs [33] again confirmed this model by showing that at 5 K the NCs have an open hysteresis loop with large H_c = 1573 Oe; the authors pointed out that the FC magnetization increase below 150 K is related to the freezing of the surface spins in an FM-like shell. In their pioneering article, Kan et al. [18] also studied the magnetic properties of spherical α-MnS NCs with size 20, 40, and 80 nm. They found a generally AFM behavior with size-dependent T_N (50, 97, and 116 K) and Θ (−59, −86, and −174 K) for 20, 40, and 80 nm NCs, respectively. Again, this indicates that the AFM magnetic ordering is less effective at small size. These Θ values are, however, high when compared to the size trend displayed by octahedral NCs [32], and their hysteresis loop is closed. This suggests that these spherical NCs actually are agglomerates of smaller NCs, a conclusion also supported by the TEM images.

Two facts about the magnetic properties of 0D α-MnS NCs are generally agreed upon. First, the NC core undergoes a low-temperature AFM transition characterized by negative Curie-Weiss temperature Θ that approaches the bulk value (−465 K) as the NC size increases. Second, the NC surface spins form an FM-like layer, which is clearly observed in the low-temperature hysteresis loops. The observed coercivity ranges from the hundreds of Oe for 20–30 nm octahedral NCs to the kOe range for 40 nm nanocubes and 45 nm nanostars. Exchange coupling between the AFM core and the FM shell has been detected as a shift (exchange bias) of the field-cooled hysteresis of octahedral α-MnS NCs [32].

The magnetic properties of 1D α-MnS nanosystems have also been studied. α-MnS nanowires, slightly doped with Cd, have been grown by CVD to about 20 μm length and average diameter 70 nm [49]. The growth direction is along [110]. They generally behave as an AFM material both at low temperature (where they display closed isothermal magnetization curves) and at high temperature ($T > 100$ K) where they follow the Curie-Weiss law with $\Theta = -903$ K. This value is much higher than the bulk value, suggesting a remarkably stronger AFM correlation in the nanowires than in the bulk. The authors hypothesized that the stronger AFM could be caused by the intrinsic geometrical anisotropy of the nanowires but did not exclude that the Cd^{2+} impurity played a role. We also remark the absence of magnetic signatures indicating the presence of an FM-like phase at the nanowires surface. The magnetism of α-MnS nanobelts has also been characterized [37]. These nanobelts, obtained by an autoclave solvothermal technique and grown along the [100] direction, are longer than 1 μm, 50-150 nm wide and 25 nm thick. An FM-like shell since was present since the hysteresis loop at 5 K is open with coercivity H_c = 4013 Oe, a large value that might be due to the relatively high surface/volume ratio of the nanobelts with respect to the 0D NCs. It remains unclear why these nanobelts have an FM-like shell while the above nanowires do not. We can tentatively propose that the reason is that the nanowires prepared by CVD are free from organic coating on the surface (the nanobelts are stabilized by dodecylamine) and perhaps have a better lattice quality in the near-surface region.

The magnetic properties of γ-MnS nanosystems have been rarely reported. However, a detailed study of 0D γ-MnS NCs has been described. Spherical γ-MnS NCs with size 3–11 nm were prepared by intrapore synthesis inside mesoporous silica [43]. The high-temperature part of the NC magnetization versus temperature curve follows the Curie-Weiss law with Θ ranging from about −500 K (11 nm) to about −100 K (6 nm). It is noteworthy that 3 nm γ-MnS NCs closely follow the Curie-Weiss law down to 2 K, indicating that AFM ordering is not present even at such low temperature. Therefore, the behavior of γ-MnS NCs agrees with the conclusion drawn for α-MnS NCs, i.e., the onset of AFM ordering occurs at lower temperature for smaller NCs. Deeper insight into the magnetism of the γ-MnS NCs was obtained by studying their electron paramagnetic resonance (EPR) spectrum. EPR is a local probe of the magnetic state of transition ions but, because of the strong interactions between Mn^{2+} ions, the EPR spectrum appears as a broad line without any resolved feature from zero-field or hyperfine interaction. Careful analysis of the temperature-dependent EPR spectra showed that they are actually composed of a broad and a narrow line, the latter following the purely PM Curie law. The broad line follows the Curie-Weiss law with negative Θ typical of AFM materials and the EPR-derived Θ values closely agree with those derived from magnetization measurements, evidencing that the broad EPR line corresponds to the AFM ordered core of the NCs. Since optical measurements showed that the Mn^{2+} exchange constant does not depend on size, the strong decrease in the AFM onset with size was attributed to the presence of a disordered PM shell. By analyzing the relative intensity of the broad (AFM) and narrow (PM) EPR lines, the authors could conclude that the surface/volume ratio is critical to determine the onset of the long-range ordered AFM phase and that the minimum diameter to observe AFM ordering of γ-MnS NCs at non-zero temperature is between 3 and 6 nm, corresponding to 1000–4000 Mn^{2+} ions.

Heterostructures consisting of a 16 × 16 nm γ-MnS nanocylinder epitaxially grown on a 16 × 4 nm $Cu_{1.94}S$ nanoplate were synthesized by a solvothermal technique [50]. The isothermal magnetization curve is purely PM but no conclusion can be drawn since the measurement temperature is not specified. The magnetic properties of solvothermally prepared γ-MnS nanorods (2.4 × 20 nm) were measured [17]. The isothermal magnetization at 5 K displayed an open loop with large coercivity (H_c = 1020 Oe) and the ZFC magnetization showed a maximum at 30 K. These observations support the presence of an FM-like shell surrounding an AFM core, as occurs in α-MnS nanosystems. The large coercivity is expected on the basis of the large surface/volume ratio of 1D nanosystems.

No magnetic characterization of β-MnS NCs seems to have been reported except for a brief EPR study. The RT EPR spectrum of solvothermally prepared β-MnS nanowires with 25 nm diameter was described [27] as a single line at g = 2.0064 with linewidth about 1000 Oe, typical of strongly coupled high-spin Mn^{2+} ions. The line shape is asymmetric reflecting the morphological anisotropy and inhomogeneity of the sample. The linewidth is larger than that observed for 0D γ-MnS NCs (about 500 Oe for 11 nm NCs) [43] as expected on the basis of the nanowire diameter. Finally, a mixture of α-, β-, and γ-MnS NCs (size 3.5–5.1 nm) within the 7 nm pores of mesoporous P123 silica displayed ZFC/FC curves similar to those observed for α-MnS NCs, with a maximum at T = 42 K attributed to the FM-like surface shell of the NCs. The presence of such an FM-like region was confirmed by recording isothermal hysteresis loops between 10 and 50 K. The loops up to 35 K are open and support the presence of an FM-like component while the loop at 50 K is purely reversible. Coercivity values were not given.

In conclusion, the magnetic properties of MnS nanosystems can mostly be understood on the basis of the magnetic core-shell model where just the nanosystem core undergoes a size-dependent AFM transition (the smallest the size, the lowest the transition temperature) whereas the nanosystem surface shell form a FM-like (α- and γ-MnS) or PM (γ-MnS) layer. We recall that the observation of exchange bias shifted hysteresis loops in octahedral α-MnS NCs [32] is a strong evidence of the presence of an FM/AFM interface.

As a very promising application of the magnetism of MnS, we mention a report demonstrating very good performance of α-MnS NCs as a T_1 contrast agent for MRI [9]. The usefulness of α-MnS microparticles as a T_1 contrast agent had been demonstrated as early as 1985 [8] and very recently, α-MnS NCs were shown to have larger T_1 relaxivity than a commercial gadolinium-based contrast agent. Solvothermally prepared (5 ± 2) nm α-MnS NCs were coated with citrate to obtain a stable NC dispersion in water. It was shown that they have T_1 relaxivity = 3.4 mM^{-1} s^{-1} in water and 8.7 mM^{-1} s^{-1} when bovine serum albumin (BSA) was added to the aqueous dispersion to simulate the protein content of blood. These NCs have larger T_1 relaxivity than most reported values for the widely studied Mn oxide (MnO and Mn_3O_4) NCs and, more importantly, than the commercial T_1 contrast agent Gd-(diethylenetriaminepentacetic acid) complex. This means that 5 nm α-MnS NCs can provide equal image contrast with a ca. tenfold lower dose. In vitro viability assays showed that these NCs are not cytotoxic to HepG2 cells up to a dose of 70 μg/ml. In vivo T_1-weighted images of a nude mouse at 0.47 T showed good contrast in liver and kidneys 1 h after injection. Thus, α-MnS NCs undergo the usual fate of IV injected NCs, that is capture by the liver MPS cells. Thanks to their very small size, the NCs are partially excreted through the renal pathway. The uptake and excretion rate, both for liver and kidneys, are similar to that of the commercial Gd-based T_1 agent.

6. Other

The mechanical properties of individual heterogeneous γ-MnS@(graphitic carbon) nanowires synthesized by CVD have been measured using an AFM holder in a TEM chamber [11]. The nanowires have hexagonal section (60–100 nm) and are over 100 μm long in the [0001] direction; the inner γ-MnS core has a diameter of 20–50 nm. Under uniaxial compression, the nanowires first underwent a weak plastic deformation followed by fracture in brittle mode at a critical force of 1330 nN. The fracture occurs perpendicular to the nanowires axis along the [0001] direction. The calculated Young modulus (65 GPa) is comparable to that of 1D ZnS nanosystems and three orders of magnitude lower than that of carbon fibers, suggesting that the modulus is dominated by the γ-MnS core. This was supported by bending and axial stress experiments, which showed that the strain is mostly confined to the γ-MnS core.

As a last application, we mention that 3–5 nm MnS NCs (a mixture of all three polymorphs) prepared inside mesoporous silica by impregnation/calcination were successfully used as a catalyst for the growth of carbon nanocages when a flow of methane was passed on the NC-containing silica [55]. The latter have non-uniform size but their framework elements are 3–7 nm thick, consistent with the size of the MnS NCs. This suggests that the framework elements grow starting from the MnS NCs in the silica pores. A growth mechanism has been proposed involving the formation of reduced metal sites on the MnS NC surface where methane cracking occurs. The deposited carbon cannot form carbides and nucleates the growth of the nanocages. The authors claim that this is the first example of using a transition metal sulfide for the synthesis of carbon nanostructures.

7. Conclusions

We have reviewed the literature concerning the synthesis, properties, and application of MnS nanosystems having at least one dimension smaller than 100 nm, with a few exceptions regarding MnS sub-microparticles with interesting properties and applications. The wide variety of successful synthetic techniques for the preparation of 0D and 1D MnS nanosystems (both homogeneous and heterogeneous) with size, shape, and polymorphism control is astounding. However, the issue of independently controlling morphology and crystal structure is not yet completely solved. Further research in this direction is needed to foster application of MnS nanosystems in the above described fields. Recently, excellent results on the performance of MnS nanomaterials have been reported that we believe will further stimulate basic and applicative research in the exciting field of MnS nanosystems.

Author details

Anna M. Ferretti, Sara Mondini and Alessandro Ponti*

*Address all correspondence to: alessandro.ponti@istm.cnr.it

Laboratorio di Nanotecnologie, Istituto di Scienze e Tecnologie Molecolari, Consiglio Nazionale delle Ricerche, Milano, Italy

References

[1] Lokhande CD, Ennaoui A, Patil PS, Giersig M, Diesner K, Tributsch H. Process and characterisation of chemical bath deposited manganese sulphide (MnS) thin films. Thin Sol Films. 1998;330:70–75. doi: 10.1016/s0040-6090(98)00500-8

[2] Danielian A, Stevens KWH. Exchange interactions in the polymorphic forms of MnS. Proceedings of the Physical Society. 1961;77:124–128. doi: 10.1088/0370-1328/77/1/315

[3] Tappero R, Darco P, Lichanot A. Electronic structure of α-MnS (alabandite): An ab initio study. Chem Phys Lett. 1997;273:83–90. doi: 10.1016/s0009-2614(97)00591-5

[4] Hobbs D, Hafner J. Magnetism and magneto-structural effects in transition-metal sulphides. J Phys Condens Matter. 1999;11:8197–8222. doi: 10.1088/0953-8984/11/42/303

[5] Kravtsova AN, Stekhin IE, Soldatov AV, Liu X, Fleet ME. Electronic structure of MS (M=Ca,Mg,Fe,Mn): X-ray absorption analysis. Phys Rev B. 2004;69:134109, 12p. doi: 10.1103/PhysRevB.69.134109

[6] Viswanath R, Naik HSB, Kumar GSY, Kumar PNP, Harish KN, Prabhakara MC. Luminescence properties of blue-red emitting multilayer coated single structure ZnS/MnS/ZnS nanocomposites. Spectrochim Acta A. 2014;125:222–227. doi: 10.1016/j.saa.2014.01.022

[7] Zhang XV, Martin ST, Friend CM, Schoonen MAA, Holland HD. Mineral-assisted pathways in prebiotic synthesis: Photoelectrochemical reduction of carbon(+IV) by manganese sulfide. J Am Chem Soc. 2004;126:11247–11253. doi: 10.1021/ja0476415

[8] Chilton HM, Jackets SC, Hinson WH, Ekstrand K. Use of a paramagnetic substance, colloidal manganese sulfide, as an NMR contrast material in rats. J Nucl Med. 1984;25:604–607.

[9] Meng J, Zhao Y, Li Z, Wang L, Tian Y. Phase transfer preparation of ultrasmall MnS nanocrystals with a high performance MRI contrast agent. RSC Adv. 2016;6:6878–6887. doi: 10.1039/c5ra24775f

[10] Beltran-Huarac J, Palomino J, Resto O, Wang J, Jadwisienczak WM, Weiner BR, Morell G. Highly-crystalline γ-MnS nanosaws. RSC Adv. 2014;4:38103–38110. doi: 10.1039/c4ra05561f

[11] Beltran-Huarac J, Resto O, Carpena-Nunez J, Jadwisienczak WM, Fonseca LF, Weiner BR, Morell G. Single-crystal γ-MnS nanowires conformally coated with carbon. ACS Appl Mater Interfaces. 2014;6:1180–1186. doi: 10.1021/am404746k

[12] Lee SM, Lee J-K, Kang YC. Electrochemical properties of hollow-structured MnS-carbon nanocomposite powders prepared by a one-pot spray pyrolysis process. Chem. Asian J. 2014;9:590–595. doi: 10.1002/asia.201301261

[13] Ha D-H, Ly T, Caron JM, Zhang H, Fritz KE, Robinson RD. A general method for high-performance li-ion battery electrodes from colloidal nanoparticles without the introduction of binders or conductive-carbon additives: The cases of MnS, $Cu_{2-x}S$, and Ge. ACS Appl Mater Interfaces. 2015;7:25053–25060. doi: 10.1021/acsami.5b03398

[14] Tang Y, Chen T, Yu S. Morphology controlled synthesis of monodispersed manganese sulfide nanocrystals and their primary application in supercapacitors with high performances. Chem Commun. 2015;51:9018–9021. doi: 10.1039/c5cc01700a

[15] Chen T, Tang Y, Qiao Y, Liu Z, Guo W, Song J, Mu S, Yu S, Zhao Y, Gao F. All-solid-state high performance asymmetric supercapacitors based on novel MnS nanocrystal and activated carbon materials. Scientific Reports. 2016;6:23289, 9p. doi: 10.1038/srep23289

[16] Lu J, Qi PF, Peng YY, Meng ZY, Yang ZP, Yu WC, Qian YT. Metastable MnS crystallites through solvothermal synthesis. Chem Mater. 2001;13:2169–2172. doi: 10.1021/cm010049j

[17] Jun YW, Jung YY, Cheon J. Architectural control of magnetic semiconductor nanocrystals. J Am Chem Soc. 2002;124:615–619. doi: 10.1021/ja016887w

[18] Kan SH, Felner I, Banin U. Synthesis, characterization, and magnetic properties of α-MnS nanocrystals. Isr J Chem. 2001;41:55–61. doi: 10.1560/1FB3-1PF4-72JQ-0AQC

[19] Corliss L, Elliott N, Hastings J. Magnetic structures of the polymorphic forms of manganese sulfide. Phys Rev. 1956;104:924–928. doi: 10.1103/PhysRev.104.924

[20] Papp G. History of minerals, rocks and fossil resins discovered in the Carpathian region. Budapest: Hungarian Natural History Museum; 2004.

[21] Eriksson L, Kalinowski MP. $Mn_{1-x}Fe_xS$, $x \cong 0.05$, an example of an anti-wurtzite structure. Acta Crystallogr Sect E Struct Rep Online. 2001;57:I92–I93.

[22] Ma C, Beckett JR, Rossman GR. Browneite, MnS, a new sphalerite-group mineral from the Zaklodzie meteorite. Am Mineral. 2012;97:2056–2059.

[23] Pradhan N, Katz B, Efrima S. Synthesis of high-quality metal sulfide nanoparticles from alkyl xanthate single precursors in alkylamine solvents. J Phys Chem B. 2003;107:13843–13854.

[24] Tian L, Yep LY, Ong TT, Yi J, Ding J, Vittal JJ. Synthesis of NiS and MnS nanocrystals from the molecular precursors (TMEDA)M(SC{O}C_6H_5)$_{(2)}$ (M = Ni, Mn). Cryst Growth Des. 2009;9:352–357. doi: 10.1021/cg800536w

[25] Barry L, Copley M, Holmes JD, Otway DJ, Kazakova O, Morris MA. Synthesis and characterization of nanoparticulate MnS within the pores of mesoporous silica. J Solid State Chem. 2007;180:3443–3449. doi: 10.1016/j.jssc.2007.10.004

[26] Peng L, Shen S, Zhang Y, Xu H, Wang Q. Controllable synthesis of MnS nanocrystals from a single-source precursor. J Colloid Interface Sci. 2012;377:13-17. doi: 10.1016/j.jcis.2012.03.052

[27] Moloto N, Moloto MJ, Kalenga M, Govindraju S, Airo M. Synthesis and characterization of MnS and MnSe nanoparticles: Morphology, optical and magnetic properties. Opt Mater. 2014;36:31–35. doi: 10.1016/j.optmat.2013.06.023

[28] Li Z, Ji Y, Xie R, Grisham YS, Peng X. Correlation of CdS nanocrystal formation with elemental sulfur activation and its implication in synthetic development. J Am Chem Soc. 2011;133:17248–17256. doi: 10.1021/ja204538f

[29] Thomson JW, Nagashima K, Macdonald PM, Ozin GA. From sulfur-amine solutions to metal sulfide nanocrystals: Peering into the oleylamine-sulfur black box. J Am Chem Soc. 2011;133:5036–5041.

[30] Joo J, Na HB, Yu T, Yu JH, Kim YW, Wu FX, Zhang JZ, Hyeon T. Generalized and facile synthesis of semiconducting metal sulfide nanocrystals. J Am Chem Soc. 2003;125:11100–11105. doi: 10.1021/ja0357902

[31] Capetti E, Ferretti AM, Dal Santo V, Ponti A. Surfactant-controlled composition and crystal structure of manganese(II) sulfide nanocrystals prepared by solvothermal synthesis. Beilstein J Nanotech. 2015;6:2319–2329. doi: 10.3762/bjnano.6.238

[32] Puglisi A, Mondini S, Cenedese S, Ferretti AM, Santo N, Ponti A. Monodisperse octahedral α-MnS and MnO nanoparticles by the decomposition of manganese oleate in the presence of sulfur. Chem Mater. 2010;22:2804–2813. doi: 10.1021/cm903735e

[33] Tian Q, Tang M, Jiang F, Liu Y, Wu J, Zou R, Sun Y, Chen Z, Li R, Hu J. Large-scaled star-shaped α-MnS nanocrystals with novel magnetic properties. Chem Commun. 2011;47:8100–8102. doi: 10.1039/c1cc11621e

[34] Wang DS, Zheng W, Hao CH, Peng Q, Li YD. A synthetic method for transition-metal chalcogenide nanocrystals. Chem Eur J. 2009;15:1870–1875.

[35] An CH, Tang KB, Liu XM, Li FQ, Zhou G, Qian YT. Hydrothermal preparation of α-MnS nanorods from elements. J Cryst Growth. 2003;252:575–580. doi: 10.1016/s0022-0248(03)01001-7

[36] Zhang HT, Hyun BR, Wise FW, Robinson RD. A generic method for rational scalable synthesis of monodisperse metal sulfide nanocrystals. Nano Lett. 2012;12:5856–5860. doi: 10.1021/nl303207s

[37] Ma D, Huang S, Zhang L. One-pot synthesis and magnetic, electrical properties of single-crystalline α-MnS nanobelts. Chem Phys Lett. 2008;462:96–99. doi: 10.1016/j.cplett.2008.07.078

[38] Mu J, Gu Z, Wang L, Zhang Z, Sun H, Kang S-Z. Phase and shape controlling of MnS nanocrystals in the solvothermal process. J Nanopart Res. 2008;10:197–201. doi: 10.1007/s11051-007-9216-8

[39] Zhang C, Tao F, Liu GQ, Yao LZ, Cai WL. Hydrothermal synthesis of oriented MnS nanorods on anodized aluminum oxide template. Mater Lett. 2008;62:246–248. doi: 10.1016/j.matlet.2007.05.008

[40] Yang X, Wang Y, Wang K, Sui Y, Zhang M, Li B, Ma Y, Liu B, Zou G, Zou B. Polymorphism and formation mechanism of nanobipods in manganese sulfide nanocrystals induced by temperature or pressure. J Phys Chem C. 2012;116:3292–3297. doi: 10.1021/jp209591r

[41] Powell AE, Hodges JM, Schaak RE. Preserving both anion and cation sublattice features during a nanocrystal cation-exchange reaction: Synthesis of metastable wurtzite-type CoS and MnS. J Am Chem Soc. 2016;138:471–474. doi: 10.1021/jacs.5b10624

[42] Michel FM, Schoonen MAA, Zhang XV, Martin ST, Parise JB. Hydrothermal synthesis of pure α-phase manganese(II) sulfide without the use of organic reagents. Chem Mater. 2006;18:1726–1736. doi: 10.1021/cm048320v

[43] Kurz T, Chen L, Brieler FJ, Klar PJ, von Nidda HAK, Froeba M, Heimbrodt W, Loidl A. Minimal number of atoms to constitute a magnet: Suppression of magnetic order in spherical MnS nanoparticles. Phys Rev B. 2008;78:132408, 124p. doi: 13240810.1103/PhysRevB.78.132408

[44] Brieler FJ, Grundmann P, Froba M, Chen LM, Klar PJ, Heimbrodt W, von Nidda HAK, Kurz T, Loidl A. Formation of $Zn_{1-x}Mn_xS$ nanowires within mesoporous silica of different pore sizes. J Am Chem Soc. 2004;126:797–807. doi: 10.1021/ja038960j

[45] Zheng J, Yuan X, Ikezawa M, Jing P, Liu X, Zheng Z, Kong X, Zhao J, Masumoto Y. Efficient photoluminescence of Mn^{2+}ions in MnS/ZnS core/shell quantum dots. J Phys Chem C. 2009;113:16969–16974. doi: 10.1021/jp906390y

[46] Zheng J, Ji W, Wang X, Ikezawa M, Jing P, Liu X, Li H, Zhao J, Masumoto Y. Improved photoluminescence of MnS/ZnS Core/shell nanocrystals by controlling diffusion of Mn Ions into the ZnS shell. J Phys Chem C. 2010;114:15331–15336. doi: 10.1021/jp104513k

[47] Ge JP, Li YD. Controllable CVD route to CoS and MnS single-crystal nanowires. Chem Commun. 2003:2498–2499. doi: 10.1039/b307452h

[48] Ge JP, Wang J, Zhang HX, Li YD. A general atmospheric pressure chemical vapor deposition synthesis and crystallographic study of transition-metal sulfide one-dimensional nanostructures. Chem Eur J. 2004;10:3525–3530.

[49] Kim DS, Lee JY, Na CW, Yoon SW, Kim SY, Park J, Jo Y, Jung MH. Synthesis and photoluminescence of Cd-doped α-MnS nanowires. J Phys Chem B. 2006;110:18262–18266. doi: 10.1021/jp063965z

[50] Zhou J, Huang F, Xu J, Wang Y. $Cu_{1.94}S$-MnS dimeric nanoheterostructures with bifunctions: Localized surface plasmon resonance and magnetism. CrystEngComm. 2013;15:4217–4220. doi: 10.1039/c3ce00015j

[51] Gui YC, Qian LW, Qian XF. Hydrothermal synthesis of uniform rock salt (α-) MnS transformation from wurtzite (γ-) MnS. Mater Chem Phys. 2011;125:698–703. doi: 10.1016/j.matchemphys.2010.09.071

[52] Yang X, Wang Y, Sui Y, Huang X, Cui T, Wang C, Liu B, Zou G, Zou B. Size-controlled synthesis of bifunctional magnetic and ultraviolet optical rock-salt MnS nanocube superlattices. Langmuir. 2012;28:17811–17816. doi: 10.1021/la304228w

[53] Zhang N, Yi R, Wang Z, Shi R, Wang H, Qiu G, Liu X. Hydrothermal synthesis and electrochemical properties of α-manganese sulfide submicrocrystals as an attractive

electrode material for lithium-ion batteries. Mater Chem Phys. 2008;111:13–16. doi: 10.1016/j.matchemphys.2008.03.040

[54] Pearce CI, Pattrick RAD, Vaughan DJ. Electrical and magnetic properties of sulfides. Sulfide Mineral Geochem. 2006;61:127–180. doi: 10.2138/rmg.2006.61.3

[55] Barry L, Tobin J, Copley M, Holmes JD, Otway DJ, Morris MA. MnS doped mesoporous silica catalysts for the generation of novel carbon nanocages. Appl Catal A. 2008;341:8–11. doi: 10.1016/j.apcata.2007.12.008

[56] Xu G, Zhu YL, Ma XL. Cu2S nanowires and MnS/Cu2S nanojunctions derived from γ-MnS nanowires via selective cation-exchange reaction. Phys Status Solidi A. 2011;208:123–128. doi: 10.1002/pssa.201026124

Chapter 7

Optical Excitations of Colloidal Core/Shell Semiconductor Quantum Dots

Tiberius O. Cheche

Additional information is available at the end of the chapter

http://dx.doi.org/10.5772/65091

Abstract

The energy structure of multi-layer core/shell semiconductor quantum dots (QDs) is simulated and the optical absorption is described within the linear response theory. The lattice-mismatch strain field of the multi-layer nanostructures of spherical symmetry is modelled by a linear continuum elasticity treatment. The excitonic effect is estimated by a configuration interaction approach. Multi-layer core/shell QDs of II-VI semiconductors with heterostructures of type I and II are discussed. Localization of the photo-excited carriers induced by coating explains the optical stability of these multi-layer nanostructures.

Keywords: strain, semiconductor, quantum dot, optical absorption

1. Introduction

Semiconductor quantum dots (QDs) are nanometre-scale objects that are different to the usual optical materials, have size-tuneable and narrow fluorescence and broad absorption spectra. The large interest for these materials comes from their technical applications and theoretical openings. For example, the colloidal multi-shell QDs have led to the development of high-efficiency solar cells [1] or laser applications [2], and probably the most important in the immediate future, to molecular diagnostics and pathology [3]. At theoretical level, the study of magnetism [4] or photon entanglement [5] is an example in which the theoretical studies emerge in promising new properties of electronic devices for spintronics or computer microchips. The main factors used in tuning the physical properties of QDs are the shape and size confinement, and the lattice-mismatch-induced strain. 'Giant' core/multi-shell of 18–19 monolayers (MLs) shell thickness can be prepared with low cost by chemical synthesis [6, 7]. The theoretical predictions of electronic structures and optical properties are important in the core/shell quantum dots (CSQDs) engineering. Modelling of the lattice-mismatch strain is a key factor in obtaining an accurate physical description of CSQDs. Widely used for analysing the linear elasticity of epitaxial-strained heterointerfaces, the valence force field method (see, e.g. Ref. [8]) has some limitations as the predictions are dependent of *a priori* information regarding the interface structure and surface passivation. There are several theoretical studies of multi-component nanocrystals, in which the role of the strain is considered by *ab-initio* calculations [9–11]. Structural limitations of these first-principle calculations (e.g. band gap underestimation) and also more important, the computational cost for larger QDs, make difficult comparison between theoretical predictions and the experiment. The continuum elasticity approach in the limits of homogeneous and isotropic materials has been shown to be in good agreement with the valence force field models for semiconductor QDs of spherical shape and cubic symmetry (see, e.g. Ref. [12]).

In this context, in this chapter, we discuss optical properties of CSQDs. In Section 2, we introduce the theoretical modelling. In Section 2.1, we describe the lattice-mismatch strain field in core/multi-shell nanostructures by a continuum elasticity approach. In Sections 2.2 and 2.3, we describe the electronic structures and obtain the single-particle states (SPSs) by an effective two-band model and by an eight-band model within the **k·p** theory, respectively. In Section 2.4, we discuss the excitonic effect in QDs. In Section 2.5, the optical absorption coefficient is obtained by taking into account the excitonic effect. In Section 3, we apply the theory presented in the previous section to several types of semiconductor CSQDs with heterostructures of type I and II. Conclusions are presented in Section 4.

2. Theoretical modelling of core/shell semiconductor quantum dots

The theoretical models introduced in this section combine knowledge from classical mechanics, solid state physics and quantum optics for description of elasticity, electronic structure and optical spectra in CSQDs.

2.1. Strain field and the band lineup in the presence of the strain

In the elasticity theory, description of the strain field in finite-domain elastic bodies is a difficult problem. Elaborate solutions for such a problem are obtained in few particular cases for Eshelby-type inclusion of a finite elastic body (see Ref. [13]). In what follows, we introduce an efficient simple method to obtain the strain field in concentric spherical domains of different elastic parameters, which satisfies the necessary accuracy level for our problem. We start by observing that for an elastic spherical core/shell structure, the displacement field, **u**, has radial symmetry and consequently the field is *irrotational*. Within the continuum elasticity approach, the equilibrium equation is grad $div\, \mathbf{u} = 0$ [14]. To obtain the strain field, the linear stress-strain tensor ($\sigma_{ij} - \varepsilon_{ij}$) relation is used. For a structure core/shell/shell (CSS QD) with radii r_1 (for the core), r_2 (for core + middle shell) and R (for the total radius of core + middle shell + outer shell), we consider the following boundary conditions: (a) continuous stress at the interfaces,

(b) zero pressure outside nanostructure and (c) *shrink-fit* induced by the lattice-mismatch. Transposed in algebraic equations, we have

$$\sigma_{rr}^{A}(r_1) = \sigma_{rr}^{B}(r_1), \sigma_{rr}^{B}(r_2) = \sigma_{rr}^{C}(r_2), \tag{1a}$$

$$\sigma_{rr}^{C}(R) = 0 \tag{1b}$$

$$u_r^A(r_1) - u_r^B(r_1) = \varepsilon_1 r_1, u_r^B(r_2) - u_r^C(r_2) = \varepsilon_2 r_2, \tag{1c}$$

where $\varepsilon_1 = (a_B - a_A)/a_A$ and $\varepsilon_2 = (a_C - a_B)/a_B$ are relative lattice-mismatches, and A, B and C denote the core, middle and outer shell, respectively. In Eq. (1c), which makes connection between the continuum elastic and the discrete crystalline approaches, a_A, a_B, a_C represent the lattice constants of the crystals. The solutions for the radial displacement in domain X ($X = A, B, C$) should respect the above-mentioned equilibrium equation, that in spherical coordinates r, θ, φ are of the form $u_r^X(r) = X_1 r + X_2/r^2$ (with X_1, X_2 as constants). Then, using the spherical coordinates, we derive the strain tensor components from their definition with respect to the displacement. Then, we obtain the stress tensor σ_{ij} in each domain by applying, as mentioned above, the linear stress-strain tensor Hooke's law, $\sigma_{ij} = E(1+\nu)^{-1}\left[\varepsilon_{ij} + \nu(1-2\nu)^{-1}\varepsilon_{ll}\delta_{ij}\right]$ (where E is the Young modulus and ν is Poisson ratio). Finally, the constants X_1, X_2 for each domain are found by solving Eqs. (1) and the strain tensor components emerge from their definitions. The values of the hydrostatic strain, $\varepsilon_{hyd} = \varepsilon_{rr} + \varepsilon_{\theta\theta} + \varepsilon_{\varphi\varphi}$ for the case of similar elastic parameters of the three materials, $\nu_A = \nu_B = \nu_C = \nu$, $E_A = E_B = E_C$ are as follows:

$e_{hyd}^{A} = \alpha\left[\varepsilon_1\left(1 - r_1^3 R^{-3}\right) + \varepsilon_2\left(1 - r_2^3 R^{-3}\right)\right]$, $e_{hyd}^{B} = \alpha\left[-\varepsilon_1 r_1^3 R^{-3} + \varepsilon_2\left(1 - r_2^3 R^{-3}\right)\right]$, $e_{hyd}^{C} = -\alpha\left(\varepsilon_1 r_1^3 R^{-3} + \varepsilon_2 r_2^3 R^{-3}\right)$, with $\alpha = 2(1-2\nu)(1-\nu)^{-1}$. One can observe that in the limit of thick outer shell, the outer shell hydrostatic strain vanishes, which is in accordance with the physics anticipation. Complete expressions of the strain tensor components are presented in Ref. [15]. Our expressions for the strain tensor recover the results obtained in Ref. [16] for one shell or those obtained in Ref. [17] for core embedded in an infinite matrix. The method we described can be applied to spherical multi-shell nanostructures and to cylindrical multi-layer structures as well.

The band offset is an important parameter in modelling the physical properties of semiconductor heterostructures. Simple relations are obtained for the band lineup in the presence of the strain for a two-band model within the effective mass approximation. Thus, the energy values at the Γ point of the valence band (VB) and conduction band (CB) for direct band gap semiconductors are given by the equation [18]:

$$E_{v,c} = E_{v,c}^{u} + a_{v,c}\varepsilon_{hyd}, \tag{2}$$

where the unstrained (bulk) values are related by $E_c^u = E_v^u + E_g$ (with E_g, the unstrained (bulk) band gap), and $a_{v(c)}$ is the volume deformation potential (subscript $v(c)$ stands for VB(CB)). Within **k·p** theory, in a multi-band approach, the kinetic and strain Hamiltonians are obtained according to the point group the semiconductors belong. By using a perturbation theory or a group theory-based method (method of invariants), the expressions of the kinetic and strain Hamiltonians are written for the desired number of bands. For example, based on the work of Kane [19] and Bir and Pikus [20], expressions of the eight-band kinetic and strain bulk Hamiltonians, including the spin-orbit interaction for zinc blende crystals, are obtained in Ref. [21]. The multi-band kinetic plus strain Hamiltonians are adapted to limited size structures by adding a cut-off potential (corresponding to a buffer shell), which allows the electron to leak more or less into the environment, outside the structure.

2.2. Single-particle states by two-band model within the effective mass approach

The electron and hole SPSs are obtained by solving the Schrödinger equation for the envelope wave function $H\psi(\mathbf{r}) = E\psi(\mathbf{r})$. The one-band effective Hamiltonian is

$$\left[\frac{p^2}{2\mu(r)} + V(r)\right]\psi(\mathbf{r}) = E\psi(\mathbf{r}), \tag{3}$$

where $\mu(r) = \mu_i$ is the photoexcited (electron or hole) carrier r-dependent effective mass, $V(r) = V_i$ is the step confinement potential generated by the band lineup of the materials in the presence of the strain, as described in Section 2.1, and $i = A, B, C$ for core, middle shell outer shell, respectively. The solution $\psi(\mathbf{r})$ separates in a product of a radial function and spherical harmonics $\psi_{nlm}(\mathbf{r}) = R_l(r)Y_l^m(\theta, \varphi)$, and with $\rho = k_i r$, $R_l(r) = v_l(\rho)$ and $k_i^2 = 2\mu_i|E - V_i|/\hbar^2$ for the radial one-particle Schrödinger equation, one obtains the spherical Bessel differential equation:

$$\rho^2 \frac{d^2 v_l(\rho)}{d\rho^2} + 2\rho \frac{dv_l(\rho)}{d\rho} \pm \left[\rho^2 \mp l(l+1)\right] v_l(\rho) = 0, \tag{4}$$

where the upper sign corresponds to $E > V_i$ and the lower to $E < V_i$. Solutions of Eq. (7) are spherical Bessel functions $j_l(\rho), y_l(\rho)$ (upper sign) or modified spherical Bessel functions $i_l(\rho), k_l(\rho)$ (lower sign). The radial wave function for the three regions, A, B, C, are linear combinations of these functions as follows:

$$\begin{cases} R_l^A(r) = A_1^l\, f_{1l}^A(k_A r), \; 0 \le r < r_1 \\ R_l^B(r) = B_1^l\, f_{1l}^B(k_B r) + B_2^l\, f_{2l}^B(k_B r), \; r_1 \le r < r_2 \\ R_l^C(r) = C_1^l\, f_{1l}^C(k_C r) + C_2^l\, f_{2l}^C(k_C r), \; r_2 \le r < R \end{cases}, \tag{5}$$

http://dx.doi.org/10.5772/65091

where A_1^l, $B_{1,2}^l$, $C_{1,2}^l$ are constants and $f_{1l,2l}^{\eta}$ are spherical Bessel functions with the argument dependent of $k_\eta = \hbar^{-1}\sqrt{2m_\eta|E - V_\eta|}$ $(\eta = A, B, C)$. The functions *f* are chosen according to the resulting band lineup (with taken into account the strain). Imposing the physical conditions of continuity, $R_l^A(r_1) = R_l^B(r_1)$, $R_l^B(r_2) = R_l^C(r_2)$ and $R_l^C(R) = 0$, and conservation of the probability current, $m_A^{-1}\left(dR_l^A/dr\right)_{r\to r_1} = m_B^{-1}\left(dR_l^B/dr\right)_{r\to r_1}$ and $m_B^{-1}\left(dR_l^B/dr\right)_{r\to r_2} = m_C^{-1}\left(dR_l^B/dr\right)_{r\to r_2}$, we obtain the transcendental equation valid for *any* relative position of the band lineup for the three regions:

$$\frac{m_C f_{1l}^{B\,\prime}(r_2) - m_B F_l^C f_{1l}^B(r_2)}{m_C f_{2l}^{B\,\prime}(r_2) - m_B F_l^C f_{2l}^B(r_2)} = \frac{m_A f_{1l}^A(r_1) f_{1l}^{B\,\prime}(r_1) - m_B f_{1l}^{A\,\prime}(r_1) f_{1l}^B(r_1)}{m_A f_{1l}^A(r_1) f_{2l}^{B\,\prime}(r_1) - m_B f_{l1}^{A\,\prime}(r_1) f_{2l}^B(r_1)}, \tag{6}$$

where $F_l^C = [f_{1l}^{C\,\prime}(r_2) f_{2l}^C(R) - f_{l1}^C(R) f_{2l}^{C\,\prime}(r_2)]/[f_{1l}^C(r_2) f_{2l}^C(R) - f_{1l}^C(R) f_{2l}^C(r_2)]$ and the prime are used to denote the first radial derivative. From the condition of normalization $\int_0^R R_l^2(r)\, r^2\, dr = 1$ and the explicit form of functions *f*, we find A_1^l, $B_{1,2}^l$, $C_{1,2}^l$ and consequently the explicit analytical expressions for the normalized eigenfunctions. Explicitly values of functions *f* are given in Ref. [15].

2.3. Single-particle states by multi-band model within the k·p approach

The expansion of the envelope functions in terms of plane waves (see, e.g. Refs. [22, 23]) is one of the most used techniques within the **k·p** theory to obtain the energy structure of the nanostructures. By this method, Schrödinger's equation is solved by using Fourier transform of a superlattice formed by periodic boxes containing the nanostructure that is analysed. To avoid the artefacts induced by the artificial periodic boundary conditions imposed by the array of boxes, which can affect the accuracy of the results especially when Coulomb and/or strain field interactions are present, we consider a single nanostructure whose Hamiltonian is solved in a localized basis set defined over the volume of the nanostructure. Regarding the inner (core/shell interface) boundary conditions, we adopt the *weak* solution method (see, e.g. Lassen et al. [24]) and integrate over the whole volume of nanostructure to obtain matrix elements between basis states (formed by the products of envelope functions and zone-centre Bloch functions) for the Hamiltonian. This method is a convenient alternative to the method of imposing continuity of the envelope function and probability current at the core/shell interface to satisfy the boundary conditions. Details of computing the SPS energy structure by an eight-band model for CSQDs are given elsewhere [25].

2.4. Excitonic effect

To obtain an accurate description of the optical properties, in addition to the geometrical confinement and strain, the excitonic effect should be considered. In order of their contribution to the excitonic effect, the interactions that are commonly considered are as follows: the Coulomb attractive interaction of the photoexcited electron-hole pair, the polarization interaction of the charges induced at the interfaces, the electron-hole exchange interactions and the correlation interactions [26, 27]. The polarization interaction, according to calculus of Brus [28], is about 3–5 times lower than the absolute value of Coulomb interaction and is smaller than the nanostructure size for ZnO, CdS, GaAs and InSb. The electron-hole exchange interaction is smaller. For example, in spherical InAs QDs of radius 3 nm, it is of 2.093 meV comparatively to the Coulomb interaction of 60.6 meV [26] and of order 0.1 meV in CdSe/CdS QD with thick shell [27]. The correlation interaction is also of order 1–2 meV or smaller [29]. For the level of accuracy of the two-band model introduced in Section 2.1, the Coulomb electron-hole interaction yields a reliable estimation of the excitonic effect. The muti-band **k·p** method is more accurate and, within this approach, calculus of all the above-mentioned interactions is possible at the desired accuracy.

Next, we limit discussion at modelling the excitonic effect by the Coulomb electron-hole interaction mediated by a *homogenized* dielectric constant (see below). We write the exciton state α as a configuration interaction expansion, $\left|\psi^{(\alpha)}\right\rangle = \sum_{i,j=1} c_{ij}^{(\alpha)} c_i^+ h_j^+ |0\rangle$, with $|0\rangle$ the ground state (filled VB) and c_i^+ (h_i^+) as creation operator of the electron (hole) state $\psi_i^e(\mathbf{r}) = \langle \mathbf{r}|c_i^+|0\rangle$ ($\psi_i^h(\mathbf{r}) = \langle \mathbf{r}|h_i^+|0\rangle$). The excitonic spinless QD Hamiltonian formed by the kinetic part and Coulomb electron-hole interaction in the second quantization is written as [30]

$$H_D = \sum_m E_m^e c_m^+ c_m + \sum_m E_m^h h_m^+ h_m + \sum_{m,n,p,q} V_{mnpq}^{eh} c_m^+ h_n^+ h_p c_q, \tag{7}$$

where the first and second term stand for electron and hole kinetic energy, and the third for electron-hole Coulomb interaction, respectively. According to Maxwell Garnett (MG) formalism [31, 32] for spherical structures, the non-local dielectric constant may be replaced by a homogenized value, and will adopt this approach in our calculus. With the algebra of the second quantization, one obtains the secular equation:

$$\sum_{i,j,k,l=1} \left[\left(E_i^e + E_j^h - E \right) \delta_{ik} \delta_{jl} + V_{ijlk}^{eh} \right] C_{kl} = 0, \tag{8}$$

where $V_{i,j,k,l}^{eh} = -e^2 \left(4\pi\varepsilon_0 \bar{\varepsilon}_r\right)^{-1} \iint_V d\mathbf{r}_e d\mathbf{r}_h \psi_i^e(\mathbf{r}_e)^* \psi_j^h(\mathbf{r}_h)^* \left|\mathbf{r}_e - \mathbf{r}_h\right|^{-1} \psi_k^h(\mathbf{r}_h) \psi_l^e(\mathbf{r}_e)$ are Coulomb matrix elements and $\bar{\varepsilon}_r$ is the homogenized screened relative dielectric constant. In the concrete calculus of the Coulomb matrix elements, we used the series expansion of the Coulomb Green function in spherical harmonics

$$\frac{1}{\left|\mathbf{r}_e - \mathbf{r}_h\right|} = 4\pi \sum_{L=0}^{\infty} \sum_{m=-L}^{L} \frac{1}{2L+1} \frac{r_<^L}{r_>^{L+1}} Y_{Lm}^*(\hat{\mathbf{r}}_e) Y_{Lm}(\hat{\mathbf{r}}_h) \tag{9}$$

http://dx.doi.org/10.5772/65091

and the angular integrals separately factorize for electron and hole and they are computed analytically by using Gaunt's formula.

2.5. Optical absorption

In this section, we derive an expression of the absorption coefficient, which is valid in the limit of low-light irradiation power. We approximate the absorptive material as being formed by QDs in contact and assume that each QD is an absorber of volume V_{QD}. At the low-light irradiation power, the assumption of a linear relation between the polarization and the electric field is a common approximation. By treating the QD-field interaction as a perturbation, absorption coefficient expression can be obtained using either first-order complex susceptibility or Fermi's golden rule. Here, in the proof, we consider Fermi's golden rule as follows. For a monochromatic electromagnetic wave $\mathbf{E} = \mathbf{E}_0\cos(\omega t - kz)$, propagating in z direction in a macroscopic non-conducting media, the gradient of the energy density is (see, e.g. Ref. [33]) $dw/dz = nc^{-1}dw/dt$. The loss of energy per time unit is the loss of energy in a *single* QD that is, $dw/dt = -\hbar\omega V_{QD}^{-1}R$, where R is the rate of change of the number of photons. From the last two equations, one obtains $dw/dz = -\hbar\omega R V_{QD}^{-1} nc^{-1}$. With the energy density $w = \varepsilon_r\varepsilon_0 E_0^2/2$, the intensity of the electromagnetic wave is (see Eqs. (7.13) (7.14) in Ref. [34]) $I = \sqrt{\varepsilon_r\varepsilon_0/(\mu_r\mu_0)}E_0^2/2 \approx c\varepsilon_0\varepsilon_r E_0^2/(2n) = wc/n$, where $n = \sqrt{\mu_r\varepsilon_r} \cong \sqrt{\varepsilon_r}$ (Eq. (7.5) in Ref. [34]). Next, by combining the expressions of dw/dz, the derivative dI/dz and the Beer-Lambert law, $dI/dz = -\alpha_{QD}I$, we obtain the single QD absorption coefficient.

$$\alpha_{QD} = \frac{2\hbar\omega}{cn\varepsilon_0 E_0^2 V_{QD}} R. \tag{10}$$

Expression of the probability rate for absorption in the first order of approximation, R, can be obtained following the standard textbook derivation of Fermi's golden rule as

$$R = \frac{\pi e^2 E_0^2}{2m_0^2\hbar\omega^2}\sum_{i,f}|M_{if}|^2\,\delta\left(E_{if} - \hbar\omega\right), \tag{11}$$

where $M_{if} = \langle\Psi_i|\mathbf{e}\cdot\mathbf{P}|\Psi_f\rangle$ is the optical matrix element between the initial $|\Psi_i\rangle$ and the final $|\Psi_f\rangle$ states, $E_{if} = \hbar\left(\omega_i - \omega_f\right)$, and δ is the Dirac function. To mimic the homogenous broadening that takes into account the relaxation phenomena of the carriers, one replaces the Dirac function by a Lorentzian of width γ, and the single QD absorption coefficient valid for low temperatures becomes.

$$\alpha_{QD} = \frac{e^2}{cn\varepsilon_0 m_0^2 \omega V_{QD}} \sum_{i,f} \frac{\gamma |M_{if}|^2}{\gamma^2 + \left(E_{if} - \hbar\omega\right)^2}. \tag{12}$$

Eq. (12) can be applied to obtain the excitonic absorption by computing the optical matrix element $M_{0\alpha} = \langle 0|\mathbf{e} \cdot \mathbf{P}|\psi^{(\alpha)}\rangle$, where **e** is the polarization light unit vector. Thus, by writing the transition linear momentum operator in the basis set $\{|0\rangle, |\psi^{(\alpha)}\rangle\}$ as $\mathbf{P} = \sum_\alpha |0\rangle\langle 0| \sum_{i=1}^N \mathbf{p}_i |\psi^{(\alpha)}\rangle\langle\psi^{(\alpha)}| + h.c.$, one obtains

$$M_{0\alpha} = \mathbf{e} \cdot \sum_{i,j,k=1} C_{ij}^{(\alpha)} \langle 0|\mathbf{p}_k c_i^+ h_j^+|0\rangle = \mathbf{e} \cdot \sum_{i,j=1} C_{ij}^{(\alpha)} \langle \psi_j^h u_v|\mathbf{p}|\psi_i^e u_c\rangle = \mathbf{e} \cdot \mathbf{p}_{vc} \sum_{i,j=1} C_{ij}^{(\alpha)} \langle \psi_j^h|\psi_i^e\rangle, \tag{13}$$

where $\mathbf{p}_{vc} = \langle u_v|\mathbf{p}|u_c\rangle$; the last equality is obtained by making use of the slow spatial variation of the envelope wave functions over regions of the unit cell size and the orthonormality of the Bloch cell wave functions. By introducing the Kane momentum matrix element, $P = -i(\hbar/m_0)\langle s|p_z|z\rangle = -i(\hbar/m_0)\langle u_c|p_z|u_v\rangle = -i(\hbar/m_0)p_z^{cv}$, with $E_P = 2m_0|P|^2/\hbar^2$ and considering the polarization unit vector, **e**, parallel to the quantization axis, *z*, for example, one obtains

$$|M_{0\alpha}|^2 = \frac{E_P m_0}{2} \left| \sum_{i,j=1} C_{ij}^{(\alpha)} \langle \psi_j^h|\psi_i^e\rangle \right|^2. \tag{14}$$

Then, expression of the single QD absorption coefficient from Eq. (12) becomes

$$\alpha_{QD}(\omega) = \frac{\alpha_0}{\omega} \sum_\tau \left| \sum_{i,j=1} C_{ij}^{(\tau)} \langle \psi_i^e|\psi_j^h\rangle \right|^2 \frac{\gamma}{(E_\tau - \hbar\omega)^2 + \gamma^2}, \tag{15}$$

where $\alpha_0 = e^2 E_P / \left(2nc\varepsilon_0 m_0 V_{QD}\right)$. The multi-shell character is brought into the excitonic optical matrix element,

$$f_{0\tau} = \left|\sum_{i,j=1} C_{ij}^{(\tau)} \langle \psi_i^e|\psi_j^h\rangle\right|^2, \tag{16}$$

by the domain dependence of the electron and hole single SPSs. The quantity expressed by Eq. (15) is the exciton oscillator strength. For the exciton state at resonance, one obtains that the absorption coefficient is proportional to the exciton oscillator strength. As the parameters entering Eq. (15) characterizes a single QD, we name it *single* QD absorption coefficient. Neglecting light scattering, a good approximation for QD of small size comparatively to the light wavelength, we introduce a *colloidal* absorption coefficient for the measured absorption coefficient in experiment:

$$\alpha_{sol} = -c_{QD}^{1/3} \ln\left[1 - 4R^2 c_{QD}^{2/3}\left(1 - e^{-2R\,\alpha_{QD}}\right)\right], \tag{17}$$

where c_{QD} is the concentration of QDs in solution and ln is the natural logarithm function. The model is based on a probabilistic hitting QD as a target and details of the derivation can be found in Ref. [15]. For $\alpha_{QD}R < < 1$ (that is QDs with radius smaller than approximately 10 nm) and $Rc_{QD}^{1/3} < < 1$ (that is for dilute solutions), we obtain $\alpha_{sol} = 8R^3 c_{QD}\alpha_{QD}$.

3. Application of type I and II semiconductor heterostructures

In this section, we apply the theory developed in Section 2 to heterostructures of type I and II of direct band gap semiconductors. In accordance with our continuum model for strain, we consider QD structures with thicker shells. On the other hand, as a two-band model neglects both VB-CB and VB sub-band mixing, we assess its validity in our modelling. As a first requirement in neglecting VB-CB mixing, the semiconductors should be of wide enough band gap. Regarding neglecting of the VB sub-band mixing, things are more complicated, and to assess to what extent a two-band model respects a specific accuracy, a more advanced theoretical treatment is necessary. For this purpose, we compare the SPSs generated by the two- and eight-band approaches for CSQDs of wide band gap semiconductors. For the modelling, we consider ZnTe/ZnSe CSQD (a type II heterostructure), ZnTe/ZnSe/ZnS CSS QD (a type II ZnS-coated heterostructure) and CdSe/CdS/ZnS CSS QD (a type I ZnS-coated heterostructure).

3.1. Comparison of two- and eight-band approaches for wide band gap CSQDs

Before focusing discussion on quantum mechanics calculation of SPSs, we apply our method of computing the strain in lattice-mismatch heterostructures. The results obtained by solving Eqs. (1) for the general case of different elastic parameters presented in **Figure 1** for CdSe/11CdS/ZnS (CSS QDI) and ZnTe/16ZnSe/ZnS (CSS QDII); the numbers before the chemical symbols hold for the number of MLs and the core radii are of 2 nm and 2.2 nm, respectively.

The parameters we use are listed in **Table 1**.

Once the hydrostatic strain and the bulk band offset are known, we can obtain the band lineup in the presence of the strain, by using Eq. (2). The bulk energy structure of the heterostructures (band offset) we discuss is schematically shown in **Figure 2** for CdSe/CdS/ZnS and ZnTe/ZnSe/ZnS (in what follows we take as zero reference the bulk VB edge of the core semiconductor).

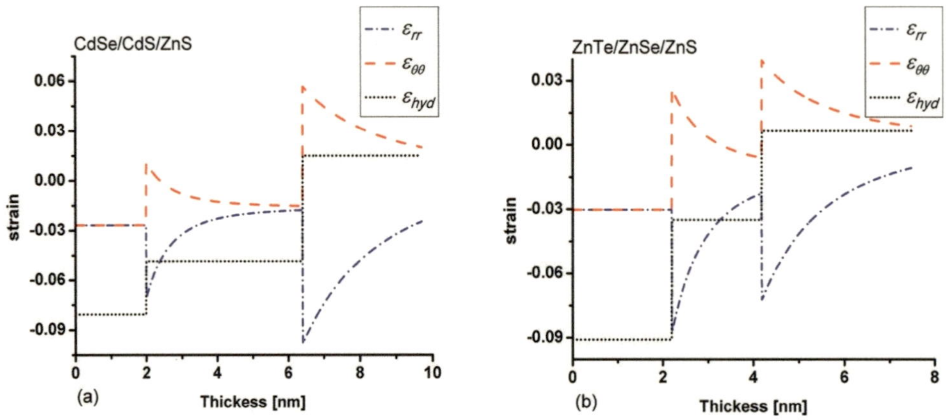

Figure 1. Diagonal spherical tensor components, ε_{rr}, $\varepsilon_{\theta\theta}$, $\varepsilon_{\varphi\varphi}$ for (a) CdSe/11CdS/ZnS; (b) ZnTe/16ZnSe/ZnS for outer shell thickness of 3.3 nm.

	ZnTe	ZnSe	CdSe	CdS	ZnS
a (Å)	6.08[a]	5.65[a]	6.05[b]	5.82[b]	5.40[c]
E (10^{10} N m^{-2})	4.17[a]	4.51[a]	2.87[d]	3.26[d]	5.55[c]
ν	0.363[a]	0.376[a]	0.408[d]	0.410[d]	0.384[c]
E_{gap} (eV)	2.25 [e]	2.69 [e]	1.74[e]	2.49[e]	3.61 [e]
E_v (eV)	-5.34 [e]	-6.07 [e]	-6.00 [e]	-6.42 [e]	-6.6 [e]
a_v (eV)	0.79[f]	1.65[f]	0.9[b]	0.4[b]	2.31[f]
a_c (eV)	-5.83[f]	-4.17[f]	-2.00[b]	-2.54[b]	-4.09[f]
γ_1	3.74[g]	3.77[g]	3.33[h]	4.11[d]	2.54[g]
γ_2	1.07[g]	1.24[g]	1.11[h]	0.77[d]	0.75[g]
γ_3	1.64[g]	1.67[g]	1.11[h]	1.53[d]	1.09[g]
m^{lh_i}	0.152	0.148	0.18	0.15	0.225
m^{hh_i}	1.092	1.292	0.90	0.60	1.582
m^{el}	0.20[j]	0.21[j]	0.15[k]	0.22[k]	0.34[l]
ε^m	7.4	9.1	10	8.9	9

[a]Ref. [35].
[b]Ref. [36].
[c]Ref. [37].
[d]Ref. [38].
[e]Ref. [39].

[f]Ref. [40].
[g]Ref. [41].
[h]Ref. [42].
[i]Calculated with $m^{hh/lh} = m_0\gamma_1^{-1}\left[1 \pm \left(6\gamma_3 + 4\gamma_2\right)/\left(5\gamma_1\right)\right]^{-1}$ from Ref. [43].
[j]Ref. [44].
[k]Ref. [45].
[l]Ref. [46].
[m]Ref. [47].
For 1 ML of ZnSe, we considered the thickness of 0.33 nm [48], and for 1 ML of ZnS, we approximated it as 0.33 nm also.

Table 1. Material parameters used in the work

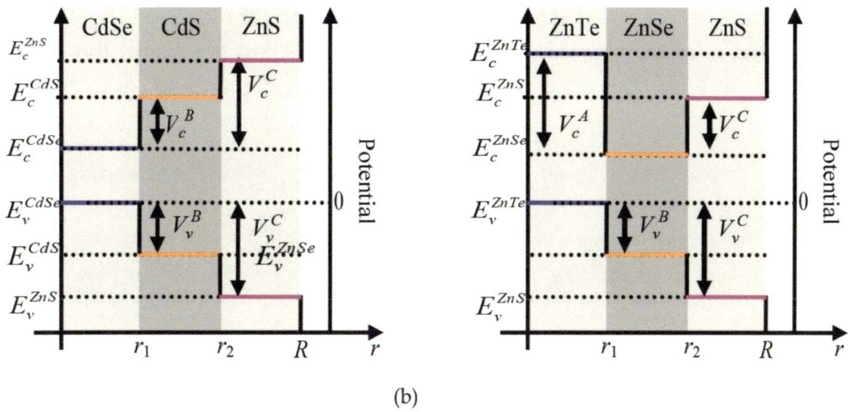

Figure 2. Schematic band lineups in CSS QDs for (a) CdSe/CdS/ZnS; (b) for ZnTe/ZnSe/ZnS heterostructures.

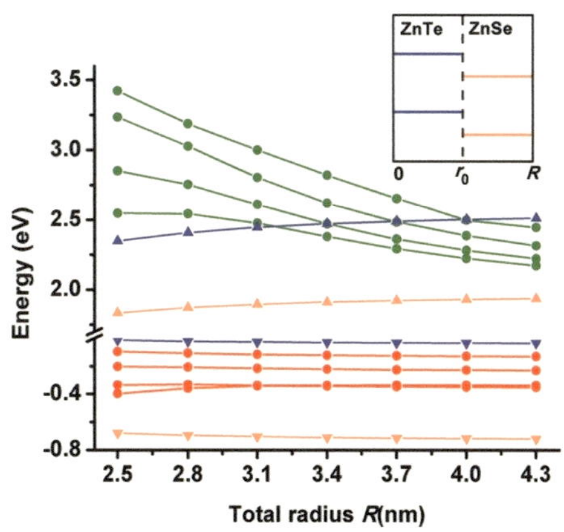

Figure 3. Energy of the first four electron (green colour) and hole (red colour) SPSs in ZnTe/ZnSe CSQDs of core radius 2.2 nm and variable total radius R in a two-band model. Continuum lines with up (down) triangle symbols show the band lineup in presence of lattice-mismatch strain of electron (hole) states for ZnTe (orange colour) and ZnSe (blue colour). The inset is for the lineup guidance. Zero reference is the bulk ZnTe VB edge.

Next, we discuss the SPSs obtained for ZnTe/ZnSe CSQDs by the two-band model introduced in Section 2. After calculus of the hydrostatic strain (by the method we mentioned above), we calculate the band lineup in the presence of the strain. For this, the bulk band offset assumed by Eq. (2) is take from Ref. [39]. To obtain the SPS structures, we take the version of Eqs. (5) and (6) for core/shell structure (see, for example, Ref. [16]). The eigenvalues of several electron and hole SPSs are presented in **Figure 3**. For a core radius of 2.2 nm, beginning with shell thicker than 3.1 nm, the lowest electron states start to be located in the shell while the highest hole states are practically insensitive to the shell thickness and located in the core. Our calculus shows that the first two or three (for shell thinner than 3.1 nm) hole SPSs are of heavy hole type.

To get a more complete image, in **Figure 4**, the probability density (PD) $|\psi_{nlm}(\mathbf{r})|^2$ is presented for a shell thickness of 4.3 nm. The SPSs energetically predicted location by the calculus of the eigenvalues is confirmed (checked) by the geometrical location. The degeneracy of the states is dictated by the angular momentum quantum number L and the PD shape by the values of n, L and $|m|$ (which results from the $Y_l^m(\theta,\varphi)$ dependence of the state).

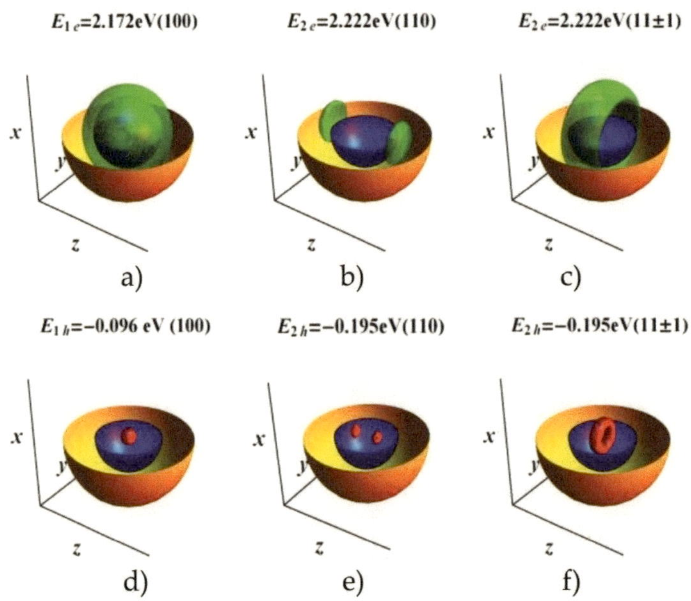

Figure 4. Two-band electron (green colour) and hole (red colour) SPS probability density $0.75 \times \max(|\psi_{nlm}(\mathbf{r})|^2)$ for ZnTe/ZnSe with core radius 2.2 nm and shell thickness 4.3 nm. The quantum numbers shown in the parentheses are, in order, n, L and m.

For comparison, we make several remarks on the SPS structure we obtained (see Ref. [25]) by the **k·p** method. According to the irreducible representations of T_d point group, the maximum value of degeneracy is four. Absent in the two-band model, the time-reversal symmetry is present in the eight-band model, and all SPSs are at least two-fold degenerate (Kramers degeneracy). The cubic symmetry of the zinc blende structures can also be observed in the PD shape. The shape of the PD, except the mentioned cubic symmetry, is similar in the two approaches. To have a more complete image of the density of states in the two approaches, we collect and compare the eigenvalue for CSQDs of different shell thicknesses. One obtains that the density of states is larger in the eight-band model, and for these wide band semiconductors, this is mainly caused by the VB sub-band mixing. However, when the shell thickness increase, the CB SPSs become denser and the discrepancy decreases. Consistently, in both approaches, one obtains the red shift of the SPS fundamental inter-band transition with the shell thickness. The conclusion of the comparison is that if only several excited states are sufficient to be taken into account in modelling, then CSQDs with thicker shells and larger cores can be approximately characterized by a two-band model. Generally, the VB SPSs are a mixture of heavy, light and split-off hole states. The two-band calculus shows, as expected, that the close to the VB edge, the SPSs are heavy hole state type, and this is another approximation that can be used in a two-band modelling of this type of CSQDs. On another hand, in case higher excited states are necessary in modelling, then a two-band model becomes inadequate. We can compare our two-band model prediction with the experimental measurement of the fundamental absorption for this type of CSQDs with shell thickness of 5 ML. Thus, for core radius of 2.2 nm and total radius of 3.85 nm (corresponding approximately to 5ML ZnSe shell thickness), we obtain a transition of 517 nm (2.397 eV) comparatively to approximately 550 nm (2.254 eV) reported in Ref. [49]. Energy overestimation by our prediction is mainly due to neglecting of the excitonic effect. If we estimate the excitonic effect (binding energy) by an attractive interaction of 0.13 eV, then our prediction becomes 547 nm, close to the experiment. In the next section, we obtain an estimation of the binding energy from the calculus of the Coulomb electron-hole attraction.

3.2. Excitonic effect and optical absorption

In this section, we evaluate the excitonic effect for CSS QDI and CSS QDII with thick outer shell. First, following the model from the previous section, by using Eqs. (1), (2), (5) and (6), we obtain the SPS structure. In the calculations, for the both nanostructure types, we find that the first four VB SPS states are heavy hole states and the fifth is the first light hole. Since the density of states is higher for heavy hole states, according to discussion from the previous section, we can appreciate that a two-level approach is at least satisfactory if only the first several hole levels are taken into account in the modelling. The characteristics of the SPS structure are as follows. The red shift of the fundamental inter-band transition with the middle shell thickness is found. The optical stability effect of the outer shell as reported by experiment [7, 49] is explained by the slight variation of the SPS energy with the ZnS shell thickness. Both electron and hole are located in the core for CSS QDI while for CSS QDII, the hole is located in the core and the electron in the shell, revealing the type I and type II character of the heterostructures, respectively. These observations are shown in **Figure 5** (reproduced from Ref. [15]). The PDs for the SPSs can be seen in Ref. [15]. With the SPS structure obtained, we estimate the exciton binding energy (or in an equivalent formulation, the fundamental excitonic absorption (FEA)) by computing the Coulomb electron-hole attractive interaction by a configuration interaction method as introduced in Section 2.4. With the first four energy levels we consider, we form for both modelled nanostructures the set of configurations as follows. The SPSs have the quantum numbers (n, L) = (1, 0), (2, 1), (3, 2) and (4, 3) for electron and (4, 0) for hole, and each state is $2L+1$ degenerated. Then, we build the configurations by replacing one from the set of (1 + 3 + 5 + 1) VB states with one of the (1 + 3 + 5 + 7) CB states, and obtain a total of 160 configurations.

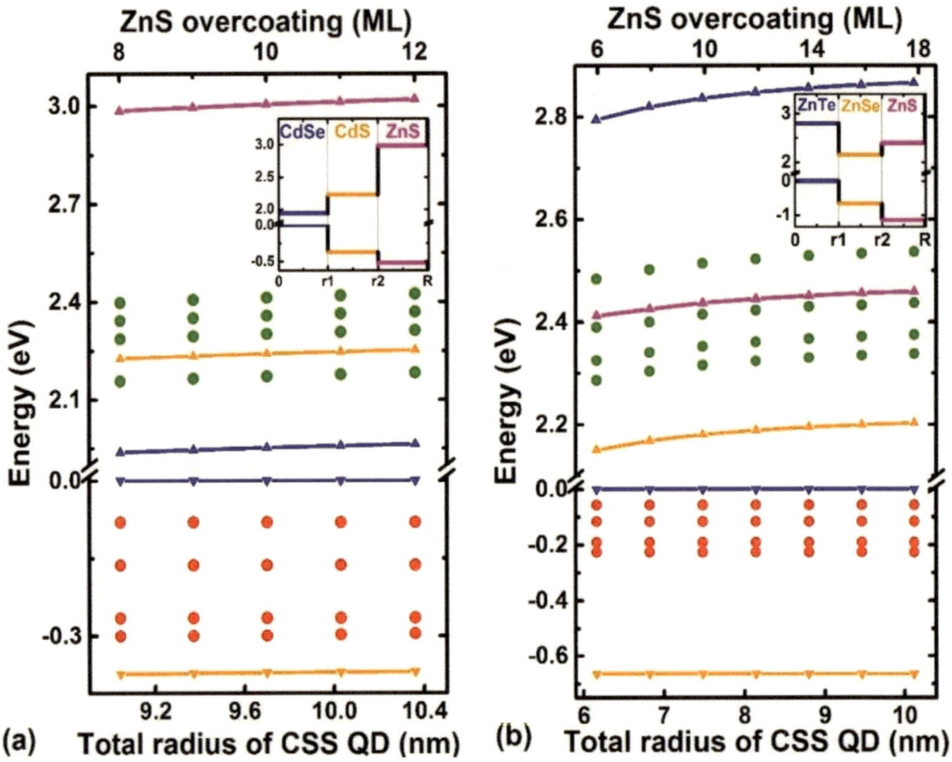

Figure 5. Energy of the first four electron (green colour) and hole (red colour) SPSs in (a) CSS QDI; (b) CSS QDII of total radius R. Continuum lines with up (down) triangle symbols show the band lineup in the presence of lattice-mismatch strain of electron (hole) states for ZnTe (blue colour), ZnSe (orange colour), ZnS (violet colour) in figure (a) and for CdSe (blue colour), CdS (orange colour), ZnS (violet colour) in figure (b). The insets are for the lineup guidance. Zero reference is core VB edge.

In **Figure 6** (reproduced from Ref. [15]), we show the energy value of FEA, E_{FEA}, the single QD absorption coefficient for the corresponding FEA (obtained by Eq. (15)), $\alpha_{FEA}(x)$ and the colloidal absorption coefficient for the corresponding FEA (obtained by Eq. (17)) $\alpha_{sol}(x)$; the argument x represents the number of ZnS-coating MLs. For the Kane energy, which is a necessary ingredient, we estimate E_P = 20.4eV for CSS QDI and E_P = 19.1eV for CSS QDII from Ref. [36]. Another necessary parameter is the CdS ML thickness, which we take as 0.4 nm, in accord with Ref. [27]. For **Figure 6(a)**, first, we setup the screened dielectric constant that is used to obtain the Coulomb matrix elements. Thus, we fit the experimental E_{FEA} of 620 nm from experiment [7] for CdSe/19CdS (CSQD with core radius of 2 nm and CdS shell thickness of 19

ML). Then, we analyse the effect of the ZnS coating and obtain the following results. E_{FEA} is slightly blue shifted with ZnS thickness and asymptotically blue shifted with the ZnS overcoating. Thus, a relative weak change of E_{FEA} with the ZnS overcoating as reported by the experiment [7] is obtained. This behaviour is primarily the result of the lattice-mismatch strain and secondly of the excitonic effect. $\alpha_{FEA}(x)$ obtained is asymptotically decreasing with ZnS coating. On another hand, the value obtained by the experiment, $\alpha_{sol}(x)$, is weakly changed by ZnS coating for 100 μM concentration of the colloidal QD solution. This is in agreement with the reported optical stability of such overcoated QDs [7] and with the slow variation of SPSs mentioned above. Also, from Eq. (15) we obtain that for $0 \rightarrow X_g$ transition corresponding to FEA, f_{0X_g} is very slightly varying with the ZnS coating. In a similar manner, we obtained **Figure 6(b)** as follows. First, to set-up the screened dielectric constant, we fit the experimental E_{FEA} for ZnTe/6ZnSe (CSQD with core radius of 2.2 nm and ZnSe shell thickness of 6ML) of 570 nm from experiment [49]. Then, we analyse the effect of the ZnS coating and obtain the following results.E_{FEA} is weakly blue shifted with the ZnS overcoating as reported by the experiment [49]. Differently from the CSS QDI, the single QD absorption coefficient is decreasing by one order of magnitude with ZnS coating. This is the result of the orbitals overlap decreasing with the ZnS coating, the electron is more pronouncedly shifted to the outer shell. We obtain that α_{sol} also strongly decreases with the ZnS coating but as in the type I case, it is almost constant with the ZnS overcoating. This dependence can be related to the protecting effect of the ZnS overcoating reported by experiment [49]. From Eq. (15), we obtain f_{0X_g} decreases about five times by ZnS coating and is weakly sensitive with the ZnSovercoating.

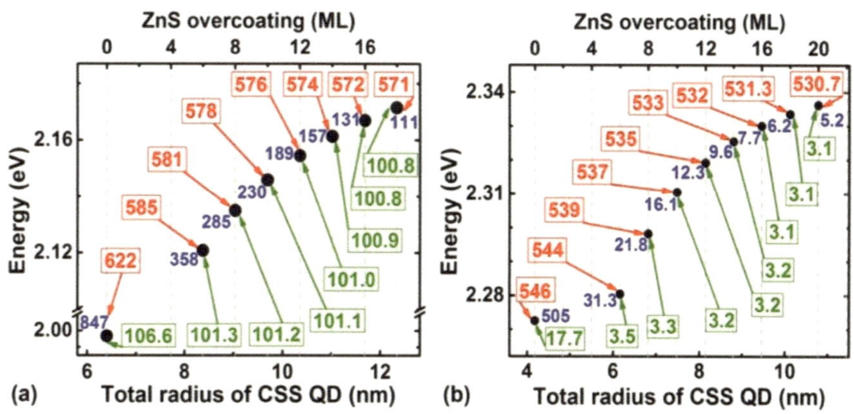

Figure 6. FEA (black circle and red-bordered label for the energy expressed in units of nm), α_{FEA} (blue label, units of 10cm^{-1}), α_{sol} (green, bordered label, units of 0.01cm^{-1}) for (a) CdSe/11CdS/xZnS QD; (b) ZnTe/6ZnSe/xZnS QD. α_{sol} is obtained for 100 μM concentration of the colloidal QD solution.

Regarding the location of the photoexcited charges, we introduce the radial probability density $D_{\alpha}^{(X_g)}(r) = \sum_{i,j}\left|C_{ij}^{(X_g)}\right|^2 r^2 \int \left|\psi_i^{\alpha}(\mathbf{r})\right|^2 d\Omega$, where $\alpha = e,\ h$ for electron and hole, respectively. One observes that $\int_0^R D_{\alpha}^{(X_g)}(r)\, dr = 1$ since $\sum_{i,\ j}\left|C_{ij}^{(X_g)}\right|^2 = 1$ from the orthonormalization condition. The radius expectation value of the photoexcited electron or hole is obtained with $\overline{r_{\alpha}^{(X_g)}} = \sum_{i,j}\left|C_{ij}^{(X_g)}\right|^2 \langle\psi_i^{\alpha}|r|\psi_i^{\alpha}\rangle$. The results are remarkable by revealing the ZnS coating effect. Thus, coating is not significant for CSS QDI, the electron and hole remain located in the core not close to the core/shell interface. On the other hand, for CSS QDII, the hole radius expectation value is practically not affected by the ZnS coating and the hole remains close to the centre but the ZnS coating has a strong effect on the electron localization, namely, the electron moves to the middle of the ZnSe shell. Numerical details are given in (Ref. [15]). Thus, according to our modelling in both CSS QDI and CSS QDII, the electron and hole are not confined in the proximity of the interfaces and surface. Consequently, QD coating induces larger photoexcited carrier-trap (interface defects and impurities close to the core/shell interface or to the outer-shell-environment interface) separation. This might have an important effect regarding the competition between recombination and trapping. Thus, one of the interesting effects of the QD coating is the blinking suppression observed, for example, in ultra-thick-shell CdSe/CdS [50]. According to our results, the model assuming a tunnelling barrier between the photoexcited carriers and the trap states, similar to that proposed in Ref. [51] can simplely explain the non-blinking as follows. Coating induces increasing of the barrier thickness and consequently a lowering of the trapping probability. Then, under a continuous photoexcitation, (stimulated) recombination is more probable than the trapping and a continuous luminescence (non-blinking) is observed from coated QDs. Given the low trapping probability heuristically concluded by our modelling, for the non-bliking in coated QDs, we rather advocate the barrier model than the model of Auger recombination of the photoexcited carriers on the trapping states as proposed in Ref. [50].

4. Conclusions

We described optical properties of CSQDs of type I and II with thick shells by a two-band model, which first is justified by a comparison with an eight-band **k·p** model. To obtain the lattice-mismatch strain effect on the energy structure, we developed a continuum elasticity model for isotropic, homogeneous and finite size multi-layer structures. When implemented in the specific quantum mechanics of the multi-shell semiconductor nanocrystals, the strain model we introduce is able to predict the main characteristics of fundamental absorption in thick shell CSQDs. By taking into account the excitonic effect, we can explain the optical stability of the overcoated core/shell QDs. According to our estimations, the measured absorption coefficient in colloidal QD solutions is practically not sensitive to the overcoating. The most important finding regarding the coating effect is related to the photoexcited carrier localization. According to our predictions, the photoexcited carriers are moved away from the surface and interfaces by overcoating. We believe that this work is an argument for improving the modelling. An eight-band treatment able to describe the optical absorption of ZnTe/ZnSe CSQDs which considers high excited states is already done (see Ref. [25]). In this context, the analysis of the multi-exciton generation, electron-phonon interaction and relaxation dynamics are promising successful directions of research in finding responses to the complex optical processes that take place in multi-layer QDs.

Author details

Tiberius O. Cheche

Address all correspondence to: cheche@gate.sinica.edu.tw

Faculty of Physics, University of Bucharest, Bucharest, Romania

References

[1] Kamat PV. Quantum dot solar cells. The next big thing in photovoltaics. The Journal of Physical Chemistry Letters. 2012;4:908–918.

[2] Chen Y, Herrnsdorf J, Guilhabert B, Zhang Y, Watson IM, Gu E, Laurand N, Dawson MD. Colloidal quantum dot random laser. Optics Express. 2011;19(4):2996–3003.

[3] Kairdolf BA, Smith AM, Stokes TH, Wang MD, Young AN, Nie S. Semiconductor quantum dots for bioimaging and biodiagnostic applications. Annual Review of Analytical Chemistry (Palo Alto Calif). 2013;6:143–162.

[4] Dietl T. Exchange interactions and nanoscale phase separations in magnetically doped semiconductors. In: Dietl T, Awschalom DD, Kaminska M, Ohno H, editors. Spintronics, Semiconductors and Semimetals. Vol 82. 1st ed. USA: Academic Press; 2008. pp. 371–412.

[5] Trotta R, Zallo E, Ortix C, Atkinson P, Plumhof D, van den Brink J, Rastelli A, Schmidt OG. Universal recovery of the energy-level degeneracy of bright excitons in InGaAs quantum dots without a structure symmetry. Physical Review Letters. 2012;109(14): 147401(1–4).

[6] García-Santamaría F, Chen Y, Vela J, Schaller RD, Jennifer A, Hollingsworth JA, Klimov V. Suppressed auger recombination in "giant" nanocrystals boosts optical gain performance. Nano Letters. 2009;9(10):3482–3488.

[7] Chen Y, Vela J, Htoon H, Casson JL, Donald J, Werder DJ et al. "Giant" multishell CdSe nanocrystal quantum dots with suppressed blinking. Journal of the American Chemical Society. 2008;130(15):5026–5027.

[8] Gronqvist J, Sondergaard N, Boxberg F, Guhr T, Aberg S, Xu HQ. Strain fields in core-shell nanowires. Journal of Applied Physics. 2009;106:053508.

[9] Sarkar P, Springborg M, Seifert G. A theoretical study of the structural and electronic properties of CdSe/CdS and CdS/CdSe core/shell nanoparticles. Chemical Physics Letters. 2005;405:103–107.

[10] Li JB, Wang LW. First principle study of core/shell structure quantum dots. Applied Physics Letters. 2004;84(18):3648–3650.

[11] Yang SY, Prendergast D, Neaton JB. Strain-induced band gap modification in coherent core/shell nanostructures. Nano Letters. 2010;10:3156–3162.

[12] Pryor C, Kim J, Wang LW, Williamson J, Zunger A. Comparison of two methods for describing the strain profiles in quantum dots. Journal of Applied Physics. 2009;106(5): 2548–2554.

[13] Ma HM, Gao XL. Strain gradient solution for a finite-domain Eshelby-type plane strain inclusion problem and Eshelby's tensor for a cylindrical inclusion in a finite elastic matrix. International Journal of Solids and Structures. 2011;48:44–55.

[14] Landau LD, Lifshitz EM. Theory of elasticity. Bristol. Pergamon. 1970.

[15] Pahomi TE, Cheche TO. Strain influence on optical absorption of giant semiconductor colloidal quantum dots. Chemical Physics Letters. 2014;612:33–38.

[16] Cheche TO, Barna V, Chang YC. Analytical approach for type-II semiconductor spherical core–shell quantum dots heterostructures with wide band gaps. Superlattice and Microstructures. 2013;60:475–486.

[17] Grundmann M, Stier O, Bimberg D. InAs/GaAs pyramidal quantum dots: Strain distribution, optical phonons, and electronic structure. Physical Review B. 1995;52(16): 11969–11981.

[18] Van de Walle CG. Band structure of indium antimonide. Journal of Physics and Chemistry of Solids. 1989;39(4):1871–1883.

[19] Kane EO. Band structure of indium antimonide. Journal of Physics and Chemistry of Solids. 1957;1(4):249–261.

[20] Bir GL, Pikus GE. Symmetry and Strain-Induced Effects in Semiconductors. New York: Wiley; 1975.

[21] Bahder TB. Eight-band k.p model of strained zinc-blende crystals. Physical Review B. 1990;41(17):11992–12001.

[22] Baraff GA, Gershoni D. Eigenfunction-expansion method for solving quantum-wire problem: Formulation. Physical Review B. 1991;43(5):4011–4022.

[23] Gershoni D, Henry CH, Baraff GA. Calculating the optical properties of multidimensional heterostructures: application to the modeling of quatemary quantum well lasers. IEEE Journal of Quantum Electronics. 1993;29(9):2433–2450.

[24] Lassen B, Lew Yan Voon LC, Willatzen M, Melnik R. Exact envelope-function theory versus symmetrized Hamiltonian for quantum wires: a comparison. Solid State Communications. 2004;132:141–149.

[25] Cheche TO, Chang YC, Barna V. CMD-24, Abstract Book. Discretized total angular momentum basis within eight-band k.p theory: application to heteroepitaxial core-shell quantum dots. [Internet]. September 2012 . Available from: http://www.cmd-24.org/home.

[26] Luo JW, Bester G, Zunger A. Long- and short-range electron–hole exchange interaction in different types of quantum dots. New Journal of Physics. 2009;11:123024.

[27] Brovelli S, Schaller RD, Crooker SA, Garcia-Santamaria G, Chen Y, Viswanatha R et al. Nano-engineered electron–hole exchange interaction controls exciton dynamics in core–shell semiconductor nanocrystals. Nature Communications. 2011;2(280):1-8. DOI: 10.1038/ncomms1281

[28] Brus LE. Electron–electron and electron–hole interactions in small semiconductor crystallites: the size dependence of the lowest excited electronic state. Journal of Chemical Physics. 1989;80(9):4403–4409.

[29] Shumway J, Franceschetti A, Zunger A. Correlation versus mean-field contributions to excitons, multiexcitons, and charging energies in semiconductor quantum dots. Physical Review B. 2001;63:155316 1–13.

[30] Hawrylak P. Excitonic artificial atoms: Engineering optical properties of quantum dots. Physical Review B. 1999;60:5597–5608.

[31] Garnett JCM. Colours in metal glasses and in metallic films. Philosophical Transactions of the Royal Society A. 1904;203:385–420.

[32] Holmström P, Thylén L, Bratkovsky A. Dielectric function of quantum dots in the strong confinement regime. Journal of Applied Physics. 2010;107:064307(1–7).

[33] Mitin V, Kochelap V, Stroscio MA. Quantum Heterostructures: Microelectronics and Optoelectronics. 1st ed. Cambridge: Cambridge University Press; 1999.

[34] Jackson JD. Classical Electrodynamics. 3rd ed. New York: John Wiley and Sons; 1999.

[35] Berlincourt D, Jaffe H, Shiozawa LR. Electroelastic properties of the sulfides, selenides, and tellurides of zinc and cadmium. Physical Review. 1963;129:1009.

[36] Li YH, Gong XG, Wei SH. Ab initio all-electron calculation of absolute volume deformation potentials of IV–IV, III–V, and II–VI semiconductors: the chemical trends. Physical Review B. 2006;73:245206(1–5).

[37] Hall RB, Meakin JD. The design and fabrication of high efficiency thin film CdS/Cu2S solar cells. Thin Solid Films. 1979;63(1):203–211.

[38] Adachi S. Properties of Group-IV, III-V and II-VI Semiconductors. Chichester: John Wiley and Sons; 2005.

[39] Lo SS, Mirkovic T, Chuang CH, Burda C, Scholes GD. Emergent properties resulting from type-II band. Advanced Materials. 2010;XX:1–18.

[40] Van de Walle CG. Band lineups and deformation potentials in the model-solid theory. Physical Review B. 1989;39(3):1871.

[41] Lawaetz P. Valence-band parameters in cubic semiconductors. Physical Review B. 1971;4(19):3460–3467.

[42] Mourad D, Richters JP, Gerard L, Andre R, Bleuse J, Mariette H. Determination of the valence band offset at cubic CdSe/ZnTe type II heterojunctions: a combined experimental and theoretical approach [Internet]. Available from: arXiv:1208.2188v2

[43] Baldereschi A, Lipari NO. Spherical model of shallow acceptor states in semiconductors. Physical Review B. 1979;8(6):2697–2709.

[44] Singh J. Physics of Semiconductors and Their Heterostructures. New York: McGraw-Hill; 1993.

[45] Singh J. Semiconductor Optoelectronics: Physics and Technology. New York: McGraw-Hill; 1995.

[46] Barman B, Sarma KC. Luminescence properties of ZnS quantum dots embedded in polymer matrix.Chalcogenide Letters. 2011;8(3):171–176.

[47] Palmer DW. The Semiconductors-information [Internet]. 2008 03 . Available from: www.semiconductors.co.uk

[48] Smith AM, Mohs AM, Nie S. Tuning the optical and electronic properties of colloidal nanocrystals by lattice strain. Nature Nanotechnology. 2009;4(1):56–63.

[49] Bang J, Park J, Lee JH, Won N, Nam J. ZnTe/ZnSe (core/shell) type-II quantum dots: Their optical and photovoltaic properties. Chemistry of Materials. 2010;22(1):233–240.

[50] Galland C, Ghosh Y, Steinbruck A, Hollingsworth JA, Htoon H, Klimov VI. Lifetime blinking in nonblinking nanocrystal quantum dots. Nature Communications. 2012;3(908):1–7. DOI: 10.1038/ncomms1916

[51] Kuno M, Fromm DP, Hamann HF, Gallagher A, Nesbitt DJ. Nonexponential "blinking" kinetics of single CdSe quantum dots: a universal power law behavior. Journal of Chemical Physics. 2000;112(7):3117–3120.

Section 3

State-of-the-Art Nanocomposites

Chapter 8

Colloidal Behaviors of Conducting Polymer/Chitosan Composite Particles

Mehmet Çabuk

Additional information is available at the end of the chapter

http://dx.doi.org/10.5772/65125

Abstract

In this study, colloidal behaviours of conducting polymer (CP) and chitosan (CS) multifunctional composite particles were investigated by using electrokinetic measurements. The effects of pH, electrolytes, surfactants and temperature on electrokinetic properties were determined as a function of ζ-potential of the colloidal dispersions. The obtained result showed that the zeta (ζ)-potential of CP increases to the higher positive region due to interaction with polycationic CS. Isoelectric points (IEP) of the CP/CS composite are also shifted to higher pH values. The positive ζ-potential of the composite shifts to more positive regions when the electrolyte valence and cationic surfactant concentration are increased. Also, these ζ-potential value shifts to negative region with the addition of anionic surfactant. Temperature affects the ζ-potential of the CP/CS dispersion. As a conclusion, colloidal behaviours of CP/CS composite particles can be altered by using different charged molecules. Therefore, CP/CS composite particles could be a good adsorbent for separation process.

Keywords: zeta potential, isoelectric point, conducting polymer, chitosan, colloidal particle

1. Introduction

1.1. Electrokinetic properties

Electrokinetic measurements are very useful techniques to present the interactions between dispersed particles in an aqueous dispersion. Electrokinetic properties of materials are also important to prepare sedimentation problems of solved materials. Zeta (ζ)-potential value is a surface charge of colloidal particles in an aqueous solution [1–3] and gives information about surface groups of colloidal particles and interaction between solid and liquid interfaces. It can be used for determination of some physical properties such as colloidal properties, Particle interactions and synthesis of charged polymeric composites. ζ-Potential cannot be measured directly. It can be determined using electrokinetic techniques, such as streaming current or potential, electric conductivity and electrophoretic mobility [4, 5].

The electrokinetic potential on surface of a solid particle in contact with a polar medium (i.e. H_2O) is controlled by the dissociation of surface groups, the preferential adsorption of cations or anions, the adsorption of polyelectrolytes and surfactants, the isomorphic substitution of cations and anions [6]. Generally, descriptions of the charge distribution at solid/liquid interface assume the presence of an electrical double layer consisting of a fixed (Stern) layer and a diffuse layer [7]. Externally applied electrical force causes a relative movement between these layers (**Figure 1**).

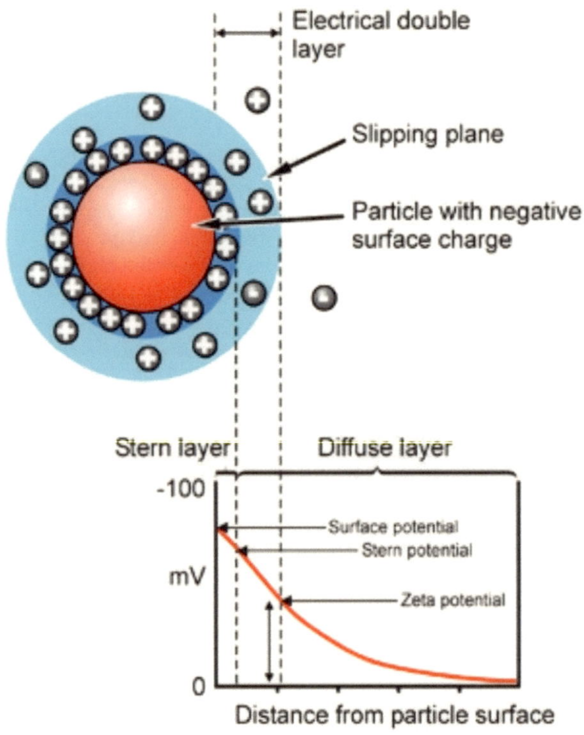

Figure 1. Zeta potential on a charged particle.

The interface between a solid and a solution may be treated as a semi-permeable membrane, which allows only the charged species common to both the solid and the pass through. These species are called potential determining ions and have the ability to affect the sign of the ζ-potential. They are the major ions responsible for establishment of the surface charge of particle. On the other hand, some ions called indifferent ions are adsorbed by electrostatic attraction only thereby affecting the magnitude of ζ-potential and not by the surface potential

http://dx.doi.org/10.5772/65125

[8–11]. The indifferent ions remain within outer (diffuse) part of the electrical double layer and do not strongly adsorb at the solid particle's wall.

The potential at this surface is called the ζ-potential. IEP of dispersed particles and potential determining ions in a solution have important role to understand the adsorption mechanism of various ionic species onto the dispersed particles [12]. Also, they help for flotation or coagulation in various colloidal dispersions.

When the ζ-potential is low, attraction of suspended particles to each other exceeds repulsion and the particles tend to aggregate. The ζ-potential can be determined from measured electrophoretic mobility and Henry's function, $f(\kappa\alpha)$ value is either 1.5 or 1.0. For the Smoluchowski approximation, $f(\kappa\alpha)$ is 1.5 in aqueous (polar) solutions. For the Hückel approximation, $f(\kappa\alpha)$ is 1.0 in non-polar medium [13–15].

As shown in **Figure 2**, when ζ-potential value of a dispersion is more positive than +30 mV or less negative than -30 mV, the dispersed particles will tend to be stable. Some parameters such as time, pH, ion and surfactant concentrations and types, and temperature may affect the value of ζ-potential [16, 17].

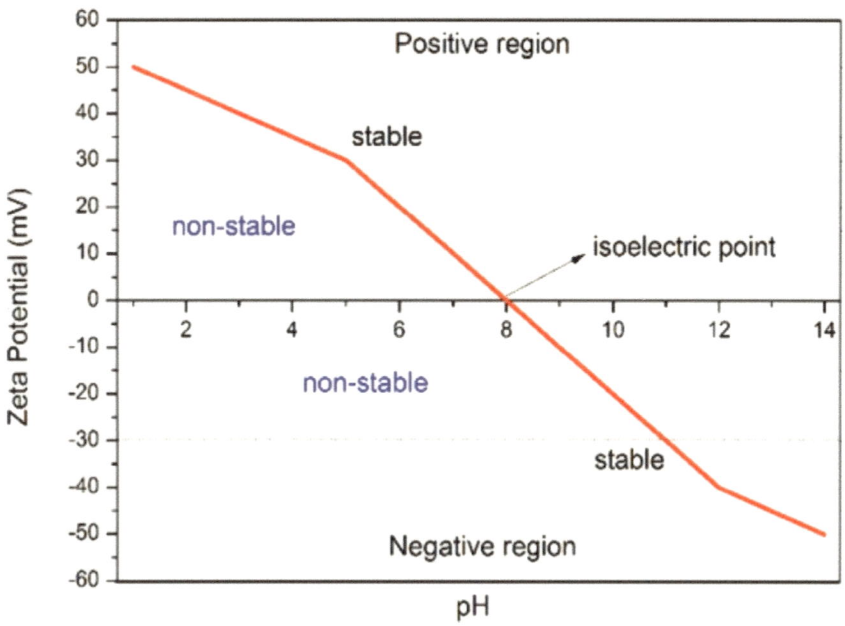

Figure 2. Schematic illustration of change of zeta (ζ) potential with pH.

1.2. Conducting polymers

Conducting polymers have unique properties such as electrical stability, ease of synthesis and durability leading to a variety of potential applications. Conjugated polymers have attracted enormous attentions from both science and technology as semiconductors and electroactive materials for their applications in batteries, molecular electronic devices, sensors, solid phase extractions and light emitting diodes [18–21]. The most common conducting polymers are polypyrrole, polythiophene and polyaniline, as shown in **Figure 3**. The biggest advantage of these polymers is their processability, mainly by dispersion.

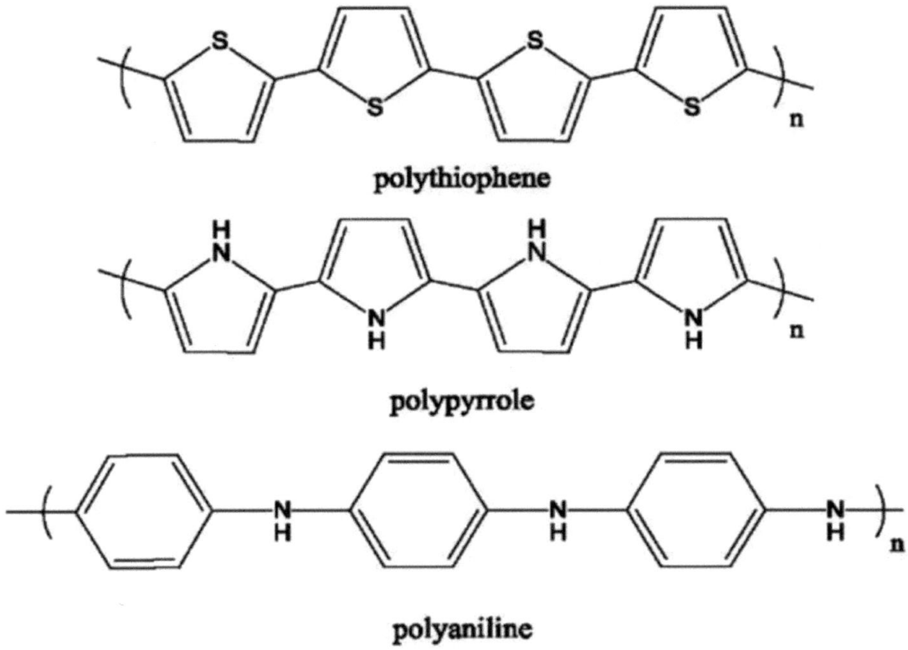

Figure 3. Chemical structures of some conjugated conducting polymers.

1.3. Chitosan as a biodegradable polymer

Polysaccharides are a major class of biomaterials, such as chitosan, chitin, starch, cellulose and alginate. Chitin is extracted from the shells of crustaceans like crabs and shrimps [22]. Chitosan [poly (β-(1-4)-2-amino-2-deoxy-D-glucan] is *N*-deacetylated derivative of chitin (**Figure 4**). CS is a polycationic and insulating biopolymer and contains free amino groups (-NH2) at neutral and alkaline pH values, but they are protonated (-NH3+) under acidic conditions, thus making ionic interactions possible upon the reaction with anions [23–26]. CS has interesting characteristics such as biodegradability, chemical inertness, biocompatibility and low cost. Therefore, chemical derivatives of CS with conducting polymers are important topic for production of multi-functional materials. CS as a biodebredable polymer have been using in many application area especially in the food and medical industry. Chitosan is easily enzymatically biodegradable in nature [27].

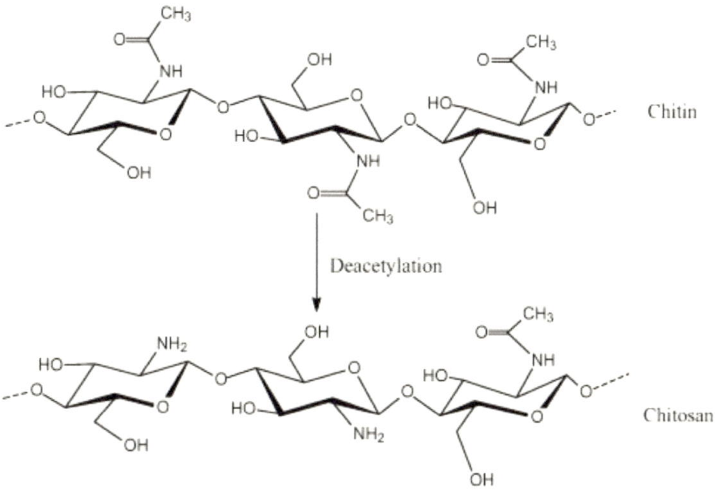

Figure 4. Structures of chitin and chitosan.

2. Electrokinetic measurements of colloidal particles

The incorporation of conducting polymers into natural polymers can enhance the performance of both the 'host' and the 'guest' thereby leading to interesting physical and chemical properties [28]. Generally, fabrication of CP/CS composite is carried out by chemical polymerization of monomer such as aniline, pyrrole and thiophene using as initiator ($FeCl_3$, APS, etc.) in the presence of 1% CS solution. The ζ-potential measurements are performed with a Malvern Nano ZS which works with Laser Doppler Electrophoresis technique using folded capillary cell equipped with gold electrodes (**Figure 5**). Note that 0.1 g/L dispersions is taken and pH value is adjusted with 0.1 M $HCl_{(aq)}$ and 0.1 M $NaOH_{(aq)}$ solution.

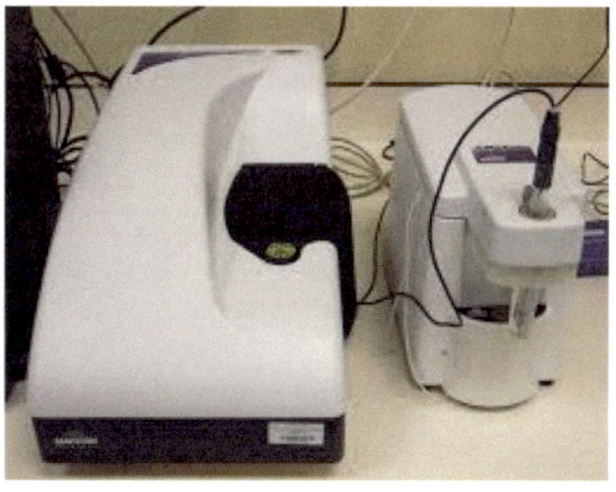

Figure 5. Image of the Malvern Nano Zeta Sizer.

Zeta-Sizer software is used for the ζ-potential measurements of the dispersed particles. In aqueous medium, the ζ-potential is determined by using the Smoluchowski model. The thickness of the electrical double layer, EDL (κ^{-1}) is accepted to be smaller in comparison with the particle size. First, to measure surface potential, solid particles are dispersed in deionized water as 0.1 g/L concentration. This dispersion is subjected to ultrasonication for 30–60 min. And then the dispersion is left to establish equilibrium for 2–3 h. Finally, the supernatant liquid is taken for the ζ-potential measurements. pH of the dispersion is adjusted with diluted $HCl_{(aq)}$ and $NaOH_{(aq)}$ solution by MPT-2 autotitrator unit at 25°C. The effects of pH, electrolytes, surfactants and temperature on the ζ-potentials of the colloidal dispersions can be investigated.

2.1. Effect of pH on colloidal behaviour of dispersions

Electrokinetic properties of polyaniline, polypyrrole and polythiophene-chitosan composites are given as a function of pH in **Figure 6**. It is shown that ζ-potential values of the conducting polymers increase and shift to positive region due to positive surface charge of polycationic chitosan molecules in the composite structure. At the same time, it causes to an increase in colloidal stabilities of the dispersed particles. With the addition of $HCl_{(aq)}$ solution into the dispersion medium, concentration of hydronium cations increases in the aqueous dispersion, and the particle surfaces are surrounded by the hydronium ions. This causes to the protonation of the amino groups of CS or the selective adsorption of the hydronium cations on the surfaces of the dispersed colloidal particles. As a result, the ζ-potential values of the particles gradually increase with decreasing pH values, as shown in **Figure 6**. ζ-potential of CS molecules is related with number of $-NH_3^+$ ions [29]. With the addition of $NaOH_{(aq)}$ solution into the dispersion medium, number of hydronium ions reduces and the number of OH^- anions increases. And then, deprotonation of protonated $-NH_3^+$ ions to $-NH_2$ and selective adsorption of OH^- anions on the surfaces of the dispersed particles occur. As a result, ζ-potential of the colloidal dispersions decreases and shifts to more negative values due to the overcompensation of the positive surface charge with the negatively charged counter ions.

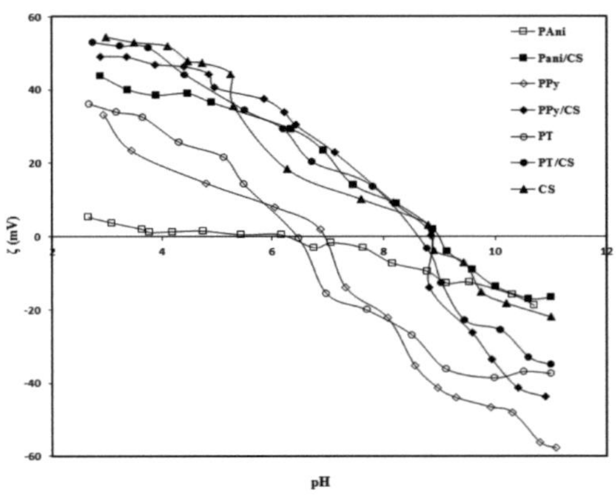

Figure 6. Change of ζ-potential with pH, 10^{-3} M NaCl, c = 0.1 g/L, T = 25°C.

Dispersed particles have zero surface charge at their isoelectric points. At this point, number of positive charges is equal to the number of negative charges. Therefore, the particles in a solution are only stabilized by steric effect. CS is deprotonated around pH 8.5 due to the zwitterionic structure of carboxyl groups in CS molecules. To signify the unique zwitterionic properties, the CS derivatives are named as zwitterionic chitosans. Conducting polymers may act as anion or cation exchangers [30–33].

2.2. Effect of electrolytes on colloidal behaviour of dispersions

The effect of cationic and anionic electrolytes on ζ-potentials of conducting polymers and their CS composites is presented in **Figure 7**. As can be seen, monovalent (Na^+) and divalent (Ba^{2+}) cations and divalent (SO_4^{2-}) anions are indifferent ions which only compressed the electrical double layer. As shown in **Figure 7**, ζ-potential values of colloidal particles shift to the positive regions in 1 M $BaCl_2$ solution due to increasing valency of the cations. It can be attributed to the larger ionic radius of the Ba^{2+} ions, which makes difficult to exchange with the other ions such as Na^+ present in the dispersion. On the other hand, with increasing valency of the anion, ζ-potential of the colloidal dispersions shifts to more negative region. It can be attributed the presence of a Stern layer formed by negatively charged divalent counter ions (SO_4^{2-}), over compensating the positive surface charge of the colloidal particles [34, 35]. Also, it was observed that IEP values of the dispersions shift to higher or lower pH values with increasing ionic strength of the cations and anions present in the dispersion medium, respectively.

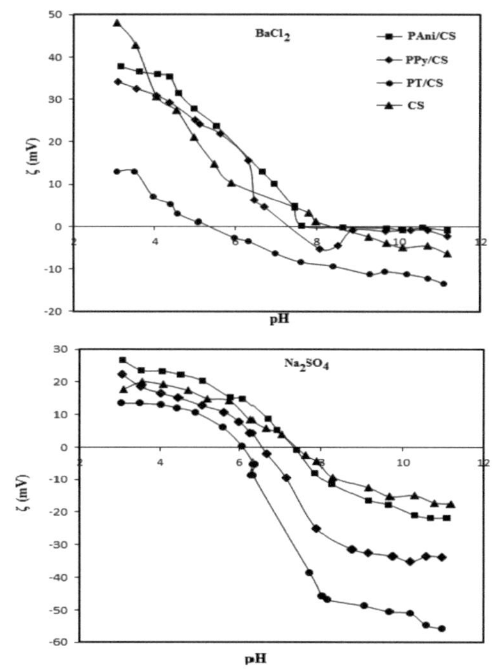

Figure 7. Change of ζ-potential with pH, 10^{-3} M $BaCl_2$ and Na_2SO_4, $c = 0.1$ g/L, $T = 25°C$.

2.3. Effect of various surfactants on the ζ-potential of colloidal dispersions

A small amount of surfactants could be adsorbed easily on the surfaces of the dispersed particles due to their hydrophobic surfaces and cause rapid ζ-potential change. The changes in ζ-potentials of the colloidal dispersions as a function of cationic (CTAB), anionic (SDS) and nonionic (Triton X-100) surfactant concentrations are given in **Figure 8**. As shown, CTAB is added into the colloidal dispersion, ζ-potential of the dispersions increases, without passing through an IEP. After the surface coverage is completed, ζ-potential value slowly decreases and then similarly charged CTAB ions repulsive each other [36, 37]. With the addition of the anionic SDS molecules, the surface coverage of the particles increases with an increasing number of negatively charged SDS molecules. As a result, ζ-potential of the particles in the dispersion shifts to negative values with passing through an IEP. It may be caused by the electrostatic interactions between the negatively charged SDS tails and the positively charged CP/CS surfaces [38]. On the other hand, with the addition of non-ionic surfactant (Triton X-100), ζ-potential value of the colloidal dispersion is almost similar, as shown in **Figure 8**.

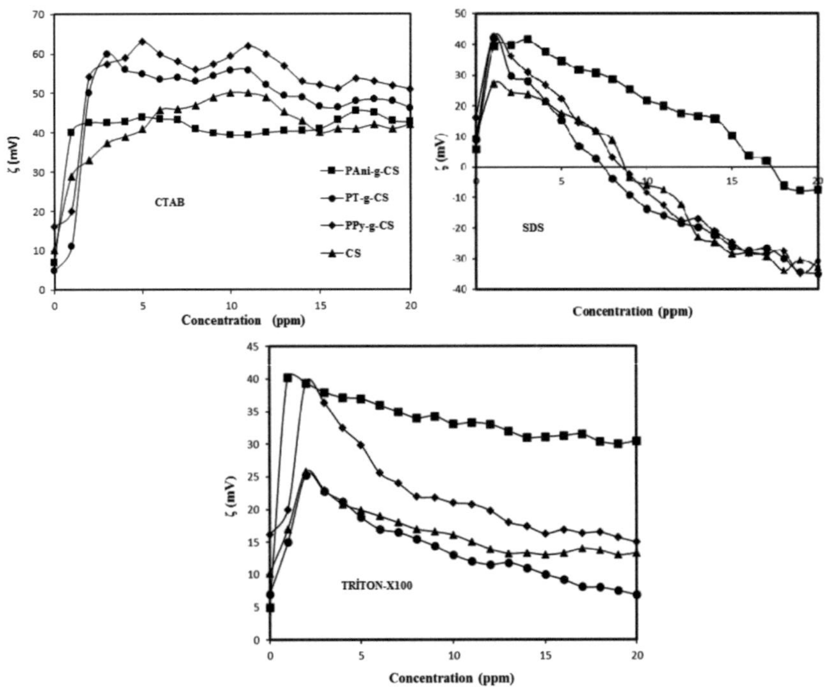

Figure 8. Effect of surfactant type on ζ-potential of colloidal dispersions.

It is known that mobility of the dispersed particles increases with increasing temperature. When the ionic mobility increases, viscosity of dispersion decreases at higher temperatures [39, 40]. Therefore, the establishment of diffuse layer leading to the formation of ζ-potential around the particles becomes difficult. Therefore, ζ-potential value of the CP/CS dispersions decreases with increasing temperature.

http://dx.doi.org/10.5772/65125

3. Conclusion

These results showed that the ζ-potential of the both CP and CS can be changed with the synthesis of their CP/CS composites. Besides, the ζ-potential of CP/CS colloidal dispersions can be altered by using different electrolytes, cationic or anionic surfactants and their colloidal stability can be improved. It is an important result from their industrial applications point of view.

Author details

Mehmet Çabuk

Address all correspondence to: mehmetcabuk@sdu.edu.tr; mhmtcbk@gmail.com

Chemistry Department, Süleyman Demirel University, Isparta, Turkey

References

[1] Gestel TV, Vandecasteele C, Buekenhoudt A, Dotremont C, Luyten J, Leysen R, Bruggen BV, Maes G. Salt retention in nanofiltration with multiplayer ceramic TiO_2 membranes. J. Membr. Sci. 2002; 209: 379–389.

[2] El-Gholabzouri O, Cabrerizo-Vilchez MA, Hidalgo-Alverez R. Zeta-potential of polystyrene latex determined using different electrokinetic techniques in binary liquid mixtures. Colloids Surf. A 2006; 291: 30–37.

[3] Alkan M, Demirbas O, Dogan M. Electrokinetic properties of sepiolite suspensions in different electrolyte media. J. Colloid Interf. Sci. 2005; 281: 240–248.

[4] Davis JA, James R, Leckie JO. Surface ionization and complexation at the oxide/water interface: I. Computation of electrical double layer properties in simple electrolytes. J. Colloid Interf. Sci. 1978; 63: 480–499.

[5] Duman O, Tunc S. Electrokinetic and rheological properties of Na-bentonite in some electrolyte solutions. Micropor. Mesopor. Mater. 2009; 117: 331–338.

[6] Gur A. Dissolution mechanism of colemanite in sulphuric acid solutions. Korean J. Chem. Eng. 2007; 24(4): 588–591.

[7] Hang JZ, Zhang YF, Shi L, Feng Y. Electrokinetic properties of barite nanoparticles suspensions in different electrolyte media. J. Mater. Sci., 2007; 42: 9611–9616.

[8] Ozdemir O, Celik MS. Surface properties and flotation characteristics of boron minerals. Open Min. Proces. J. 2010; 3: 2–13.

[9] Stamm M. Polymer Surface and Interface Characterization Techniques. Polymer Surfaces and Interfaces, (Stamm M, Editor) Springer-Verlag, Berlin Heidelberg, 2008.

[10] Rao SR. Surface Chemistry of Forth Flotation, (Rao, S R, Editor) Luwer Academic/ Plenum Publishers, New York, 2004, p. 223.

[11] Unal HI, Sahan B, Erol O. Investigation of electrokinetic and electrorheological properties of polyindole prepared in the presence of a surfactant. Mater. Chem. Phys. 2012; 134: 382–391.

[12] Wan Y, Creber KAM, Peppley B, Bui VT. Synthesis, characterization and ionic conductive properties of phosphorylated chitosan membranes. Macromol. Chem. Phys. 2003; 204: 850–858.

[13] Jin X, Wang J, Bai J. Synthesis and antimicrobial activity of the Schiff base from chitosan and citral. Carbohydr. Res. 2009; 344: 825–829.

[14] Li X, Wang Y, YangX, Chen J, Fu H, Cheng T. Conducting polymers in environmental analysis. Trends Analyt. Chem. 2012; 39: 163–179.

[15] Ma Y, Hou S, Ji B, Yao Y, Feng X. A novel temperature-responsive polymer as a gene vector. Macromol. Biosci. 2010; 10: 202–210.

[16] Fessi H, Elaissari A. Polymer-based nanocapsules for drug delivery. Int. J. Pharm. 2010; 385: 113–142.

[17] Koyuncu K, Unal HI, Gumus OY, Erol O, Sari B, Ergin T. Electrokinetic and electrorheological properties of poly(vinyl chloride)/polyindole conducting composites. Polym. Advanc. Technol. 2011; 23: 1464–1472.

[18] Cabuk TZ, Sari B, Unal HI. Preparation of novel polyindene/polyoxymethylene blends and investigation of their properties. J. Appl. Polym. Sci. 2010; 117: 3659–3664.

[19] Xu MJ, Zhang J, Wang SR, Guo XZ, Xia HJ, Wang Y, ZhangSM, Huang WP, Wu SH. Gas sensing properties of SnO_2 hollow pheres/polythiophene inorganic-organic hybrids. Sens. Actuator B-Chem. 2010; 146: 8–13.

[20] Gumus OY, Unal HI, Erol O, Sari B. Synthesis, characterization, and colloidal properties of polythiophene/borax conducting composite. Polym. Compos. 2011; 32: 418–426.

[21] Zhang X, Bai R. Surface electric properties of polypyrrole in aqueous solutions. Langmuir 2003; 19: 10703–10709.

[22] Sashiwa H, Aiba S. Chemically modified chitin and chitosan as biomaterials. Prog. Polym. Sci. 2004; 29; 887–908.

[23] Du WL, Xu YL, Xu ZR, Fan CL. Preparation, characterization and antibacterial properties against *E. coli* K88 of chitosan nanoparticle loaded copper ions. Nanotechnology 2008; 19: 85707–85711.

[24] Sabaa MW, Mohamed NA, Mohamed RR, Khalil NM, Abd El Latif SM. Synthesis, characterization and antimicrobial activity of poly(N-vinyl imidazole) grafted carboxymethyl chitosan. Carbohydr. Polym. 2010; 79: 998–1005.

[25] Wang QZ, Chen XG, Liu N, Wang SX, Liu CS, Meng XH, Liu CG. Protonation constants of chitosan with different molecular weight and degree of deacetylation, Carbohydr. Polym. 2006; 65: 194–201.

[26] Cabuk M, Yavuz M, Unal HI, Erol O. Synthesis, characterization and electrorheological properties of biodegradable chitosan/bentonite composites. Clay Min. 2013; 48: 129–141.

[27] Liu Y, Naidu R, Ming H. Surface electrochemical properties of red mud (bauxite residue): zeta potential and surface charge density. J. Colloid Interface Sci. 2013; 394: 451–457.

[28] Bower MJD, Bank TL, Geise RF, VanOss CJ. Nanoscale forces of interaction between glass in aqueous and non-aqueous media: a theoretical and empirical study. Colloids Surf. A 2010; 362: 90–96.

[29] Xu P, Bajaj G, Shugg T, Van Alstine WG, Yeo Y. Zwitterionic chitosan derivatives for pH-sensitive stealth coating, Biomacromolecules 2010; 11: 2352–2358.

[30] Guzel S, Unal HI, Erol O, Sari B. Polyindene/organo-montmorillonite conducting nanocomposites I: synthesis, characterization and electrokinetic properties. J. Appl. Polym. Sci. 2011; 123: 2911–2922.

[31] Cabuk M, Yavuz M, Unal HI, Alan Y, Synthesis, characterization, and enhanced antibacterial activity of chitosan-based biodegradable conducting graft copolymers. Polym. Compos. 2014; 36: 497–503.

[32] Parks GA, The isoelectric points of solid oxides, solid hydroxides and aqueous hydroxo complex systems. Chem. Rev. 1965; 65: 177–198.

[33] Boomi P, Prabu HG, Mathiyarasu J. Synthesis and characterization of polyaniline/Ag-Pt nanocomposite for improved antibacterial activity. Colloid Surf. B 2013; 103: 9–14.

[34] Ishikawa Y, Katoh Y, Ohshima H. Colloidal stability of aqueous polymeric dispersions: effect of pH and salt concentration. Colloid Surf. B 2005; 42: 53–58.

[35] Zhao Y, Xing YZW, Xu N, Wong FS. Effects of inorganic electrolytes on zeta potentials of ceramic microfiltration membranes. Sep. Purif. Technol. 2005; 42: 117–121.

[36] Ruso JM, Deo N, Somasundaran P. Complexation between dodecyl sulfate surfactant and zein protein in solution. Langmuir 2004; 20: 8988–8991.

[37] Blanco E, Ruso JM, Prieto G, Sarmiento F. On relationships between surfactant type and globular proteins interactions in solution. J. Colloid Interf. Sci. 2007; 316: 37–42.

[38] Kirby BJ, Hasselbrink EF. Zeta potential of microfluidic substrates: 1. Theory, experimental techniques, and effects on separations. Electrophoresis 2004; 25: 187–202.

[39] Garcia-Garcia S, Jonsson M, Wold S. Temperature effect on the stability of bentonite colloids in water. J Colloid Interf. Sci. 2004; 298: 694–702.

[40] Hsu JP, Tai YH, Yeh LH. Importance of temperature effect on the electrophoretic behavior of charge-regulated particles. Langmuir 2012; 28: 1013–1019.

Chapter 9

Colloids: Applications in Sperm Preparation for Assisted Reproduction

Jane Morrell

Additional information is available at the end of the chapter

http://dx.doi.org/10.5772/64898

Abstract

Colloids have been used for several decades to prepare spermatozoa for assisted reproduction, initially for *in vitro* fertilization but, with the development of scaled-up techniques, increasingly for artificial insemination and cryopreservation as well. The colloids usually consist of coated silica particles. Using colloid centrifugation, it is possible to select sperm subpopulations consisting of motile spermatozoa with intact membranes, stable DNA and normal morphology and to separate them from the rest of the ejaculate. This review explains why different protocols for colloid centrifugation are needed for different species, as well as species-specific colloid formulations, to match the physical characteristics of the semen. The advantages and disadvantages of sperm preparation by this technique will be outlined. An emerging area of interest is the ability to separate spermatozoa from the bacteria that contaminate semen during collection. Thus, colloid centrifugation represents an alternative to using antibiotics in semen extenders. Since there is a worldwide movement to restrict the use of antibiotics, the possibility of physically removing the bacteria is of considerable interest. Moreover, it may be possible to use colloids to reduce viruses in semen. Transmission of viruses through semen is an emerging problem as more and more viruses are being identified that can potentially be spread in this manner.

Keywords: sperm quality, single layer centrifugation, antimicrobial resistance

1. Introduction

Assisted reproduction is the creation of a pregnancy by a means other than natural mating [1]. The most widely used form of assisted reproduction in animals is artificial insemination (AI); *in vitro* fertilization (IVF) and intracytoplasmic sperm injection (ICSI) are confined to specialist laboratories, for example, for bovine IVF, or are used occasionally for research or for conservation breeding. In contrast, human fertility treatments usually involve either IVF or ICSI. However, there are many differences between species regarding sperm numbers needed for AI, the type of semen used (fresh, stored or cryopreserved), the timing relative to ovulation, and site of deposition of the semen. These differences create a range of challenges for sperm preparation according to the species concerned. Despite the differences, one feature common to all species is that the chance of a successful outcome, that is, a pregnancy, is related to the quality of the sperm sample used. Colloid centrifugation can be used to select good quality spermatozoa for assisted reproduction [2].

The purpose of this review is to describe colloid centrifugation through species-specific colloids, particularly single layer centrifugation (SLC), to select functional spermatozoa. In this review, the history of single layer centrifugation (SLC) will be presented, the method of preparing semen will be explained and its applications in the semen industry will be proposed. Apart from describing the necessity for species-specific protocols, this review will also look at species-specific applications, particularly for the equine, porcine and bovine breeding industries, and the possibilities for germ-plasm conservation.

2. Why is it necessary to select spermatozoa for assisted reproduction?

Spermatozoa to be used for assisted reproduction technologies (ART) must be able to function normally, that is, to find and fertilize an oocyte and to direct the future development of the zygote [2]. Whichever form of assisted reproduction is used, good quality gametes are needed to ensure a successful outcome. Semen samples usually contain a heterogeneous population of spermatozoa of different stages of maturity and functional ability [3]. It is possible to select the subpopulations consisting of motile spermatozoa with intact membranes, stable DNA and normal morphology and to separate them from the rest of the ejaculate, using colloid centrifugation. These spermatozoa are most likely to be functional, to be able to fertilize and activate oocytes and to contribute to the zygote's future development.

3. Different types of semen

Semen for ART is available either fresh or cryopreserved, depending on the species [1]. Most bull spermatozoa are cryopreserved in small quantities, for example, 0.25 mL. However, the semen of some species or individuals may not freeze well; thus, boar and stallion semen is usually used fresh or after cooled storage. As a result, different protocols are needed to deal with the vast range of semen volumes and sperm concentration seen in different species, ranging from a few microliters (cat ejaculate or frozen semen of many species), to several hundred mL for a fresh boar ejaculate. Human semen in fertility clinics is used either fresh or frozen, if from the patient; if the semen has come from a sperm bank, it is usually frozen.

4. Colloid centrifugation

Colloid centrifugation is a sperm preparation technique whereby a sub-population of motile spermatozoa with normal morphology, intact membranes and good chromatin integrity are separated from the rest of the ejaculate, including the seminal plasma, by centrifugation through a colloid [2]. These robust spermatozoa pass through the colloid during gentle centrifugation and form a pellet in the bottom of the conical centrifuge tube. The seminal plasma, with its content of non-sperm cells, bacteria, etc., and semen extender are retained above the colloid. At the interface between the colloid and the seminal plasma, spermatozoa that have not been able to pass into the colloid are retained [4]. Typically, the spermatozoa at the interface are immotile spermatozoa, abnormal spermatozoa and those with damaged chromatin. If too many spermatozoa are loaded on to the colloid or there are very high numbers of abnormal spermatozoa, some normal spermatozoa may be prevented from entering the colloid and will be found in the interface population. Typically, the sperm pellet will have better sperm quality than the unselected samples [5, 6] and the resulting sperm sample will retain motility [7] and fertilizing ability [8] for longer than the unselected samples.

The colloids typically used to prepare spermatozoa are coated silica colloids in a mixture of salts, sugars and buffer. The composition of the mixture, and the pH, osmolarity and density of the colloid all affect the result in terms of number of spermatozoa in the pellet and their quality [9]. Although commercial colloid products are sterile, they do not contain antibiotics. Care should be taken to use aseptic techniques when opening the bottle to avoid contaminating the contents.

5. History of colloid centrifugation

Colloid centrifugation is a technique for separating different types of cells. The first application of this technique was as a density gradient, involving two or more layers of colloids of different densities [10]. During centrifugation on the density gradient, the cells move to a point in the gradient that matches their own density, which is called the isopycnic point. The degree of separation achieved depends on the number of layers of different densities used in the preparation of the density gradient and the differences in densities of the cell populations to be separated. The first and most widely used colloid was Percoll™ (now sold by GE Healthcare, Uppsala, Sweden), based on polyvinylpyrrolidone-coated silica particles. Since the mid-1990s, other coated silica colloids have been used, for example, silane-coated silica such as Redigrad® (originally sold by Amersham Biosciences, now GE Healthcare, Uppsala, Sweden) and Iodixanol (OptiPrep™, Progen, Heidelberg, Germany).

Semen samples consist of a heterogeneous population of spermatozoa of different ages and stages of maturity. Before spermatozoa can fertilize an oocyte, they must undergo a series of membrane changes over a period of time, known as capacitation, followed by the acrosome reaction just before they bind to the oocyte. If they do not bind to an oocyte after undergoing the acrosome reaction, they will die. The purpose of proving a mixture of spermatozoa of different ages is to ensure that spermatozoa capable of fertilization will be available in sequence over a period of time. However, if the sperm sample is to be stored prior to being inseminated into a female, the presence of dead or dying spermatozoa, or cellular debris, may adversely affect the remaining spermatozoa. Separation of the viable spermatozoa from the rest of the ejaculate therefore enables them to remain functional over a longer period of time. A further advantage is that spermatozoa can be removed from seminal plasma (which contains decapacitating factors) and semen extender before use [11].

Early in the 1980s, researchers began to use density gradients for selecting viable spermatozoa [12, 13]. The spermatozoa that passed through the colloid and could be retrieved from the sperm pellet were mostly those that were motile, with normal morphology, intact sperm membranes and good chromatin integrity. Since these parameters of sperm quality have been correlated with pregnancy rate after artificial insemination in a number of species, it was proposed that the spermatozoa in the pellet were more likely to generate pregnancies when used for assisted reproduction than the uncentrifuged sample. Since the seminal plasma and extender were retained on the top of the uppermost layer of colloid, the method was ideally suited for preparing frozen spermatozoa for IVF or ICSI, where separation of the spermatozoa from seminal plasma was necessary. Immotile and immature spermatozoa or morphologically abnormal spermatozoa and those with damaged chromatin are mostly retained at the interface between the seminal plasma/extender and the colloid.

Density gradient centrifugation (DGC) was used for preparing human semen in fertility clinics, since the volume of semen to be processed was generally small and the sperm concentration low. Typically, the volume of colloid in each layer was 1–2 mL, depending on the number of layers used, and the volume of semen to be processed was up to approximately 1.5 mL. Thus, the technique had severe limitations when trying to process animal semen, which typically has much larger volumes and higher sperm concentrations than human semen. Nevertheless, frozen animal semen (mostly bovine) was processed for IVF; there were some reports of animal semen being processed by density gradient in the laboratory to improve sperm quality, but the consensus of opinion was that it was too impractical for processing whole ejaculates for AI in the field. Since AI is by far the most widely used biotechnology in animal breeding, there was little use for sperm preparation by DGC in this area.

In the mid-1990s, there was some debate as to whether Percoll™ itself was detrimental to spermatozoa, with some researchers claiming adverse effects from the use of the PVP-coated silica (e.g. [14]). Other researchers reported conflicting results with bovine and murine samples [15, 16]. This issue has never been fully resolved but factors other than the type of colloid may also be implicated, such as the centrifugation force and time used in the protocol [17], the osmolarity and chemical composition of the buffered salt solution used to dilute the colloid [9], the semen extender used for the semen [7], and even the sperm concentration of the sperm suspension [7].

Towards the end of the 1990s, a modification of DCG was developed in which only one layer of Percoll™ was used to prepare human spermatozoa [18]. These authors concluded that the single layer gave good results with oligospermic ejaculates but did not improve the sperm quality of normospermic semen. Thus, the technique was not universally adopted in fertility clinics at that time, since there was a preference to have one standard protocol for all sperm preparations regardless of the quality of the original semen sample. Similar findings were later reported using one layer of Percoll™ for stallion semen [19], in that sperm quality was only improved in poor ejaculates.

Commercially available colloid formulations for use as a density gradient became available early in the twenty-first century. The first of these colloids for density gradients, EquiPure (Nidacon International AB, Mölndal, Sweden) was reported to improve stallion sperm quality [17], and it was used to prepare small numbers of spermatozoa from problem ejaculates for low dose insemination [20]. However, sperm yields were too low and the method too impractical for routine application when preparing spermatozoa for conventional insemination in the field.

Single layer centrifugation (SLC) using only one layer of a species-specific colloid formulation (Androcoll-E, now known as Equicoll) for preparing stallion spermatozoa was reported by Morrell et al. [21]. They described studies in which "normal" stallion semen, that is, from stallions with no known fertility problems, was split and processed by DGC and SLC. Sperm quality was improved in both types of selected samples compared to the uncentrifuged samples, and there was no difference in sperm quality between the two colloid centrifugation groups [5, 6]. The results from 250 ejaculates processed by SLC in three breeding seasons showed that the technique was repeatable and reliable under a variety of field conditions for stallions at commercial stud farms [7]. In addition, the development of a different colloid formulation (Androcoll-E Large) for use in 50-mL tubes enabled SLC to be scaled up to process whole stallion ejaculates [22].

It is also possible to use SLC process 25 mL of extended semen in 100-mL tubes or even 150-mL extended semen in 500-mL tubes [23]. The latter is particularly useful for processing boar ejaculates, which tend to be voluminous. However, the centrifuges commonly found on stud farms have rotors for 50-mL tubes; most stallion ejaculates can be processed in 4–6 of these 50-mL tubes.

Recently, the original protocol (4-mL colloid in a 15-mL centrifuge tube—designated Small SLC— was compared with protocols using 1 mL colloid, either in a 15-mL centrifuge tube (Mini-SLC) or in a microcentrifuge tube (Mini-EP), for preparing bovine spermatozoa for IVF [24]. The highest sperm yield was obtained from the Mini-SLC, that is, from 1 mL colloid in a 15-mL tube, although sperm quality was good in all three variants. An IVF trial using spermatozoa processed by Mini-SLC compared to swim-up did not show significant differences between the swim-up control and Mini-SLC, but since the chromatin integrity of the Mini SLC sperm preparation was significantly higher than the control, it would be expected that differences in embryo development might occur at a later stage in vivo than studied in vitro.

Thus, protocols have been developed for a range of tube sizes, different types of semen and various species as shown in **Table 1**. For optimum results, the specific colloid formulation for a particular size of tube should be used. However, the formulation developed for use in 50-mL tubes has also given good results when used in 15-mL centrifuge tubes (L Barros, personal communication). Protocols based on other colloids have also been tried, such as Iodixanol (OptiPrep™, Progen, Heidelberg, Germany) [25, 26].

Type of semen/species	Tube size (mL)	Volume colloid (mL)	Volume extended semen* (mL)
Thawed: Red deer [64] Bull [24]	12–15	1.0	0.25
Fresh or thawed: Stallion [21] Bull [48] Boar [65]	12–15	4.0	4.5
Fresh: Stallion [22] Boar [42]	50	15	15–18
Fresh: Stallion [23]	100	20	25
Fresh: Stallion [22] Boar [23]	200	60	60
Fresh: Boar [23, 43]	500	150	150

Table 1. Modifications of single layer centrifugation according to volume of semen to be processed.

6. The method of single layer centrifugation

Preparing spermatozoa by SLC is a straightforward procedure (**Figure 1**). The following instructions refer to SLC in a 15 mL tube; the modifications required for other sizes of tube are shown below.

1. Equilibrate colloid and extended semen to room temperature to avoid cold-shocking the spermatozoa.
2. Adjust sperm concentration to not more than 100×10^6/mL for stallions and boars and not more than 50×10^6/mL for bulls.
3. Carefully layer up to 4.5 mL of extended semen on top of 4-mL Equicoll using a Pasteur pipette.
4. Use a centrifuge with a swing-out rotor; centrifuge at 300 g for 20 min.
5. Remove supernatant (seminal plasma, semen extender and most of the colloid) using a Pasteur pipette, taking care not to agitate the sperm pellet. Leave approximately 2 mm colloid above the sperm pellet.

6. Using a clean pipette, aspirate the sperm pellet from underneath the colloid.

7. Add extender to the sperm pellet to achieve the desired sperm concentration.

The same procedure is used for the other variants of SLC, differing only in the size of the tube, the type and volume of colloid, and the volume of semen to be used. For 50-mL tubes, 15-mL colloid and up to 20-mL extended semen are used. For 500-mL tubes, 150-mL colloid and 150-mL extended semen are used.

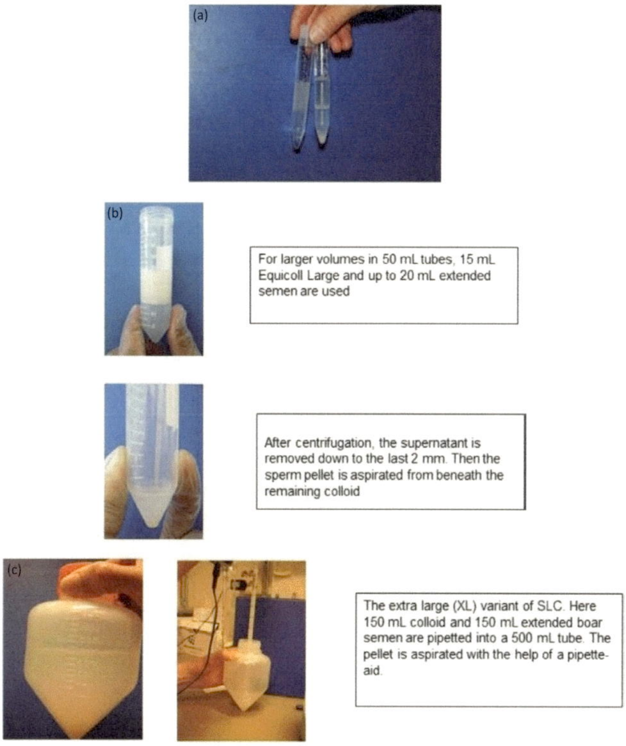

Figure 1. (a) Single layer centrifugation to process up to 4.5-mL semen ("Small"). (b) Single layer centrifugation to process up to 20-mL semen ("Large"). (c) Single layer centrifugation to process up to 150-mL semen ("XL") [23].

7. Specific applications of SLC

7.1. Applications in equine breeding

The equine breeding industry is extremely diverse ranging from studs with one stallion and a few mares to complex operations sending out hundreds of semen doses all over the country and beyond, and also large-scale embryo transfer facilities, not to mention stud farms with natural mating. The scale of the industry and the geographical distribution of the stud farms present specific challenges in the distribution of good quality semen doses, which is where SLC could be of particular benefit.

7.1.1. Improving sperm quality for artificial insemination

Sometimes sperm quality is not as good in some ejaculates as desired. Since sperm morphology and chromatin integrity are correlated with pregnancy rates after AI, improving these parameters can lead to an increase in the pregnancy rate [27, 28]. Both normal morphology and chromatin integrity are higher in SLC-selected samples than in uncentrifuged control semen, indicating that pregnancy rates can be improved using SLC to select the spermatozoa. This has been shown to be the case both for problem stallions [27] and for those with no known fertility problem [28]. However, since the poor quality spermatozoa are removed, the number of spermatozoa remaining after SLC is less than in the original sample.

7.1.2. Maintaining sperm quality during transport

The sperm quality of the semen doses must be maintained during transport so that good pregnancy rates can be obtained after AI. Colloid centrifugation is ideal for this purpose, since the robust spermatozoa are separated from the dead and dying spermatozoa and also from seminal plasma and any bacteria contaminating the ejaculate at collection. Dead and dying spermatozoa can release intracellular contents, such as reactive oxygen species, that are detrimental to the survival of healthy spermatozoa. Thus, separating the robust spermatozoa from sources of reactive oxygen species helps to maintain them in good condition. Separating spermatozoa from seminal plasma and bacteria also increases the length of time for which the spermatozoa remain viable [20]. The spermatozoa are not damaged by SLC, whereas other centrifugation protocols to remove seminal plasma, such as cushion centrifugation, may cause damage [29]. The SLC-selected sperm samples are capable of fertilization when inseminated at least up to 96 h after semen collection [8], considerably longer than the 24–36 usually chosen by stud personnel.

7.1.3. Dealing with ejaculates that do not tolerate cooling

Some sperm samples cannot be cooled because the spermatozoa die. Previously, the only chance to use such ejaculates was to inseminate the mare immediately after semen collection at the same stud farm. However, the risk of injury and disease transmission make long-distance transport of mares for breeding undesirable. Removing all the seminal plasma from the spermatozoa and selecting the robust spermatozoa by SLC allows them to survive being cooled, and their fertilizing capability is retained [27]. Alternatively, SLC-selected sperm samples can be maintained at around 15° for at least 48 h and still retain their motility [30] and fertilizing capacity (J. Newcombe, personal communication)

7.1.4. Extending the shelf life of a conventional semen dose

If a conventional stored semen dose cannot be used when anticipated (usually 24–36 h after semen collection), it can be processed by SLC to extract the robust spermatozoa, thus extending the time for which these spermatozoa can be stored prior to insemination [31]. The resulting sperm sample will have better sperm quality and a longer "shelf-life" than the stored semen dose, although sperm quality will not be quite as good or the shelf-life as long as if the sample had been prepared by SLC immediately after semen collection.

7.1.5. Selecting spermatozoa that survive cryopreservation

One way of avoiding the perishable nature of fresh semen would be to cryopreserve semen. However, not all stallions produce semen that can be frozen successfully. If the most robust spermatozoa are selected prior to freezing, sperm cryosurvival can be improved [29] and the thawed spermatozoa survive longer than control spermatozoa [32]. Since freezing protocols for stallion spermatozoa involve a centrifugation step to remove most of the seminal plasma, SLC can be used instead without creating extra effort and has the advantage of removing some of the sources of reactive oxygen species that will harmful to spermatozoa during cryopreservation. The longer survival time of the spermatozoa after thawing may allow more flexibility in the timing of insemination relation to ovulation [32].

7.1.6. Improving the sperm quality of poor quality frozen semen batches

SLC can be used to select robust spermatozoa after thawing [33–35]. Instead of discarding batches where post-thaw sperm motility is less than 30%, it may be possible to extract sufficient motile spermatozoa by SLC for insemination [36]. More straws may be needed than would normally be necessary to recover sufficient spermatozoa for an insemination dose but at least the poor quality batch has been "rescued." This is particularly relevant if the stallion is no longer available for semen collection for freezing.

7.1.7. Conservation of rare breeds or rare bloodlines

When the decision is reached to use conservation breeding to increase population numbers or to gene bank semen, there may be few males to choose from [37]. In such cases, it is important to preserve as many genes as possible, which in practice means using all available stallions regardless of sperm quality. Since no selection has been made for sperm quality, or those stallions that are available are elderly and with deteriorating sperm quality, SLC can be used to advantage to select motile spermatozoa with normal morphology, good membrane integrity and good chromatin integrity from immotile spermatozoa, abnormal spermatozoa and those with damaged chromatin. Since these parameters of sperm quality are linked to pregnancy rate after insemination, SLC can increase the chances of a mare becoming pregnant after insemination. In addition, improved sperm cryosurvival could lead to better success if frozen spermatozoa are used for insemination, thus increasing the likelihood that gene banking can actually help conservation breeding efforts.

7.2. Applications in pig breeding

Similarly to equine breeding, the pig breeding industry relies heavily on the insemination of fresh, cooled semen. Although cryopreservation protocols exist for boar semen, there is a perception that pregnancy rates are lower and litter sizes smaller if cryopreserved semen is used for AI instead of fresh semen [30].

7.2.1. *Improvement in sperm quality*

Better linear motility, normal morphology and viability were observed in SLC-selected boar samples compared to unselected controls [38, 39]. Since these parameters of sperm quality have been linked to increased pregnancy rates in artificially inseminated sows, SLC-selected sperm doses would be expected to produce higher pregnancy rates when used for AI. However, pregnancy rates are normally high for boar semen doses, so the main benefit of SLC would be to enhance sperm quality for genetically valuable boars.

7.2.2. *Reducing the number of spermatozoa required for an insemination dose*

By selecting the most robust spermatozoa, it might be possible to reduce the sperm numbers used for insemination to 1×10^9 or even lower. Thus, even with the loss of approximately 33% of sperm numbers, more doses would be produced from each ejaculate.

7.2.3. *Increasing the shelf life of stored semen*

In SLC-selected preparations, sperm quality declines more slowly during storage than in unselected semen samples [38].

7.2.4. *Removal of bacteria and viruses*

Spermatozoa can be separated from bacteria [40], or viruses [41] in semen by SLC (see section on alternatives to antimicrobials).

7.2.5. *Cryosurvival*

Cryosurvival can be enhanced by SLC [42, 43]. In the first of these studies [42], boar sperm samples processed by SLC before freezing showed less membrane fluidity and less generation of reactive oxygen species after thawing than controls. In the second study, SLC in 500-mL tubes improved the quality and cryosurvival of boar semen [42]. However, the authors did point out that a high recovery rate would be required in order for SLC to be included routinely in cryopreservation protocols.

7.2.6. *For IVF*

SLC can be used to select spermatozoa for IVF, particularly where low numbers of spermatozoa are available. The rate of polyspermy was elevated when SLC-selected spermatozoa were used, indicating that SLC-selected spermatozoa are highly fertile [44]. When the sperm dose was drastically reduced to avoid polyspermy, the fertilization rate and blastocyst development rate were superior to unselected spermatozoa from the same boar.

http://dx.doi.org/10.5772/64898

7.2.7. Conservation breeding

Semen quality may be poor in rare breeds since they have not undergone selection for these traits [37]. Therefore, SLC may help to improve fertility.

7.3. Applications in cattle breeding

The cattle AI industry depends almost exclusively on cryopreserved semen in many parts of the world, except in New Zealand [1]. Cryopreservation prolongs the life of the semen sample and facilitates transport to other countries for insemination. Semen can be placed in quarantine, typically for one month after semen collection, to allow time for any disease in the donor animals to manifest itself. However, cryopreservation damages spermatozoa, although bull spermatozoa seem to be reasonably resistant to cryoinjury. The following uses of SLC have been identified for bull semen.

7.3.1. Improving sperm quality in poor semen samples prior to cryopreservation

Occasionally semen quality is poor, for example, if the proportion of spermatozoa with normal morphology is low [45], or if there is blood in the semen [46]. Even though SLC had no direct effect on total and progressive motility, it appeared to have a positive influence on several other kinematic parameters that may be important for fertilization after artificial insemination. In a previous study, the kinematics straightness (STR), Linearity (LIN) and beat cross frequency (BCF) of uncentrifuged bull sperm samples were positively correlated with the pregnancy rate at day 56 [47]. Thus, SLC samples might be expected to have better fertility than uncentrifuged ones.

7.3.2. Selecting good quality spermatozoa for IVF

Good quality spermatozoa can be selected from thawed semen samples for IVF, for example [48, 24]. Chromatin integrity particularly is improved in the SLC-selected samples: sperm chromatin damage (%DFI) was lower in SLC-selected samples than in unselected control samples (1.84 and 2.99% respectively, $p = 0.036$).

7.3.3. To select spermatozoa with high metabolic activity

Mitochondrial membrane potential ($\Delta\Psi m$) is one of the assays used to evaluate sperm metabolic activity and fertility [49]. Since SLC-selected spermatozoa were found to have a higher $\Delta\Psi m$ than controls [50], it is possible that their fertility may be higher too.

8. Removal of pathogens

8.1. Removal of bacteria from semen samples

Most ejaculates are contaminated with bacteria despite being collected from healthy animals, and further contamination can occur during semen processing [51]. Bacteria can adversely affect sperm quality, by competing for nutrients in the semen extender or from the production of metabolic by-products and endotoxins. In addition, they may cause uterine disease in inseminated females. To control the growth of bacteria, antibiotics are added to semen extenders according to national and international guidelines. The choice of antimicrobials is limited since they may be toxic to spermatozoa, and thus a cocktail of agents may be chosen to try to reduce the adverse effects on sperm survival. However, antibiotic use in general can lead to the development of microbial resistance, which can be passed on to other animals and also between humans and animals. Recent reports indicate that some microorganisms may be resistant to the antibiotics used in semen extenders for livestock [52–54]. Therefore, it might be prudent to develop alternatives to antimicrobials in semen extenders [55], although a change in the legislation would be required in the event that antibiotics are not added.

Human semen samples processed by DGC with a silane-coated silica colloid showed a 10-fold reduction in bacterial content [56]. Strict aseptic technique could reduce bacterial contamination almost completely. Increasing the *g* force used for centrifugation may have been associated with an increase in the number of positive samples for some bacteria. For 10 boar semen samples processed by SLC in a laminar air flow (LAF) bench, no bacteria were cultured from six of the SLC-selected sperm samples, and the number of bacteria in the remaining samples was reduced compared to the uncentrifuged controls [40]. Less than 1% of bacteria remained after the SLC processing. A possible bacteriostatic effect of the colloid was observed, despite the fact that no antibiotics were included in the colloid formulation. Removal of bacteria appeared to be affected by the bacterial species and the length of time between semen collection and SLC.

In a similar experiment with stallion semen, using a tube insert to facilitate harvesting of the sperm pellet [57], the removal of >90% of the total bacterial load was possible if the latter was $>5 \times 10^4$ cfu/mL although for smaller bacterial loads removal was less efficient. Bacteria that tend to clump together, for example, forming palisades, chains and pairs, were more difficult to remove, possibly due to an alteration in the density of the bacterial unit by clumping [57]. Flagellated bacteria were also more difficult to remove.

Using SLC at a higher *g* force (600 g for 10 min) than in the previous studies, approximately 50% of the contaminating bacterial load was removed from stallion semen. Post-thaw sperm motility was increased if bacterial load was reduced compared to controls, confirming the beneficial effects of bacterial removal on sperm survival [53].

8.2. Reduction in viral load

Although AI was developed first as a means of reducing disease transmission between individuals, it is still possible to spread viruses in semen [58], for example, foot and mouth disease virus, porcine respiratory and reproductive syndrome virus, equine arteritis virus, human immunodeficiency virus (HIV) and, more recently, zika virus in humans. Breeding sires used for semen collection are tested routinely for serological evidence of infection, but virus may be shed in semen for several weeks before the antibody response is detected. In other cases, individuals may be permanent virus "shedders" without showing sero-conversion. Thus, a negative serological test indicates only that the male has not mounted an antibody

http://dx.doi.org/10.5772/64898

response at that point in time rather than being an absolute indicator of non-infection. Semen from shedding males represents a source of disease transmission via artificial insemination.

For several years, human fertility clinics have used a double semen processing technique of density gradient centrifugation followed by a swim-up to process semen from patients known to be shedding HIV or hepatitis C virus, where the female partner is not infected with the virus (so-called discordant couples) [59, 60]. These attempts to remove virus were apparently successfully in that there are no reports of infection of the female after the assisted pregnancy or of the birth of infected offspring. However, attempts to remove viruses from animal semen by a similar technique does not lead to the removal of all virus, possibly because unlike the human patients, the males are not treated with anti-viral drugs prior to semen collection.

In a preliminary experiment, aliquots of boar ejaculates were spiked with porcine circovirus 2 (PCV2) and processed by single layer centrifugation using Androcoll-P followed by a "swim-up" procedure. Samples before and after processing were tested for the presence of PCV2 in a tissue culture assay. The difference in virus titre corresponded to more than 99% reduction, and sperm quality was not affected by the selection procedure [41]. Other authors tested the removal of bovine viral diarrhoea virus (BVDV) from experimentally infected semen samples, comparing Percoll gradient centrifugation, swim-up, and a combination of swim-up and Percoll gradient centrifugation. They reported that no virus was detected from the combination of techniques, whereas a low level of virus was found in the swim-up preparation and a higher level in the Percoll preparation [61].

Using the "inner-tube" method described previously for some of the bacterial studies, it was possible to substantially reduce virus titre in semen samples from stallions known to be shedding equine arteritis virus [62]. However, some virus infectivity was still present, especially in stallions shedding large amounts of virus. Since it is not known how much virus in semen will cause infection in an inseminated mare, it would be unwise to assume that the use of SLC-selected samples is risk-free, unless a further modification of the procedure is developed.

9. SLC as an indicator of male fertility

The breeding industry wants to be able to identify sub-fertile individuals either to remove them from the breeding pool (bulls and boars) or to develop strategies to deal with their sub-fertility (stallions). Breeding companies would like to distinguish between highly fertile males and extremely highly fertile males. Attempts have been made to find laboratory tests that are predictive of sperm fertility, but whereas fertility is a complex multifaceted phenomenon, most assays evaluate only one parameter. However, the sperm yield after SLC is related to the quality and hence the potential fertility of the original ejaculates [63]; it is more highly correlated to fertility than the results of individual evaluations. Therefore, the suggestion was made that yield could be used as an indicator of stallion fertility in the absence of breeding data. The SLC is fast (30 min) and only requires a centrifuge, whereas assays of sperm quality require a fluorescent microscope or a flow cytometer and/or specialist skills. If the stud personnel do not wish to carry out the SLC themselves, semen doses could be transported to a laboratory for this procedure. It is possible that SLC could be used predictively in other species as well.

10. Cost-benefit of SLC

Silane-coated silica colloids are expensive, but usually the improved sperm quality and enhanced fertilizing ability of the sperm sample is worth more than the cost of the procedure. The added value of SLC-selected sample depends on many factors, such as the perceived value of offspring from a particular male. The cost of SLC is less than DGC since it uses only one layer of colloid and needs less preparation time. Other factors increase the value of the technique: the availability of sperm samples, increased cryosurvival, biosecurity, ease of transport, etc. Thus, any processing steps that improve sperm survival and fertility will have an economic benefit. The price of the colloid will depend on the volume sold, because of economies of scale. The involvement of technical help in the laboratory affects preparation costs; the final price of the semen dose may include a premium for additional quality, which will, in turn, affect the demand for SLC-selected sperm samples.

11. Conclusion

In conclusion, colloid centrifugation, especially SLC, improves sperm quality in a variety of circumstances and for different species and a range of purposes within animal breeding. The procedure is straightforward and quick; it can be used by the personnel on any stud farm or semen collection centre where there is a swing-out centrifuge.

Acknowledgements

This work was financed by grants from the Swedish Farmers' Foundation (SLF), the Swedish Foundation for Equine Research, and FORMAS.

Author details

Jane Morrell

Address all correspondence to: jane.morrell@slu.se

Department of Clinical Sciences, Swedish University of Agricultural Sciences, Uppsala, Sweden

References

[1] Morrell JM (2011) Artificial insemination: current and future trends. INTECH Open Access Publisher, Croatia. ISBN: 978-953-307-312-5, InTech, Available from: http://www.intechopen.com/articles/show/title/artificial-insemination-current-and-future-trends

[2] Morrell JM, Rodriguez-Martinez H: Biomimetic techniques for improving sperm quality in animal breeding: a review. The Open Andrology Journal. 2009;1:1-9.

[3] Suarez SS: Interactions of spermatozoa with the female reproductive tract: inspiration for assisted reproduction. Reproduction, Fertility and Development. 2007;19:103-110.

[4] Morrell JM: Stallion sperm selection: past, present and future trends. Journal of Equine Veterinary Science. 2012;32:436-440.

[5] Morrell JM, Johannisson A, Dalin A-M, Rodriguez-Martinez,H: Morphology and chromatin integrity of stallion spermatozoa prepared by density gradient and single layer centrifugation through silica colloids. Reproduction Domestic Animals. 2009;44:512-517.

[6] Morrell JM, Dalin A-M, Rodriguez-Martinez, H: Comparison of density gradient and single layer centrifugation of stallion spermatozoa: yield, motility and survival. Equine Veterinary Journal. 2009b;41:53-58.

[7] Morrell JM, Rodriguez-Martinez H, Johannisson A: Single layer centrifugation of stallion spermatozoa consistently selects the most robust spermatozoa from the rest of the ejaculate in a large sample size: data from 3 breeding seasons. Equine Veterinary Journal. 2010;42:579-585.

[8] Lindahl J, Dalin A-M, Morrell, JM: Pregnancies in mares inseminated with spermatozoa selected by single layer centrifugation and stored for 48h or 72h. Reproduction in Domestic Animals. 2011;46:P158.

[9] Morrell JM, Johannisson A, Rodriguez-Martinez H: Effect of osmolarity and density of colloid formulations on the outcome of SLC-selection of stallion spermatozoa. ISRN Veterinary Science. 2011. doi:10.5402/2011/128984.

[10] Pertoft H: Fractionation of cells and subcellular particles with Percoll. J Biochem Biophysical Methods 2000;44:1-30.

[11] Hunter RHF, Rodriguez-Martinez H: Analysing mammalian fertilisation. Reservations and potential pitfalls with an in vitro approach. Zygote. 2002;10:11-15.

[12] Bolton VN, Braude P: Preparation of human spermatozoa for in vitro fertilization by isopycnic centrifugation on self-generating density gradients. Archives Andrology. 1984;13:167-176.

[13] de Vries AC, Colenbrander B: Isolation and characterization of boar spermatozoa with and without a cytoplasmic droplet. International Journal Biochemistry. 1990;22:519-524

[14] Avery B, Greve T: Impact of Percoll on bovine spermatozoa used for in vitro insemination. Theriogenology. 1995;44:871-878.

[15] Motoishi M, Goto K, Tomita K, Ookutsu S, Nakanushi Y: Examination of the safety of intracytoplasmic injection procedures by using bovine oocytes. Human Reproduction. 1996;11:618-620.

[16] Mizuno K, Hoshi K, Huang T: Fertilization and embryo development in a mouse ICSI model using human and mouse sperm after immobilisation in polyvinlypyrrolidone. Human Reproduction. 2002;17:2350-2355

[17] Macpherson M, Blanchard TL, Love CC, Brinsko SP, Thompson JA, Varner DD: Use of a silane-coated silica particle solution to enhance the quality of ejaculated semen in stallions. Theriogenology. 2002;58:317-320

[18] Sharma RK, Seifarth K, Agarwal A: Comparison of single- and two-layer Percoll separation for selection of motile spermatozoa. International Journal Fertility. 1997;42:412-417.

[19] Sieme H, Martinsson G, Rauterberg H, Walter K, Aurich C, Patzoldt R, Klug E: Application of techniques for sperm selection in fresh and frozen-thawed stallion semen. Reproduction in Domestic Animals. 2003;38:134-140

[20] Varner DD, Love CC, Brinsko SP, Blanchard TL, Hartman DL, Bliss SB, Carroll BS, Eslick MC: Semen processing for the sub-fertile stallion. Journal Equine Veterinary Science. 2008;28: 677-685.

[21] Morrell JM, Dalin A-M, Rodriguez-Martinez H: Prolongation of stallion sperm survival by centrifugation through coated silica colloids: a preliminary study. Animal Reproduction. 2008;5: 121-126.

[22] Morrell JM, Johannisson A, Dalin A-M, Rodriguez-Martinez H: Single layer centrifugation with Androcoll™-E can be scaled-up to allow large volumes of stallion ejaculate to be processed easily. Theriogenology. 2009;72:879-884.

[23] Morrell JM, van Wienen M, Wallgren M: Single layer centrifugation can be scaled-up further to process up to 150 mL semen. ISRN Veterinary Science. 2011; doi: 10.5402/2011/183412.

[24] Abraham MC, Johannisson A, Morrell JM: Effect of sperm preparation on development of bovine blastocyst in vitro. Zygote. 2016; In press

[25] Stuhtmann G, Oldenhof H, Peters P, Klewitz J, Martinsson G, Sieme H: Iodixanol density gradient centrifugation for selecting stallion sperm for cold storage and cryopreservation. Animal Reproduction Science. 2012;133:184-190.

[26] Heutelbeck A, Oldenhof H, Rohn K, Martinsson G, Morrell JM, Sieme H: Use of density centrifugation for delayed cryopreservation of stallion sperm: perform sperm selection

directly after collection or after storage? Reproduction in Domestic Animals. 2015;50:76-83.

[27] Morrell JM, Mari G, Kutvölgyi G, Meurling S, Iacono E, Mislei B, Rodriguez-Martinez H: Spermatozoa from stallion ejaculates processed by single layer centrifugation with Androcoll™-E are capable of fertilization after artificial insemination. Reproduction Domestic Animals. 2011;46:642-645.

[28] Morrell JM, Richter J, Martinsson G, Stuhtmann G, Hoogewijs M, Roels K, Dalin A-M: Pregnancy rates are higher after artificial insemination with cooled stallion spermatozoa selected by single layer centrifugation than with control semen doses. Theriogenology. 2014;82:1102-1105

[29] Hoogewijs M, Morrell JM, Van Soom A, Govaere J, Johannisson A, Piepers P, De Schauwer C, de Kruif A, De Vliegher S: Sperm selection using single layer centrifugation prior to cryopreservation can increase post thaw sperm quality in stallions. Equine Veterinary Journal. 2011;43(Suppl 40):35-41.

[30] Morrell JM: Biomimetics in action: practical applications of single layer centrifugation for equine breeding. Veterinary Science and Technology. 2011;2:107. Doi: 10.4172/2157-7579.1000107

[31] Morrell JM, Macias Garcia B, Pena FJ, Johannisson A: Processing stored stallion semen doses by single layer centrifugation. Theriogenology. 2011;76:1424-1432.

[32] Hoogewijs M, Piepers S, Govaere J, De Schauwer C, de Kruif A, Morrell JM: Sperm longevity following pre-freeze sperm selection. Journal Equine Veterinary Science, 2012;32:489.

[33] Macias GB, Fernandez-Gonzalez L., Morrell J., Ortega Ferrusola C., Tapia JA, Rodriguez-Martinez H, Pena FJ: Single layer centrifugation through colloid positively modifies the sperm sub-population structure of frozen-thawed stallion spermatozoa. Animal Reproduction Science. 2009;114: 193-202.

[34] Macias GB, Morrell JM, Ortega Ferrusola C, Fernandez-Gonzalez L, Tapia JA, Rodriguez-Martinez H, Pena FJ: Centrifugation on a single layer of colloid selects improved quality spermatozoa from frozen-thawed stallion semen. Reproduction Domestic Animals. 2009;44:523-526.

[35] Mancill SS, Love CC, Brinsko SP, Edmond AJ. Foster ML, Teague SR, Waite JA, Varner DD: Effect of density gradient centrifugation on cryopreservation of equine spermatozoa. Animal Reproduction Science. 2010;121S:208-209.

[36] Stuhtmann G: Single layer centrifugation can improve poor quality frozen stallion ejaculates for AI. Reproduction in Domestic Animals. 2011;43:P257.

[37] Woelders H, Windig J, Hiemstra SJ: How developments in cryobiology, reproductive technologies and conservation genomics could shape gene banking strategies for (farm) animals. Reproduction in Domestic Animals. 2012;47 (Suppl. 4):264-273

[38] Morrell, JM, Saravia, F, Wallgren M, van Wienen, M & Rodriguez-Martinez, H: Selection of boar spermatozoa using centrifugation on a glycidoxypropyltrimethoxy-silane-coated silica colloid. Journal of Reproduction and Development 2009;55:547-552.

[39] van Wienen MVP, Wallgren M, Parvlevliet JM, Morrell JM: Boar spermatozoa survive freezing after single layer centrifugation with Androcoll-P. Reproduction in Domestic Animals. 2011; 46(Suppl. 2):Abst. P2.

[40] Morrell JM, Wallgren M: Removal of bacteria from boar ejaculates by single layer centrifugation can reduce the use of antibiotics in semen extenders. Animal Reproduction Science. 2011;123:64-69.

[41] Blomqvist G, Persson M, Wallgren M, Wallgren P, Morrell JM: Removal of virus from boar semen spiked with porcine circovirus type 2. Animal Reproduction Science. 2011;126:108-114.

[42] Martinez-Alborcia MJ, Morrell JM, Parrilla I, Barranco I, Vázquez JM, Martinez EA, Roca J: Improvement of boar sperm cryosurvival by using single-layer colloid centrifugation prior freezing. Theriogenology. 2012;78:1117-1125.

[43] Martinez-Alborcia MJ, Morrell JM, Barranco I, Maside C, Gil MA, Parrilla I, Vazquez JM, Martinez EA, Roca J: Suitability and effectiveness of single layer centrifugation using Androcoll-P in the cryopreservation protocol for boar spermatozoa. Animal Reproduction Science. 2013;140:173-179.

[44] Sjunnesson YC, Morrell JM, González R: Single layer centrifugation-selected boar spermatozoa are capable of fertilization in vitro. Acta Veterinaria Scandinavica. 2013;55:20-26.

[45] Morrell JM, Rodriguez-Martinez H, Andersson M: Colloid centrifugation selects normal spermatozoa from polymorphic bull ejaculates: a case study. Reproduction in Domestic Animals. 2014;49:281-284.

[46] Yusnizar Y, Abraham MC, Laskowski D, Johannisson A, Morrell JM: Changes in Bull Sperm Kinematics after single layer centrifugation. Reproduction in Domestic Animals. 2014;49:954-956.

[47] Al-Essawe E, Morrell JM: Relationship between bull sperm kinematics and fertility. Reproduciton in Domestic Animals. 2014;49:Suppl abst.3

[48] Thys M, Vandaele L, Morrell JM, Mestach J, Van Soom A, Hoogewijs M, Rodriguez-Martinez H: *In vitro* fertilising capacity of frozen-thawed bull spermatozoa selected by single-layer glycidoxypropyltrimethoxysilane-coated silica colloidal centrifugation. Reproduction in Domestic Animals. 2009;44:390-394.

[49] Oliveira BM, Arruda RP, Thomé HE, Filho MM, Oliveira G, Guimarães, Nichi M, Silva LA, Celeghini ECCC: Fertility and uterine hemodynamic in cows after artificial insemination with semen assessed by fluorescent probes Theriogenology. 2014;82:767-772.

[50] Johannisson A, Nongbua T, Edman A, Morrell JM: Effects of single layer centrifugation (SLC) on bull spermatozoa prior to freezing on post-thaw semen characteristics. Theriogenology. 2016;86:140

[51] Althouse, G.C.; Pierdon, M.S.; Lu, K.G. Thermotemporal dynamics of contaminant bacteria and antimicrobials in extended porcine semen. Theriogenology. 2008;70:1317-1323.

[52] Zampieri D, Santos VG, Braga PA, Ferreira CR, Ballottin D, Tasic L, Basso AC, Sanches BV, Pontes JH, da Silva BP, Garboggini FF, Eberlin MN, Tata A: Microorganisms in cryopreserved semen and culture media used in the in vitro production (IVP) of bovine embryos identified by matrix-assisted laser desorption ionization mass spectrometry (MALDI-MS). Theriogenology. 2013;80:337-345.

[53] Guimaraes T, Lopes G, Pinto M, Silva E, Miranda C, Correia MJ, Damásio L, Thompson G, Rocha A: Colloid centrifugation of fresh stallion semen before cryopreservation decreased microorganism load of frozen-thawed semen without affecting seminal kinetics. Theriogenology. 2015;83:186-191.

[54] Gloria A, Contri A, Wegher L, Vignola G, Dellamaria D, Carluccio A: The effects of antibiotic additions to extenders on fresh and frozen-thawed bull semen. Animal Reproduction Science 2014;150:15-23 doi:10.1016/j.anireprosci.2014.08.012.

[55] Morrell JM, Wallgren M: Alternatives to antibiotics in semen extenders: a review. Pathogens. 2014;3:934-946.

[56] Nicholson CM, Abramsson L, Holm SE, Bjurulf EX: Bacterial contamination and sperm recovery after semen preparation by density gradient centrifugation using silane-coated silica particles at different g forces. Human Reproduction. 2014;15:662-666.

[57] Morrell JM, Klein C, Lundeheim N, Erol E, Troedsson MHT: Removal of bacteria from stallion semen by colloid centrifugation. Animal Reproduction Science. 2014;145:47-53.

[58] Foote RH: The history of artificial insemination: selected notes and notables. Journal Animal Science. 2002;80:1-10.

[59] Bujan L, Daudin M, Alvarez M, Massip P, Puel J, Pasquier C: Intermittent human immunodeficiency rtype 1 virus (HIV-1) shedding in semen and efficiency of sperm processing despite high seminal HIV-1 RNA levels. Fertility Sterility. 2002;78:1321-1323.

[60] Levy R, Bourlet T, Maertens A, Salle B, Lomage J, Laurent JL, Pozzetto B, Guerin JF: Pregnancy after safe IVF with hepatitis C virus RNA-positive sperm. Human Reproduction. 2002;17:2650-2653.

[61] Galuppo AG, Junior NB, Arruda NS, Corbellini AO, Chiappetta CM, Pavão DL, D'Angelo M, Canal CW, Rodrigues JL: Evaluation of the effectiveness of semen processing techniques to remove bovine viral diarrhea virus from experimentally contaminated semen samples. Journal of Virological Methods. 2013;187:443-448.

[62] Morrell JM, Timoney P, Klein C, Shuck K, Campos J, Troedsson M: Single layer centrifugation reduces equine arteritis virus titer in the semen of shedding stallions. Reproduction in Domestic Animals. 2013;48:604-612.

[63] Morrell JM, Stuhtmann G, Meurling S, Lundgren A, Winblad C, Macias Garcia B, Johannisson A: Sperm yield after single layer centrifugation with Androcoll-E is related to the potential fertility of the original ejaculate Theriogenology. 2014;81:1005-1011.

[64] Anel-López L, Martínez-Rodríguez C, Soler AJ, Fernández-Santos MR, Garde JJ, Morrell JM: The use of Androcoll-S after thawing improves the quality of electroejaculated and epididymal sperm samples from red deer. Animal Reproduction Science. 2015;158:68-74.

[65] Morrell JM, Saravia F, Wallgren M, van Wienen M, Rodriguez-Martinez H: Selection of boar spermatozoa using centrifugation on a glycidoxypropyltrimethoxy-silane-coated silica colloid. Journal of Reproduction and Development. 2009;55:547-552.

Section 4

State-of-the-Art Nanoparticles

Chapter 10

Colloidal Aggregation Coupled with Sedimentation: A Comprehensive Overview

Agustín E. González

Additional information is available at the end of the chapter

http://dx.doi.org/10.5772/65699

Abstract

An account is made of the experimental, theoretical, and computational developments that led to our current understanding of the colloidal aggregation problem when a gravitational field is present. Starting with unaggregated colloids, a review is made of the advances that led to the founding of the barometric equation for the distribution of colloidal particles in a suspension, noticing that for large bodies, like large colloidal aggregates, their final fate *in equilibrium* is to be at the bottom of the container. Then, we briefly review the aggregation of colloids in the absence of gravity that has been amply studied by both experiments and simulations. For this purpose, the paradigmatic case of the DLVO interaction is taken as an example. Next, a brief revision is made of the seminal experimental work of C. Allain and collaborators on the colloidal aggregation problem when an external gravitational field is present, centering our study in the nongelling situations, that is, for dilute colloidal suspensions, when only sedimentation and deposition of single clusters occur. Afterward, the development of different computer simulations that treat this case of single cluster sedimentation and deposition is reviewed, and note how the different improvements of the algorithms lead to better correspondences with the experimental systems. We finally discuss further possible improvements of the algorithms and end with proposals for future work.

Keywords: colloidal particles, Brownian motion, colloidal sedimentation, colloidal aggregation, cluster sedimentation, cluster deposition

1. Introduction

Small colloidal particles—of the order from 10 nm up to 10 μm in size—immersed in a liquid experience a motion, which is the combination of a steady gravitational drift downward (assuming that their density is higher than that of the surrounding liquid, otherwise the drift would go upward) and a Brownian random movement. This is due to its translational kinetic energy coming from the principle of equipartition of energy, which, as realized by Einstein, does not distinguish the thermal motion of a solvent molecule from that of a suspended colloid or any other types of particle. The kinetic energy of a particle with mass m, translating with a speed v is

$$E_{kin} = \frac{1}{2} m\, v^2 \tag{1}$$

The equipartition principle guarantees that in thermal equilibrium, all components of a solution (solvent molecules as well as colloids or any other particles) have the same *average* translational kinetic energy, which is fixed by the absolute temperature T:

$$\langle E_{kin} \rangle = \frac{3}{2} k_B T \tag{2}$$

Thus, the root-mean-square speed of a particle is

$$\sqrt{< v^2 >} = \sqrt{\frac{3k_B T}{m}}, \tag{3}$$

showing that at a given temperature, a colloid with a large mass moves much slower than a molecule. This movement of the colloidal particles occurs in random (erratic) steps, whose finite size comes from the damping of their movement by the surrounding fluid. In the free diffusion of a particle in a liquid, far away from other particles or a wall, the diffusion coefficient D is given by Einstein's equation [1]:

$$D = \frac{k_B T}{f}, \tag{4}$$

where f is the friction coefficient of the particle. The simplest case is a spherical particle with radius a in a Newtonian liquid with viscosity η. In this case, f is the so-called Stokes friction factor: $f = 6\pi\eta a$. The combined result

$$D = \frac{k_B T}{6\pi\eta a} \tag{5}$$

http://dx.doi.org/10.5772/65699

is called the Stokes-Einstein diffusion coefficient for translational sphere diffusion. With this definition for the diffusion coefficient, Einstein reported the equation for the average quadratic displacement of a particle as follows:

$$\left\langle r^2 \right\rangle = 6Dt \tag{6}$$

If in addition to this Brownian movement a gravitational field along the vertical z direction is present in the system, the particles experience a gravitational force leading to a final equilibrium distribution of the particles which is dependent on the z coordinate. With the use of the methods of statistical mechanics, this equilibrium *average* density profile is well known to be given by

$$c(z) = c(z=0)\,exp\left[\frac{-mgz}{k_B T}\right], \tag{7}$$

where z is the vertical position measured from the free surface of the suspension fluid, considered as the origin of the *upward* z coordinate. Here c is the colloid number density. This *barometric* distribution, which only holds for noninteracting particles, has a thickness characterized by the "gravitational length": $l_g = k_B T/mg$. It is easily verified that for common gas molecules as nitrogen, l_g is of the order of 9 km, while for colloidal spheres of 0.1 μm in diameter with a density of, say, ρ = 2 g/cm^3, it is only around 1 mm, reducing its mass by the corresponding mass of the displaced fluid. For a given system of initial, homogeneous number density $c_o = N_{tot}/Ah$ inside a prismatic cell of area A and height h, $c(z = 0)$ is given by

$$c(z=0) = c_o \frac{mgh / k_B T}{exp(mgh / k_B T) - 1} \tag{8}$$

Perrin started the use of "well-defined" colloids to study the density profile on a spatial scale accessible to an optical microscope [2–6]. It should be noted that Eq. (7) assumes that all colloidal particles have the same mass m. He used fairly monodisperse spheres, obtained from laborious fractionation procedures by centrifugation of solutions of the natural gum "gamboge." In **Figure 1**, it is shown a Perrin's microscopic image of the sedimentation-diffusion equilibrium of these resin spheres of about half a micron in diameter immersed in water. His method to obtain the exponential density profile was to compare the average number of particles in a given layer of the sample to the same average number in another layer some tens of microns below. For example, from Eq. (7), we can write

$$\frac{c(z)}{c(z - 25\,\mu m)} = \exp\left[-\frac{mg}{k_B T}(25\mu m)\right] \tag{9}$$

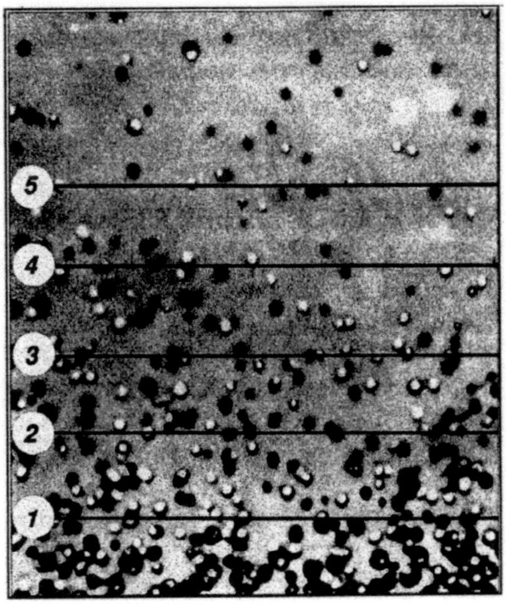

Figure 1. A micrograph taken by J. Perrin of the sedimentation-diffusion equilibrium of resin (gamboge) spheres suspended in water. From Randriamasy F. Les grandes expériences de Jean Perrin. Revue du Palais de la Découverte 1992; **20**: 18–29.

In one series of measurements, he used gamboge particles of around 0.24 μm in diameter with a density of $\rho = 1.35$ g/cm^3 (which would imply a reduced density of $\rho = 0.35$ g/cm^3). This would proportion the quantity inside the exponential with the value of −0.15, which was relatively not very difficult to measure. He used cells of a very short height (100 μm). In one of Perrin's works [5], he warned very clearly: "... One must use capillary tubing to avoid convection currents which would muddle the surface of the cloud, and place them vertically in a thermostat. These currents are produced with extreme ease in large tubes, coming from the temperature gradients."

As a final remark on the experimental work, if we combine Eqs. (7) and (8) and use the values of the density and radius of Perrin's particles given above, the quantity $\alpha = mgh/k_BT$ becomes as large as 600 for a usual laboratory vial of around 10 cm in height. Hence, *in equilibrium* the vial should be mostly empty of particles, except close to the bottom where the number density should follow the law $c = c_o\, \alpha\, exp(-\alpha\, \Delta/h)$; here Δ is the vertical distance from the bottom to the measuring point.

Subsequently, Mason and Weaver [7] investigated theoretically the manner in which the system approaches the equilibrium exponential distribution, not only for a constant uniform initial distribution c_o but for an arbitrary initial distribution $c_o = f(z)$. A consideration of the combined effect of downward drift and diffusion led them to obtain a partial differential equation for the number density of particles as a function of depth and time. Their solution showed that no matter what the initial distribution was, the final, long-time distribution was the one shown in Eq. (7). Perrin's exponential distribution was questioned by several researchers, among them

Burton and Bishop [8], since Perrin performed his measurements very close to the free surface of the suspension (of the order of a hundred microns), arguing that the distribution was merely a surface phenomena that hold for only very small distances below the free surface. They used instead a tube of a height of around 1 m [8] containing copper colloids, waiting for about 50 days for the equilibrium to be established and concluded that "… these suspensions of fine particles always reach a uniform distribution throughout the body of the liquid, and the variation expressed in Perrin's distribution law must be confined to a very small distance at the surface." The theoretical expression derived by Mason and Weaver [7] was used by Weaver [9] to try to settle the issue. He gave an estimate of the duration of the transient state for Burton's particles and tubing, before the final equilibrium distribution was reached. He found that one must wait for the time that a drifting particle sedimenting downward would travel of the order of two tube lengths, in order for the transient term in their expression to become negligible. However, he concluded that the time of 50 days spent by Burton was long enough to attain the equilibrium distribution, according to his calculations. To make a long story short, while the kinetic-molecular theory became widely accepted, the pesky colloids refused to settle in the laboratory bottles and many wondered why. Svedberg [10] pointed out that "… the discrepancy was probably due to the mixing effect of convection currents. For very small particles, the velocity of fall is extremely small and even slight temperature differences within the sol would probably disturb the equilibrium considerably." The solution to the puzzle came years later and was described in a paper by McDowell and Usher [11], using a very fine tubing and temperature-controlled apparatus, in which the temperature variations did not exceed 0.001°C. They found that the exponential law held for not only a tenth of a mm, as found by Perrin, but up to almost 1 cm, the length of their tubing apparatus, and no sensible departure from this ideal behavior was apparent. Finally, it should be mentioned that for very large particles, like the colloidal aggregates, we are going to study (sometimes consisting of up to 10,000 colloids) the quantity α becomes very large, around 6×10^6. Needless to say that *in equilibrium* those aggregates would not be found in the bulk of the suspension but at the bottom.

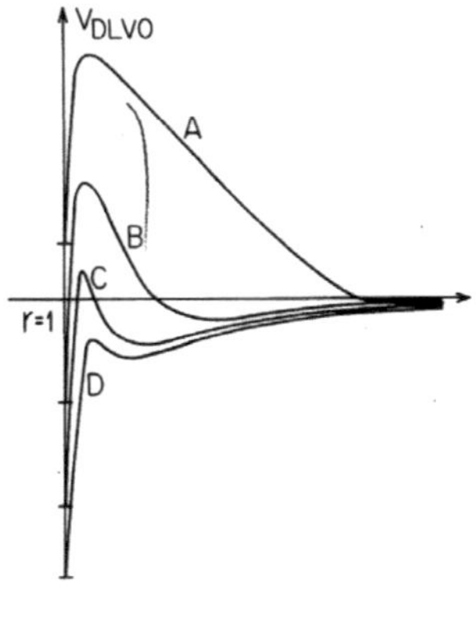

Figure 2. A sketch of the interaction potential DLVO for four electrolyte concentrations, going from a very low (curve A) to a very high (curve D) concentration.

If, furthermore, the colloids experience not only the Brownian motion and the gravitational settling but also they are allowed to aggregate or coagulate, as many colloids do under the appropriate conditions, the situation grows in complexity. Although the aggregation of colloids has been studied for over a century [12, 13], it has been in the last three decades—after Witten and Sander [14] advanced a single aggregate diffusive model now called DLA, but more appropriate for our study of colloidal aggregation, after the proposal of a diffusive model for cluster-cluster aggregation, made concurrently by Meakin [15] and Kolb et al. [16]—that the number of investigations dealing with the clustering of particles has increased considerably. This growing was enhanced by the discovery that the colloidal aggregates present a spatial (scale invariance) fractal behavior [17]. Although there are different ways in which the colloidal particles can aggregate, the paradigmatic and more common way to start the subject is with the so-called DLVO theory (after Derjaguin, Landau, Verwey, and Overbeek) of colloidal stability [18, 19]. In this theory, the colloids aggregate due to the strong attractive interactions when they are close to contact each other. This attraction comes from the London-van der Waals dipole-dipole forces between the molecules composing the two colloidal particles. Stability occurs when the repulsion of the two colloids, due to charges of the same sign on their surfaces, prevents the particles to get close enough to allow the dipole-dipole forces to take over. However, this stability can be disrupted by the addition to the colloidal suspension of a good amount of ionizable salt, which imposes the formation of a so-called double layer of ions around the macroion (colloidal particle). This double layer screens very effectively the electric repulsion between the colloids and permits their close approach, in such a way that they can get hooked by the London-van der Waals forces. In **Figure 2**, the form of the interaction potential between two of these macroions is presented, for different amounts of the added salt. Note how in some of the curves there are two minima: The first and deeper minimum (where coagulation occurs) and a secondary minimum, separated by a potential barrier. In the curve marked (D), the barrier has almost disappeared and the two particles can coagulate as soon as they collide; it is said by the computer simulators that the sticking probability after contact is one. This is the rapid coagulation regime. On the contrary in the cases (A) and (B), particularly in case (A), many collisions between the particles are necessary in order that a thermal fluctuation may help the two colliding particles to overcome this barrier. In this case, we say that the sticking probability after encounters is very low (very close to 0). We are now in the slow coagulation regime. In the olden literature on this subject (see, e.g., Ref. [20]), this is considered by the Fuchs' stability factor, which is very roughly inversely proportional to the sticking probability used by computer simulators. The first case is called diffusion-limited cluster aggregation (DLCA), while the second is termed reaction-limited cluster aggregation (RLCA). One last thing about the DLVO interaction potential: since the primary minimum where coagulation occurs is very deep and of very short range, once two particles coagulate, the formed bond stays fixed with only a *slight* possibility and range for the particles to roll over on top of each other due to the roughness of their surfaces, that is, the real interaction potential should be actually anisotropic very close to contact. This forces the formed bond to be rigid, *up to a certain degree,* and this degree allows for the elasticity and deformability of a large colloidal cluster subjected to some stresses, which depends on the nature of each particular colloid we are dealing with. This deformability (including cluster fracture or breakup) has been

http://dx.doi.org/10.5772/65699

studied in a number of works by several researchers, who considered the aggregation of clusters subjected to shear flow [21–25].

Let us call the number of aggregates in the sample containing s colloidal particles at time t as $N_s(t)$; this would proportion the number density of such clusters as $c_s(t) = N_s(t)/V$, where V is the volume of the sample. Therefore, the total number density of the monomeric units will be given by $c_o = \sum_{s=1}^{\infty} s\, c_s$. Smoluchowski [12, 13] presented the first theoretical description for the rate of change of the quantities $c_s(t)$ ($s = 1, 2, 3 \ldots$) in an equation that resembled the law of mass action in chemical reactions, named after him:

$$\frac{dc_s}{dt} = \frac{1}{2}\sum_{i=1}^{s-1} k_{i,s-i}\, c_i\, c_{s-i} - \sum_{i=1}^{\infty} k_{i,s}\, c_i\, c_s \quad s = 1,2,3,\ldots \tag{10}$$

The quantities $k_{i,j}$ are called the reaction rate constants or simply kernels. Furthermore, he was able to derive an expression for the rate constant $k_{1,1}$ that proportions the conversion of monomeric colloidal particles to dimers, on the assumption that this conversion strictly came from the collision rate of monomers diffusing (performing random walks) before the collision. His expression is $k_{1,1} = 8k_BT/3\eta$. It is customary to define the Brownian collision time as $t_B = 2/c_o k_{1,1} = 3\,\eta\,/\,4\,k_B T\, c_o$, which is the average time elapsed between two consecutive collisions of monomeric colloids in the sample, at the beginning of the aggregation. As it stands, the Smoluchowski equation with the $k_{1,1}$ given above would be valid for particles sticking at first contact after diffusing, that is, for the DLCA mechanism. However, many researchers have proposed many other forms for the rate constants that may be valid for other aggregation mechanisms, including RLCA. In the present paper, we will mainly concentrate and discuss the DLCA case coupled with sedimentation.

The colloidal aggregation problem when no gravitational field is present is now relatively well understood, profiting from many results coming from the computer simulations, for both DLCA and RLCA [26–30]. In this case, the experimentalists need to resort to a number of tricks, like flipping up and down the coagulation cell quite often during the aggregation process or considering a suspension fluid whose density closely matches that of the colloidal particles (e.g., with a mixture of water and heavy water at the correct proportions), in order to eliminate the effect of gravity. Let us now briefly review our current knowledge of the DLCA process in three dimensions [31, 32]. (a) In the dilute limit, the fractal dimension d_f of the clusters is about 1.80. For more concentrated suspensions, it increases from this value with an added term depending on the square root of the particle concentration. (b) The number-average $\left[S_n(t) \equiv \sum_s N_s(t)s \Big/ \sum_s N_s(t)\right]$ and weight-average $\left[S_w(t) \equiv \sum_s N_s(t)s^2 \Big/ \sum_s N_s(t)s\right]$ cluster sizes grow linearly with time in the dilute limit $S_n(t) \sim t^{z'}$ and $S_w(t) \sim t^z$, with $z = z' = 1$. Once again, the exponents z and z' increase from one as a square root of the particle concentration. (c) The cluster size distribution $N_s(t)$ is asymmetrically bell-shaped in the late stages of the aggregation process.

2. Colloidal aggregation coupled with sedimentation

When cluster diffusion and aggregation are coupled to the sedimentation experienced mainly by the large clusters, there appear much richer, nontrivial phenomena that proliferate in many natural and industrial processes, as opposed to the vastly investigated problem of aggregation of colloids induced merely by diffusion. Out of the numerous instances, we can mention the clarification of liquids, the aggregation and deposition of asphaltenes in crude oil, the settling of bacteria clusters in quiet water, and a number of precipitation techniques employed by the chemical industry. As quoted, studies of situations where aggregation is due to the combined action of Brownian motion and an imposed external field like gravity were scarce more than two decades ago [33, 34]. It has been only in the last 20 years that researchers have paid an increasing attention to such phenomenon. As noted before, with the addition of salts or flocculants, which screen the electrostatic repulsion between the particles or which establish bridges between them, the aggregation of particles into clusters is induced. These clusters in turn collide with other clusters, stick together, and become larger. At the beginning, the aggregates are small and essentially move by diffusion, with only a very slight sedimentation velocity. As the aggregation proceeds, the settling velocity of the large aggregates becomes significant, and the mechanism for the movement of these large aggregates by sedimentation intensifies.

2.1. The first meaningful experimental studies

A significant advance on the understanding of the process of aggregation coupled with sedimentation came after the experimental work by Allain and collaborators [35–41]. One interesting result appeared in Ref. [38], in which it was found that the clusters settling in a high cylindrical cell of length 0.8 m and 4.4 cm in diameter can grow up to a certain maximum cluster size s_{max} that cannot be exceeded, with a linear size of around $l_{max} = 0.5$ mm. This was explained in the following way: at the very beginning of the growth process, the aggregates are too small to be sensitive to hydrodynamic stresses; however, as soon as they grow beyond some limiting size and their settling velocity becomes important, the hydrodynamic force exerted by the solvent exceeds the colloidal binding force. In this way, when two large clusters come into contact, the aggregates are pulled apart and go on moving separately. In all their works, they used the same aqueous suspensions of calcium carbonate particles with a density of $\rho = 2.7$ g/cm^3 (which implies a reduced density of 1.7 g/cm^3) and radius of 0.035 μm. Enough salt was added to the suspension in order for them to be in the rapid coagulation regime (DLCA). A theoretical estimation of the l_{max} for these aggregating particles was made and found to be around $l_{max} = 0.1$ mm. Furthermore, by measuring the settling velocity of the large clusters, they found that the fractal dimension of the aggregates grows from its DLCA value of 1.8 to a value of $df \cong 2.2$–2.3. Another very interesting result was published in Ref. [37], where they reported a graph separating the different regimes of growth. This plot is shown in

Figure 3, where on the vertical axis, it showed the logarithm of the colloidal volume fraction (ϕ), while on the horizontal axis, it displayed the logarithm of the time.

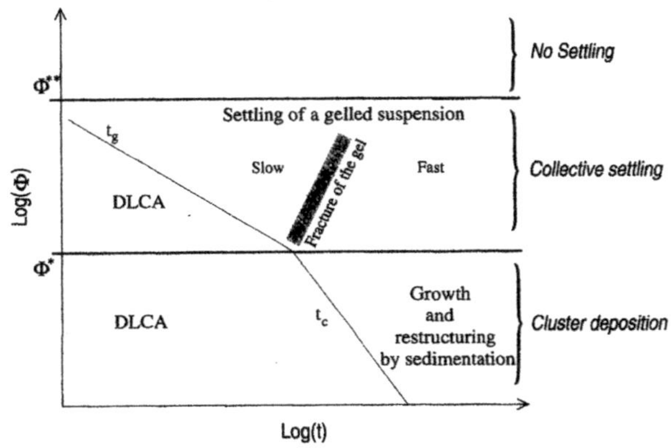

Figure 3. The phase diagram that includes the horizontal line ϕ*, which is the volume fraction that separates the single cluster sedimentation and deposition, below ϕ*, from the collective settling of a gelled suspension. Reprinted with permission from Allain C., Cloitre M., and Wafra M. Phys. Rev. Lett. **74**, 1478–1481, 1995. Copyright 1995 by the American Physical Society.

Allain et al. [37, 38] explained the large fractal dimension of the big, sedimenting clusters as coming from a restructuring mechanism due to the hydrodynamic stresses felt by those large clusters, forcing their long branches to elastically bend and form loops, which would make them more compact. Furthermore, they found experimentally for their system that ϕ* as of the order of $\phi^* \cong 3 \times 10^{-3}$. Let us try to obtain a rough, simpler estimate than the one obtained by them for ϕ*, from their data. As the cluster size distribution is bell-shaped for DLCA (sticking probability equal to one), with a typical cluster size s corresponding to the position of the maximum, we will assume *momentarily* that *all* the clusters have this size. In the following equations, we are going to make a scaling analysis, using always the sign ~ to indicate dependence and order of magnitude only, forgetting about constant factors like (4/3)π. Therefore, the results are expected to give the order of magnitude only. The *effective* volume fraction of all the clusters filling up the aggregation cell is $\phi_e \sim c_s R_s^3$, where R_s is the typical radius. By the definition of the fractal dimension d_f, we have $\left(R_s/a\right)^{d_f} \sim s$. Therefore, $\phi_e \sim c_s a^3 s\, s^{\left(3/d_f - 1\right)} \sim \phi\, s^{\left(3/d_f - 1\right)}$. To obtain the volume fraction at which the system starts to gel, we need to make $\phi_e \sim 1$, which implies $\phi^* \sim \left(R_s/a\right)^{d_f - 3}$. Now, (A) if there was no gravity involved in the problem, gelation would start when there is a single aggregate filling up the whole cell of linear size L, that is, $R_s \sim L$. Hence, $\phi^* \sim \left(L/a\right)^{d_f - 3}$. Assuming the size of their particles ($a = 0.035 \times 10^{-6}$ m) and a typical linear size of the coagulation cell of $L = 0.1$ m and taking the $d_f = 1.8$ for the DLCA case, we get $\phi^* \sim 1.8 \times 10^{-8}$. This is of the order of the very small ϕ_g concentration found in Ref. [37]. (B) If the system is under the action of the gravitational field, there is not going to be a single aggregate filling up the whole system, because the clusters cannot grow beyond an R_{max}, which would be taken as half of the linear size of the maximum size cluster they found. That is, $R_{max} \sim l_{max}/2 \sim 0.25$ mm, which would imply $\phi^* \sim \left(R_{max}/a\right)^{d_f - 3}$. Assuming in this case that the fractal dimension is $d_f = 2.2$, as found in Ref. [37], we get $\phi^* \sim 0.83 \times 10^{-3}$, a value which is close to the $\phi^* \approx 3 \times 10^{-3}$ found by the authors. In the rest of the present article, we will focus mainly on concentrations below the ϕ^*; that is, the colloidal aggregation coupled with sedimentation will be studied for single clusters diffusing and sedimenting downward and depositing at the bottom. Nevertheless, if in some cases a concentration was picked to reach gelation, the simulations were stopped at the gel point.

2.2. The first computer simulation studies

One of the first attempts to study the colloidal aggregation coupled with sedimentation was presented in Ref. [42], although with some nonessential errors but also with one essential misassumption. It started by considering the sedimentation velocity v_s experienced by a cluster of N spherical particles of radius a and mass m_o

$$v_S = \frac{m_o\left(1-\rho/\rho_o\right)gN}{f} = \frac{m_o\left(1-\rho/\rho_o\right)g}{k_BT}DN, \tag{11}$$

where ρ_o is the particle density, ρ is that of the suspension fluid, $f = 6\pi\eta R_g$ is the cluster's friction coefficient, $D(=k_BT/f)$ is its diffusion coefficient, R_g is its radius of gyration, η is the solvent viscosity, and T is the temperature. The spirit of the simulation rests on finding out, for each cluster of a certain size, by how much the cluster settles downward as compared to how much it diffuses around. Let us define t_o as the time for a cluster to diffuse a particle diameter (d = $2a$), that is, $t_o = 2a^2/D$. This was a nonessential error which will not change the nature of the results. The correct expression in the three-dimensional case is $t_o = 2a^2/3D$ that has been used in the future simulations (see below). During the same time, the cluster drifts a distance $d_S = v_S t_o = \frac{1}{3} PeNd$, where $Pe \equiv m_o\left(1 - \frac{\rho}{\rho_o}\right)ga \Big/ k_BT$ is the Peclet number of the *individual* colloidal particles in the fluid. It is found that Pe is of order unity if the particles are 1 μm in diameter, $\rho_o - \rho$ is of order 1 g/cm^3, and T is room temperature. On the other hand, if the diameter is 0.1 μm, such quantity is of the order of 10^{-4}, while if the diameter is 10 μm, Pe goes up to 10^4. As can be seen, the transition size between diffusive and drifting behavior for *individual* particles with density different from that of the medium occurs around 1 μm.

The physical system was simulated on a simple cubic lattice with periodic boundary conditions on the three spatial directions and with lattice spacing equal to d (=$2a$, the diameter of the particles). Each of the cells of the lattice can be occupied by a colloidal particle or can be empty, in which case is considered to be occupied by the suspension fluid. As the aggregation proceeds, we deal with a collection of clusters made of nearest neighbor lattice cells that are diffusing randomly and sedimenting downward. The Monte Carlo (MC) step time was picked in order to avoid a cluster to move farther than one lattice spacing. Define $\Delta t_{dif} \equiv d^2/6D_{max}$ as the time taken by the most mobile cluster to diffuse one lattice spacing. Also, Let $\Delta t_{drif} \equiv d \big/ v_s^{max}$ be the time taken by the largest cluster to sediment downward one lattice spacing. The algorithm used is as follows.

a. If $\Delta t_{drif} < \Delta t_{dif}$, then

1. pick a cluster in a cyclic way. That is, the clusters are numbered, and after having chosen a cluster, the next cluster to choose will be the following in the list, given that sedimentation is a deterministic process.

2. The time is increased by $\Delta t_{drif}/N_c(t)$ where $N_c(t)$ is the number of clusters in the sample at time t. This comes from the fact that after a cycle, when all the clusters are picked once (assuming that no merging took place), the time increase is Δt_{drif}.

3. The quantity $\Delta t_{drif} v_s/d$ is calculated, and the result is added to a real variable associated to that cluster, that is, called Z_{sed}. If the sum is greater than one, the cluster is moved downward one lattice spacing, and the new value of the variable becomes the remainder of the sum modulo one.

4. If the cluster is moved, we check for overlapping with other clusters, in which case the moved cluster is taken back to the original position and the overlapping clusters are merged. Afterward, we go back to the starting situation to calculate Δt_{dif} and Δt_{drif}.

5. The cluster now moves one lattice spacing on a random direction, with probability $\left(\frac{\Delta t_{drif}}{\Delta t_{dif}}\right)\left(\frac{D}{D_{max}}\right)$.

6. We then check for overlapping with other clusters, as in point (4).

b. If, on the other hand, $\Delta t_{dif} < \Delta t_{drif}$ we follow exactly the same procedure (a), with the exception that everywhere we find Δt_{drif} in a formula, it is replaced by Δt_{dif}

The simulations were stopped just before gelation. Only one volume fraction was considered, at the value $\phi = 0.01$, for which 50 simulations of 156,250 particles were done for each of the following Peclet numbers: Pe = 10^{-1} (the drift is felt predominantly by clusters of size ~10 or larger), 10^{-3} (the drift is felt by clusters of ~1000 or larger), and 10^{-8} (DLCA simulation, no drift). The results from those simulations were quite encouraging. For the structure of the clusters, the radius of gyration of each of the formed clusters was plotted vs. the cluster size as a point in a log-log graph. The inverse of the slope would proportion the fractal dimension of the formed clusters. In **Figure 4**, we are showing this plot for the case $Pe = 10^{-3}$. It is seen how the plotted points lie on not one but two straight lines, with a breaking point around the cluster size $s = 1000$. Two straight lines were fitted, one for the sizes below 1000 and another for the sizes above this value. For the small clusters, a fractal dimension $d_{f1} \cong 1.87$ was obtained, while for the large ones, it was found $d_{f2} \cong 2.27$. The d_{f1} is in accord with the results for the d_f in the

DLCA case for this concentration [31, 32], while the d_{f2} coincides with the value obtained by Allain et al. [37, 38]. For the other two cases, only one straight line was seen. The first, for $Pe = 10^{-8}$, proportioned the value $d_f \cong 1.89$, which again is in good accord with the known results for pure DLCA [31, 32], while the second is in very good agreement with the values obtained by Allain et al. The kinetic behavior was also studied, and it was seen that the weight-average cluster size presented the well-known behavior $S_w(t) \sim t^z$, for the DLCA case ($Pe = 10^{-8}$) only. For the other two Peclet numbers studied, the curve for $S_w(t)$ departed from the straight line in the log-log plots, this departure being more pronounced in the Pe = 0.1 case, looking more like an exponential growth rather than a power law of the time. In other words, there was a sharp speeding up of the aggregation rate. Finally, when plotting the cluster size distribution $N_s(t)$ as a function of *s* for different times during the aggregation process, it was seen a widening of this curve particularly at the later times becoming algebraically decaying, indicating the coexistence of very large clusters with smaller ones. Those three effects—the increase in the fractal dimension of the large clusters, the sharp increase in the aggregation kinetics, and the widening of the cluster size distribution—were explained by the same mechanism: there is an apparent instability in which the big clusters start to grow bigger by sweeping the smaller ones on their fast settling rate downward, increasing in this way their growth rate, capturing the smaller ones that fit in their holes and cavities (making the bigger ones more compact, with a higher fractal dimension), and forcing the very large clusters to coexist with smaller ones, not yet being trapped by the larger aggregates. It should be pointed out that it is not necessary to invoke the restructuring of the clusters—given that this is a rigid aggregation model—in order to achieve the higher values of the fractal dimension but only the sweeping of the small clusters and particles by the large drifting clusters, getting into the holes of the large ones which forces them to compactify.

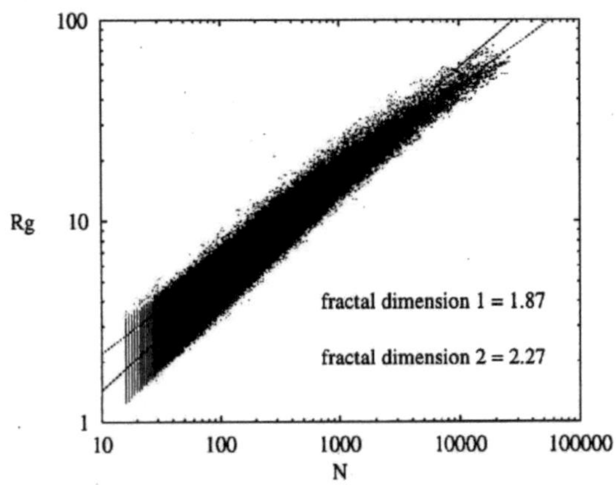

Figure 4. A log-log plot of the radius of gyration vs. the size of all the clusters formed (plotted as points) in the 50 simulations with Pe = 0.001; as it can be seen, the plot breaks into two straight lines, proportioning the values of $d_{f1} \cong 1.87$ for clusters smaller than 1000 particles and $d_{f2} \cong 2.27$ for the larger clusters, above 1000 particles. Reprinted with permission from Ref. [42]. Copyright 1997 by MRS Publishing.

One of the shortcomings of the model in Ref. [42] was that it was not possible to obtain the much higher initial volume fraction required to have gelation. This was due to the fact that there was no cluster deposition built-in in the model. We should point out that in the model, if a large cluster is rapidly settling downward, when it reaches the bottom of the simulation cell, it enters to the cell on the top of it, due to the periodic boundary conditions, but never deposits at the bottom of a real cell. To remedy this situation, an improvement of the model was made in Ref. [43]. To include the height H of any prismatic cell, a cubic lattice was considered with periodic boundary conditions on the three spatial directions, which represented *in an average way* any cubic slice of the prism of height H ($\cong 20L$ in Ref. [38]). In this sense, the model is a sort of mean field simulation. Whenever a cluster moves on the z direction a lattice spacing d, a quantity Z_s belonging to each cluster is updated that proportions the total distance in units of d which the cluster has moved downward (taking into account the movement of all its ancestors, i.e., the largest of the two quantities Z_s of the "parent" clusters is inherited to the "child" cluster after a merging). If the movement is downward, the cluster is taken out of the cubic box (which means that it has been deposited on the bottom) with probability $1/(H - Z_s)$, because of all the possible $H - Z_s$ rows where the center of the cluster can be, only at lowest row the cluster can leave the sample by a sedimentation step. Thus, the bigger clusters with the larger Z_s are taken out of the sample with higher probability. As in the previous case, the simulations were stopped just before gelation for the gelling systems; for the nongelling simulations, they were stopped after all the clusters had been deposited. One volume fraction $\phi = 0.01$ was considered, for which five values of Pe were chosen; Pe = 10^{-8} (DLCA simulation), 10^{-3}, 10^{-2}, 10^{-1}, and 10^{0}. For the first three Peclet numbers, the system was gelling, while for the last Peclet number, it was nongelling; for Pe = 0.1, some of the simulations percolated the box, while the others did not. The size of the cubic box was taken as $L = 270$—which implies 196,830 particles—and $H = 10\ L$. Another improvement made in Ref. [43] was in calculating the d_f. In the previous work [42], every cluster formed during the aggregation process was plotted as a point in the log-log plots of R_g vs. size. It was difficult to see where the breaking points of the straight lines (if any) occurred in the graph. Now it was plotted the *average* radius of gyration vs. the number of particles in the cluster, where the average was performed over all clusters formed lying on segments of constant size on the logarithmic size scale. In **Figure 5**, on the left panel, it is shown this plot for the case Pe = 0.01. We now clearly see a breaking point at around the size of about 200. On the left and right of the breaking point, there are well-defined straight lines between the arrows labeled "a" and "b" that give the two fractal dimensions: $d_{fa} = 1.868$ for the small clusters and $d_{fb} = 2.270$ for the large ones. Again, the d_{fa} is in accord with the value for DLCA at that concentration, while the d_{fb} is close to the Allain et al. value (2.2–2.3). In the right panel of **Figure 5**, we now see the average of S_w over all the simulations made for each Peclet number, except for Pe = 0.1 for which the crosses correspond to the average over the gelling simulations, while the triangles are for the nongelling simulations. As we can see, for the gelling simulations, we again find that the S_w separates from the DLCA simulation, the highest the Peclet number, the sooner the curve starts to take off. For the nongelling simulations, we now see a very interesting phenomenon: After the speeding up of the S_w, it reaches a maximum, and the average cluster size diminishes all the way to zero. This indicates that the dynamics of the aggregation accelerates and then, afterward, it slows down. This is due to the depletion of mass in the bulk as the clusters deposit on the bottom. In the same Ref. [43] and with more details in Ref. [44], it was estimated the critical concentration to have gelation for the different Peclet numbers studied, which was around the value found by Allain et al. Finally, in Ref. [44], the scaling of the cluster size distribution was studied, and it was shown that, for very low Peclet numbers, the cluster size distribution was asymmetrically bell-shaped as in DLCA. As the Peclet number increased, it was seen how that distribution was transforming into an algebraically decaying distribution (with an exponent $\tau \cong 1.73$), implying the coexistence of small and large clusters.

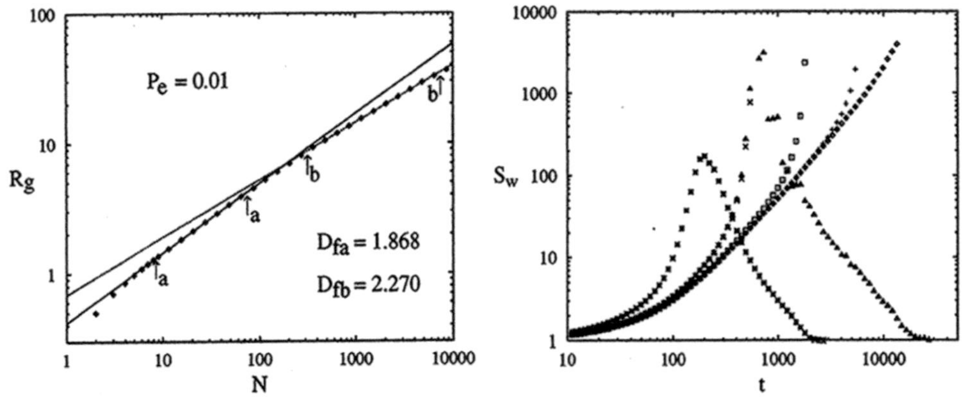

Figure 5. Left: a log-log plot of the average radius of gyration as a function of the number of particles of the clusters for Pe = 10^{-2}. Right: the weight-average cluster size S_w as a function of time for the different Peclet numbers Pe = 10^{-8} (diamonds), Pe = 10^{-3} (pluses), Pe = 10^{-2} (squares), Pe = 10^{-1} (crosses for percolating and triangles for nonpercolating systems), and Pe = 10^{0} (asterisks). Reprinted with permission from Ref. [43]. Copyright 2001 by the American Physical Society.

Subsequently Leone et al. [45] performed an off-lattice simulation with 10,000 particles inside a square prism with periodic boundary conditions on the x and y directions only, but with no periodicity on the vertical z direction, in an effort to consider the dependence of the aggregating quantities on the vertical z direction. The mechanisms for moving the particles were similar as those in Ref. [43]. However, this z dependence was obtained only for three heights: the authors divided the prismatic cell in three different equally sized regions—top, middle, and bottom. The aggregating quantities were obtained by calculating their averages separately in each of those regions. Unfortunately, due to the small number of particles and the very low concentration considered, they were not able to see very clearly the breaking point on the R_g vs. size plot that differentiates the two cluster fractal dimensions. For example, for the top region they found $d_f \cong 1.78$ for DLCA while $d_f \cong 1.88$ for Pe = 0.1, while for the middle region, they found $d_f \cong 1.74$ in the DLCA case, while for Pe = 0.1, it was obtained $d_f \cong 1.89$. As in Ref. [43], they found the slowing down of the aggregation kinetics due to the depletion of mass in the bulk, but with no speeding up before, which is probably due to the very low concentration studied which prevented the sweeping mechanism to act thoroughly. Subsequently, with this off-lattice model, the case of a sticking probability lower than one was investigated [46]. It was found as

http://dx.doi.org/10.5772/65699

expected that the fractal dimension increased to values close to the Allain et al. values of 2.2, but as a result of lowering the sticking probability, not actually due to the sedimentation effect. Finally, they also studied the nature of the deposited clusters and their fractal dimension, distinguishing two cases: for low Peclet numbers, the cluster reactions were occurring mainly in the bulk before deposition; hence, their fractal dimension was higher than that for higher Peclet numbers because, in this last case, the deposited clusters were small and their reactions were mainly occurring on the base of the prism (assuming no sticking of the clusters and particles with the base of the prism), leading to a two-dimensional aggregation that is known to produce fractal clusters with $d_f \cong 1.45$.

The low values of the d_f for the sedimenting clusters for this very low value of the volume fraction ($\phi \cong 6.7 \times 10^{-5}$), as compared to the high values obtained in Refs. [43, 44], prompted us to consider the concentration effects for the problem [47]. The two computer models, Refs. [43–45], were used to include the effects of concentration, spanning three orders of magnitude in volume fraction. Recall that for the high concentrated case [43, 44], there was a crossover from a DLCA type of regime to another regime, for larger clusters, in which the fractal dimension took a value close to the Allain et al. value of 2.2–2.3 and a speeding up followed by a slowing down of the aggregation rate that we called the sweeping scaling regime. As the concentration was lowered from this high value of $\phi = 0.01$ all the way up to $\phi = 0.0001$, it was found that the crossover to the sweeping regime remained *as long as the height of the cell was drastically increased*. For example, for $\phi = 0.01$ and $H = 2700$, there was a crossover to the sweeping scaling regime for almost all the Peclet numbers studied (except for the DLCA case, of course), with a high value of the of the fractal dimension—of the order of the Allain et al. value of 2.2–2.3—as in Refs. [43, 44]. Similarly, there was a speeding up of the aggregation rate followed by the slowing down for the nongelling simulations. For the $\phi = 0.001$ case, there was also a crossover to a sweeping scaling regime for the two heights studied, which were $H = 4500$ and 45,000. Similarly, the speeding up and, afterward, the slowing down of the aggregation rate were found. However, in this case it was found for both heights that the fractal dimension for the sweeping scaling regime did not increase as high as the Allain et al. value, but it stayed around $d_f \approx 2.05$. For the $\phi = 0.0001$ and $H = 7000$ case, it was not possible to find a sweeping scaling regime for the large clusters. In other words, in the log-log plots of R_g vs. size, only one straight line was found that proportioned the value $d_f \approx 1.83$, that is, close to the usual DLCA value. However, by increasing drastically the height (and the computer time) up to $H = 70{,}000$, the sweeping scaling regime was recovered again, with a $d_f \approx 2.05$ as in the previous $\phi = 0.001$ case. This corroborates the assertion that it is possible to find a sweeping scaling regime for low concentrations if the height of the cell is highly increased. Nevertheless, it was intriguing to find that the fractal dimension for the sweeping scaling regime decreased to the value of $d_f \approx 2.05$ for the two low concentrations studied. This could indicate that there may be a universal value for this fractal dimension, as long as the concentration is not too high. Concerning about the clusters deposited at the bottom of the prism, the new regime besides those found from the bulk was confirmed. This is the restricted quasi-two-dimensional aggregation of those clusters that have been deposited on the bottom of the container (assuming again no sticking of the clusters to the base of the prism). If the sediment clusters have been grown essentially in the bulk, which occurs basically for the more concentrated systems, one or the two fractal dimensions can be found in the sediments. However, if the system is dilute enough and/or for a high Peclet number, we may favor the arrival of not very large clusters that can aggregate in a quasi-two-dimensional way.

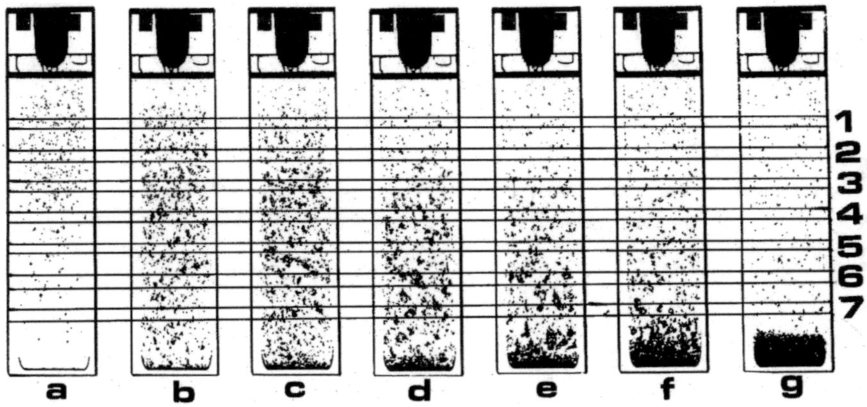

Figure 6. The time evolution of a calcium carbonate aggregating suspension of spherical particles, for the following times: (a) 5 min and 20 s, (b) 6 min and 20 s, (c) 6 min and 40 s, (d) 7 min, (e) 7 min and 20 s, (f) 7 min and 40 s, and (g) 8 min (from Ref. [36]).

2.3. Cluster restructuring vs. cavity occlusion to compact a cluster

So far, so good. Maybe not as good, since now it is necessary to explain the lower fractal dimension obtained in the simulation ($d_f \approx 2.05$) for the sweeping scaling regime at the fairly low concentrations used in the experiments, as compared to the Allain et al. value (2.2–2.3). Besides that, in Ref. [48], the authors presented a two-dimensional simulation on a rectangular lattice with periodical boundary conditions, that is, they considered the z coordinate on the direction of the gravitational field and only one perpendicular direction. They show that for aggregation combined with sedimentation, their clusters did not present spatial scale invariance, that is, they are not self-similar fractal objects. This came from the fact that the mean width perpendicular to the field direction scaled as a power law with the cluster size, while the mean height did not. The method they used to treat the mechanisms for diffusion and sedimentation of clusters differed from the previous simulations [42–47], in that they were using three probabilities for moving a cluster, $p_{\downarrow\downarrow}$ for a cluster moving downward, $p_{\uparrow\uparrow}$ for that cluster moving upward, and $p_{\perp}$ for the cluster to move on the perpendicular direction. However, these probabilities were calculated from the Stokes-Einstein formalism, and this should not present an essential problem. The only concern is about the deterministic nature of sedimentation, but it may not be very relevant. Another interesting result found in that paper was that their clusters were initially elongated along the field direction, while as time proceeds, the situation reversed and the clusters became oblate. Although this lack of spatial scaling was somewhat annoying, the decisive event that led us to consider a more realistic simulation, not the mean field one that we have shown, was the observation of a figure in Ref. [36], which is presented below in **Figure 6**.

In **Figure 6**, where the horizontal lines were drawn by the present author, we can see that the behavior of the growing and settling clusters is different in each of the layers, labeled with the numbers 1 through 7. For the top layers, we can see that the clusters grow up to a certain maximum size, and then the layer starts deplenishing of colloidal matter, which forces the mean cluster size to become smaller and smaller. For the bottom layers, a similar situation happens with the difference that this maximum size occurs at later times and, additionally, is much larger. In other words, the aggregation of colloids coupled with sedimentation is stratified, in the sense that the aggregating quantities behave differently from layer to layer. Our mean field simulation, although it gives results similar to the experimental ones, does not consider this stratification, since the simulation box represented *in an average way* any cubic slice of the aggregating prism of height H. Maybe by considering the precise, proper algorithm, it would be possible to give an answer to this lack of spatial scaling of the clusters. Furthermore, we also may be able to find a fractal dimension in the sweeping regime similar to the experimentalist value. A quick estimate to count the number of lattice cells that would be required to consider a square prism similar in volume (in units of particle volumes) to the one used by the experimentalists proportioned the number of lattice cells around 10^{18} (we recall that the long tube used in Ref. [36] was 4.4 cm in diameter, while the tube length was 80 cm and the particle diameter was 0.07 μm). To perform a simulation of that size is an impossible task, even with present-day computers. To circumvent the problem, the search for a simulation that could proportion the behavior inside a slice of arbitrary vertical thickness Δ and centered at any given depth Z (in lattice spacings) of the aggregation tube was started. In Refs. [49, 50], such simulation was presented, quite similar to the one described in Refs. [43, 44]. We considered a cubic lattice of lattice spacing d (the diameter of the particles) and lateral size $L = 650\ d$, with periodic boundary conditions on the x, y, and z directions *that represent a layer* centered at the integer depth Z (in lattice spacings), measured from the top *of a lattice prism*, and of thickness Δ (which is taken multiple of two). This differs from the previous algorithm, in that the cubic lattice there was representing the whole prism of height H. As before, whenever a cluster moves in the z direction a lattice spacing, an integer quantity Z_s is updated that proportions the total distance that the cluster has moved downward. Again, every time a cluster is picked at each Monte Carlo step of time Δt, we check whether it diffuses and/or sediments downward a lattice spacing. As we want to simulate the clusters inside the abovementioned layer of the prism, of all the $Z + \Delta/2 - Z_s \geq 1$ possible rows of cells where the center of this cluster can be, only at the lowest row—with the position $Z + \Delta/2$—will be taking out of the layer by a sedimentation step, while if it is at the row at $Z + \Delta/2$, the sedimentation step will take it inside the layer. Therefore, if Z_s has not grown enough for such cluster such that $Z - \Delta/2 - Z_s \geq 1$, the cluster is taken out of the box with probability zero (not taken out), because the cluster has the same probability to be on either of those two rows. On the other hand, if $Z - \Delta/2 - Z_s \leq 0$, the cluster cannot be at the row $Z - \Delta/2$ and is therefore taken out of the box with probability $1/(Z + \Delta/2 - Z_s)$. This is the manner used to simulate the depletion of colloidal matter inside the different layers in the aggregating prism, due to sedimentation. Finally, as before, at each diffusion or sedimentation movement that does not take the cluster out of the box, we check for overlapping with other clusters. If that happens, the cluster is taken back to its original position and the overlapping clusters are merged. The volume fraction was fixed at the value $\phi = 0.001$. Four depths were considered at $Z = 500$, 5000, 50,000, and 500,000, and one single $\Delta = 100$. For each of those depths, six Peclet numbers were considered: Pe = 0.0001, 0.001, 0.01, 0.1, 1.0, and 0.0 (DLCA simulation). In turn, for each Peclet number, a series of 10 simulations of 274,625 particles were made, in order to have enough statistics to evaluate the structural and dynamical quantities. To investigate the lack of spatial self-similarity of the formed clusters, not only the log-log plots of the radius of gyration vs. size were made but also of the x, y, and z components of the radius of gyration tensor, where $R_{gx}^2 \equiv \sum_{i=1}^{N} (x_i - x_{cm})^2 \Big/ N$, etc. In addition, the following anisotropy measures were defined: $A_{xz} \equiv \langle R_{gx}/R_g \rangle / \langle R_{gz}/R_g \rangle$ and $A_{yz} \equiv \langle R_{gy}/R_g \rangle / \langle R_{gz}/R_g \rangle$, which were also calculated as a function of size. By analyzing these quantities [50], it was found a whole variety of behaviors for the structure of the clusters in this problem, depending on the sedimentation strength (Pe), the layer depth (Z), and the region of sizes considered. Generally speaking the four radii (R_{gx}, R_{gy}, R_{gz} and R_g) scale as a power law with N, with the same scaling power, for the small, non-settling clusters, except for the cases with a high sedimentation strengths: Pe = 1.0 and 0.1 for all depths. This was called the quasi-DLCA regime, where the "quasi" means that, as there is still some sweeping of even smaller clusters, the fractal dimension is a little bit higher than the usual DLCA d_f. For the large, settling clusters, it was found cases for which the four radii scale again as a power law with N, with the same scaling power, making it possible to define a settling-cluster fractal dimension, a behavior that was called the sweeping scaling regime. There are, however, cases for which the scaling powers for the horizontal and vertical directions differ, obtaining therefore self-affine settling clusters. There are still some cases for which only the mean width or the mean height scale as a power law with N and even further cases for which no scaling as a power law of the four radii is obtained. With respect to the anisotropy measures as a function of the size, it was generally found that for the small clusters, the measure remained either close to one or diminishes somewhat from this value, making the clusters prolate along the field direction. However, as the clusters grow in size and start sweeping smaller clusters, they become oblate and in some cases quite oblate, reaching values for the anisotropy measures of up to 3. For more details about this, the reader is referred to Ref. [50]. With respect to the fractal dimension for the sweeping scaling regime for large clusters (after the quasi-DLCA regime for the smaller ones) which generally occurs for low Peclet numbers and big depths, it was found as in Ref. [47] that the d_f was generally around 2.05. This reinforces the idea that this may be a universal value *for perfectly rigid clusters*. However, this needs to be tested with more simulations for many different cases. It should be mentioned that we did not allow the clusters to rotate as in the experimental systems which, in principle, may be able to do so, canceling in this way the anisotropies found (prolate or oblate clusters). Additional anisotropies besides the prolate or oblate shape may be obtained in this problem. A moment of reflection shows that, even for intermediate drift strengths, the clusters are more compact on the bottom than at the top. This is due to the fact that the bottom is the part of the cluster that is catching up the small clusters and single particles, becoming in this way more compact. It may be conceivable that the new anisotropy may hamper the rotation of the clusters, because a higher gravitational force per unit volume would be experienced at the bottom than at the top, while the medium frictional resistance would be the same everywhere around the cluster, under a non-draining assumption. Hence, the possibility exists to find the

http://dx.doi.org/10.5772/65699

prolate or oblate shape of the experimental clusters after all, in the case that they may not be allowed to rotate; this, however, needs to be tested experimentally.

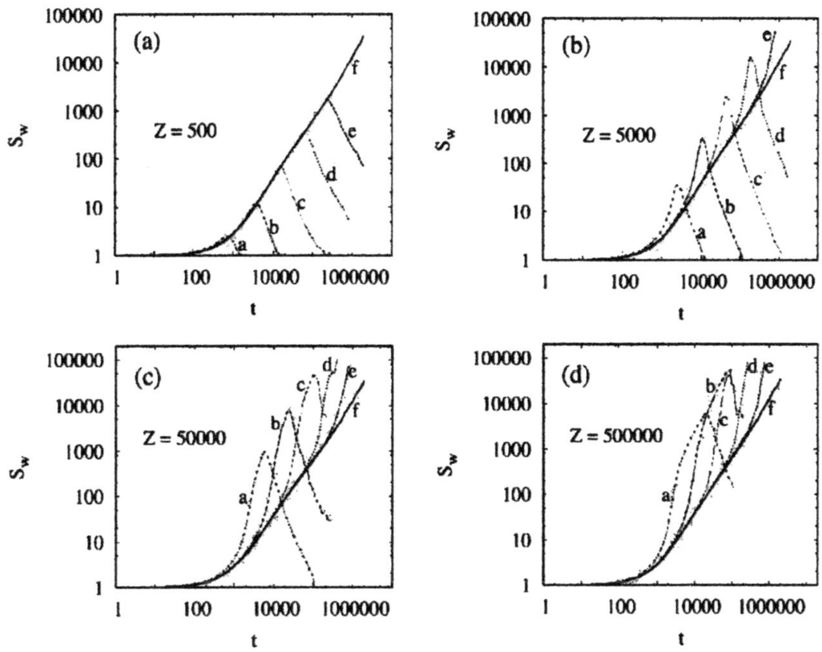

Figure 7. The logarithmic time evolution of the weight-average cluster size S_w for the layers at (a) Z = 500, (b) Z = 5000, (c) Z = 50,000, and (d) Z = 500,000. In all four figures, the labels of each of the curves a, b, c, d, e, and f correspond to the following Peclet numbers: 1.0, 0.1, 0.01, 0.001, 0.0001, and 0.0 (DLCA simulation), respectively. Reprinted with permission from Ref. [50]. Copyright 2006 by the American Physical Society.

Concerning the kinetics of the aggregation, the speeding up with respect to the DLCA case, followed by the slowing down due to the depletion of mass in the layer, was confirmed. Perhaps more interesting, the experimental behavior found in Wafra's thesis was obtained [36] that the aggregates in a layer reach a maximum size at a certain time, the deeper the layer is, the longer it takes to reach this maximum which, however, is of a much larger size. In **Figure 7**, all these kinetic behaviors are shown. Note that in those cases for which no decrease in the S_w is shown, particularly for low Peclet numbers and big depths, indicates that the system gelled at the point where the curve was cut.

As mentioned, the low value of the fractal dimension (~2.05) for the sweeping scaling regime, as compared to the Allain et al. value (2.2–2.3), still needs to be explained. There is much evidence, both theoretical [22, 25, 51, 52] and experimental [21, 23, 24, 53], that a tenuous fractal object in a shear flow would experience hydrodynamic stresses that can change its structure. At the beginning of the growth process, the aggregates are too small to be sensitive to those stresses. Nonetheless, as soon as they grow beyond some limiting size, they may undergo structural rearrangements due to the stresses felt when they sediment faster. It has been shown that the total force exerted by the fluid on the cluster is concentrated into the most peripheral particles [52]. The cluster-cluster aggregates are loopless and have an elastic behavior similar to that of a contorted chain. The elasticity is dominated by the bending elasticity which, when a pair of forces applied at the tips of the cluster bends the corresponding branches, makes the branches touch and stick. Because of this, the cluster could become more compact, with a higher fractal dimension. This is the explanation advanced by Allain et al. [38] for obtaining a high fractal dimension for the settling clusters. However, there is also much simulational evidence for obtaining a high fractal dimension of large *rigid* clusters in the sweeping scaling regime, coming from the occlusion by the small clusters and particles that enter into the holes and cavities of those large clusters. The correct explanation is probably a combination of both effects, which would make the large settling clusters to push the value of the *apparent fractal dimension* ("apparent" for the cases in which the scaling behavior is not the same along the horizontal and vertical directions) to the order of 2.2–2.3. Another factor that may increase the fractal dimension of settling clusters is the inclusion of a sticking probability less than one. As it is well known, without a gravitational field acting on the system, the d_f increases from its DLCA value of ~1.8 to an RLCA value of ~2.1. With the inclusion of sedimentation, this could increase the value to around 2.2–2.3. However, as stated by Allain et al., they believe their system belongs to a regime with a sticking probability equal to one. There is also the influence of the hydrodynamic interactions *between* clusters, whose effect on the d_f is unclear. The general belief, nonetheless, is that by working with sufficiently dilute systems ($\phi \leq 0.001$), their influence would be negligible. Needless to say that more research on this interesting subject would be worthwhile to explore, due to its practical application in several aspects of science.

3. Some other relevant simulational results

Hawick [54, 55] performed some simulations in 2D and 3D with the major emphasis in the visualization of the clusters formed during their aggregation and sedimentation. As in [48] he used rectangular in 2D or prismatic boxes in 3D, with their longest length along the field, vertical direction. Similar to [48], periodic boundary conditions were considered on the direction(s) perpendicular to the field, but not for the vertical direction with fixed vertical boundaries. For the movement of the cluster, the author used the probabilistic mechanisms proposed in Ref. [48]. In Ref. [54], for the 2D clusters, he found that the large clusters have an inverted V or W shape or more generally an inverted tree-like structure. However, for the 3D clusters, this was not the case, and they were not so anisotropic. One of the shortcomings of this work is the smallness of the systems used. For example, in Ref. [55] for the 3D case, a box of size 128×64^2 was used. One of the interesting results shown and visualized was the elongated shape along the field direction of the cluster for a Peclet number of 0.01, while by changing this number to 0.1, wider clusters were found, which is in line with the results presented here. In another work, Whitmer and Luijten [56] consider a molecular dynamics simulation for the sedimentation of diffusing and aggregating colloidal clusters. The potential of interaction between the colloids was taken as the Asakura-Oosawa-Vrij model [57, 58] of depletion-induced attraction, while the Brownian motion was controlled via a Langevin thermostat. They were mainly interested in the structure of the sediment at the bottom of the container and noticed that, for low attractions incapable of inducing cluster formation during sedimentation, this structure is similar to that expected for equilibrium colloidal suspensions. Modest attraction strengths, however, are found to enhance the growth of crystalline phases in the sediment. Furthermore, high attraction strengths induce the formation of colloidal clusters during sedimentation, which settled into loosely packed disordered structures.

4. Discussion and future work

One of the criticisms that can be made to the abovementioned simulational models is that do not proportion quantities that are usually measured by experimentalists. For example, the absence of a cluster fractal dimension due to the anisotropies found is not currently available to the experimentalists. In several experimental works, researchers are mainly focused on sedimentation rates and sedimentation halftimes, something that they can measure in their aggregation vials [59]. The structure of the sediment, as studied in Ref. [56], is another possible topic that merits a further consideration both by computer simulators and experimentalists. The sedimentation halftimes, for example, can be studied computationally by measuring the movement of the center of mass of all the colloidal particles in the clusters, as they settle downward. As it was stated above, in this review, we were mainly focused on the sedimentation and deposition of single clusters, that is, for concentrations lower than the volume fraction ϕ^* define by Allain et al. [37]. We should not be afraid to go beyond the ϕ^* and try to simulate the formation and the collective settling of a gelled suspension [39–41, 60]. For this purpose, perhaps we would need to do an off-lattice simulation with non-perfectly rigid bonds between the colloidal particles, but that can have some mild tolerance for bending and also for breaking under a high stress. As we can see, the field is wide open to exciting, future research.

Acknowledgements

The author is grateful to the Academic Supercomputing Committee at the National Autonomous University of Mexico for the computational resources assigned through DGTIC-UNAM (project SC16-1-IR-52).

Author details

Agustín E. González

Address all correspondence to: agus@fis.unam.mx

Institute for Physical Sciences, National Autonomous University of Mexico, Cuernavaca, Mexico

References

[1] Einstein A. Über die von der molekularkinetischen theorie der wärme geforderte Bewegung von in ruhenden flüssigkeiten suspendierten Teilchen. Ann. der Physik. 1905; 17: 549–560. Reproduced and translated in the book: Einstein A. Investigations on the theory of the Brownian movement. Mineola N.Y.: Dover Publications; 2000.

[2] Perrin J. The molecular agitation and the Brownian movement. Comptes Rend. Hebd. Seances Acad. Sci. 1908; 146: 967–970.

[3] Perrin J. The Stokes law and the Brownian movement. Comptes Rend. Hebd. Seances Acad. Sci. 1908; 147: 475–476.

[4] Perrin J. The origin of Brownian movement. Comptes Rend. Hebd. Seances Acad. Sci. 1908; 147: 530–532.

[5] Perrin J. Brownian movement and molecular reality. London: Taylor and Francis; 1910.

[6] Perrin J. Atoms. London: Constable & Co.; 1916.

[7] Mason M., Weaver W. The settling of small particles in a fluid. Phys. Rev. 1924; 23: 412–426.

[8] Burton E.F., Bishop E. The law of distribution of particles in colloidal solutions. Proc. R. Soc. Lond. A. 1922; 100: 414–419.

[9] Weaver W. The duration of the transient state in the settling of small particles. Phys. Rev. 1925; 27: 499–503.

[10] Svedberg T. Some recent advances in the field of colloids. Chem. Rev. 1924; 1: 249–275.

[11] McDowell C.M., Usher F.L. The distribution of suspended particles under gravity. Proc. R. Soc. Lond. A. 1932; 138: 133–146.

[12] Smoluchowski M. Drei vorträge über diffusion, brownsche molekularbewegung und koagulation von kolloidteilchen. Physik Zeitsch. 1916; 17: 557–571, 585–599.

[13] Smoluchowski M. Versuch einer mathematischen theorie der koagulationskinetic kolloider lösungen. Zeitsch. Phys. Chem. 1917; 92: 129–168.

[14] Witten T.A., Sander L.M. Diffusion-limited aggregation, a kinetic critical phenomenon. Phys. Rev. Lett. 1981; 47: 1400–1403.

[15] Meakin P. Formation of fractal clusters and networks by irreversible diffusion-limited aggregation. Phys. Rev. Lett. 1983; 51: 1119–1122.

[16] Kolb M., Botet R., Jullien R. Scaling of kinetically growing clusters. Phys. Rev. Lett. 1983; 51: 1123–1126.

[17] Mandelbrot B. The Fractal Geometry of Nature. New York: W. H. Freeman & Co.; 1983.

[18] Derjaguin B.V., Landau L. Theory of the stability of strongly charged lyophobic sols and of the adhesion of strongly charged particles in solutions of electrolytes. Acta Phys. Chim. USSR. 1941; 14: 633–662.

[19] Verwey E.J.W., Overbeek J.T-G. Theory of the stability of lyophobic colloids. Amsterdam: Elsevier; 1948.

[20] Derjaguin B.V. Main factors affecting the stability of colloids. Pure Appl. Chem. 1976; 48: 387–392.

[21] Sonntag R.C., Russel W.B. Structure and breakup of flocs subjected to fluid stresses: I. Shear experiments. J. Colloid Interface Sci. 1986; 113: 399–413.

[22] Sonntag R.C., Russel W.B. Structure and breakup of flocs subjected to fluid stresses: II. Theory J. Colloid Interface Sci. 1987; 115: 378–389.

[23] Torres F.E., Russel W.B., Schowalter F.E., Floc structure and growth kinetics for rapid shear coagulation of polystyrene colloids. J. Colloid Interface Sci. 1991; 142: 554–574.

[24] Hoekstra L.L., Vreeker R., Agterof W.G.M. Aggregation of colloidal nickel hydroxycarbonate studied by light scattering. J. Colloid Interface Sci. 1992; 151: 17–25.

[25] Wessel R., Ball, R.C. Fractal aggregates and gels in shear flow. Phys. Rev. A. 1992; 46: R3008–R3011.

[26] Jullien R., Botet R. Aggregation and Fractal Aggregates. Singapore: World Scientific; 1987.

[27] Vicsek T. Fractal Growth Phenomena. Singapore: World Scientific; 1989.

[28] Meakin P. Models for colloidal aggregation. Ann. Rev. Phys. Chem. 1988; 39: 237–267.

[29] Hidalgo-Álvarez R., Martín A., Fernández A., Bastos D., Martínez F., de las Nieves F.J. Electrokinetic properties, colloidal stability and aggregation kinetics of polymer colloids. Adv. Colloid Interface Sci. 1996; 67: 1–118.

[30] Lebovka N.I. Aggregation of charged colloidal particles. Adv. Polym. Sci. 2014; 255: 57–96.

[31] Lach-hab M., González A.E., Blaisten-Barojas E. Concentration dependence of structural and dynamical quantities in colloidal aggregation: computer simulations. Phys. Rev. E. 1996; 54: 5456–5462.

[32] González A.E., Martínez-López F., Moncho-Jordá A., Hidalgo-Álvarez H. Concentration effects on two- and three-dimensional colloidal aggregation. Physica A. 2002; 314: 235–245.

[33] Michaels A.S., Bolger J.C. Settling rates and sediment volumes of flocculated kaolin suspensions. Ind. Eng. Chem. Fundam. 1962; 1: 24–33.

[34] Couch M., Ph.D. thesis, Cambridge, UK: Cambridge University; 1993.

[35] Allain C., Cloitre M. The effect of gravity on the aggregation and the gelation of colloids. Adv. Colloid Interface Sci. 1993; 46: 129–138.

[36] Wafra M., Ph.D. Thesis, Paris: Université Paris-Nord, 1994.

[37] Allain C., Cloitre M., Wafra M. Aggregation and sedimentation in colloidal suspensions. Phys. Rev. Lett. 1995; 74: 1478–1481.

[38] Allain C., Cloitre M., Parisse F. Settling by cluster deposition in aggregating colloidal suspensions. J. Colloid Interface Sci. 1996; 178: 411–416.

[39] Senis D., Allain C. Scaling analysis of sediment equilibrium in aggregated colloidal suspensions. Phys. Rev. E. 1997; 55: 7797–7800.

[40] Senis D., Gorre-Talini L., Allain C. Settling in aggregating colloidal suspension: Analysis of cell inclination influence. Eur. Phys. J. E. 2000; 2: 31–38.

[41] Senis D., Gorre-Talini L., Allain C. Systematics study of the settling kinetics in an aggregating colloidal suspension. Eur. Phys. J. E. 2001; 4: 59–68.

[42] González A.E., Leyvraz F. Colloidal aggregation with a drift: a computer simulation. In: Proceedings of the Statistical Mechanics in Physics and Biology Symposium of the 1996 MRS Fall Meeting; 2–5 December 1996; Boston. Pittsburgh: MRS Publications; 1997.

[43] González A.E. Colloidal aggregation with sedimentation: computer simulations. Phys. Rev. Lett. 2001; 86: 1243–1246.

[44] González A.E. Colloidal aggregation in the presence of a gravitational field. J. Phys.: Condens. Matter. 2002; 14: 2335–2345.

[45] Leone R., Odriozola G., Mussio L., Schmitt A., Hidalgo-Álvarez R. Coupled aggregation and sedimentation processes: three-dimensional off-lattice simulations. Eur. Phys. J. E. 2002; 7: 153–161.

[46] Odriozola G., Leone R., Schmitt A., Moncho-Jordá A., Hidalgo-Álvarez R. Coupled aggregation and sedimentation processes: the sticking probability effects. Phys. Rev. E 2003; 67: 031401:1–031401:5.

[47] González A.E., Odriozola G., Leone R. Colloidal aggregation with sedimentation: concentration effects. Eur. Phys. J. E. 2004; 13: 165–178.

[48] Peltomäki M., Hellén E.K.O., Alava M.J. No self-similar aggregates with sedimentation. J. Stat. Mech.: Theo. Exp. 2004; P09002.

[49] González A.E. Colloidal aggregation coupled with sedimentation at a fixed depth: computer simulation. Europhys. Lett. 2006; 73: 878–884.

[50] González A.E. Stratification of colloidal aggregation coupled with sedimentation. Phys. Rev. E. 2006; 74: 061403:1–061403:14.

[51] Kantor Y., Witten T.A. Mechanical stability of tenuous objects. J. Phys. Lett. 1984; 45: 675–679.

[52] Meakin P., Deutch J.M. Properties of the fractal measure describing the hydrodynamic force distribution for fractal aggregates moving in a quiescent fluid. J. Chem. Phys. 1987; 86: 4648–4656.

[53] Oles V. Shear-induced aggregation and breakup of polystyrene latex particles. J. Colloid Interface Sci. 1992; 154: 351–358.

[54] Hawick K.A. Simulating and visualising sedimentary cluster-cluster aggregation. In: Proceedings of the International Conference on Modeling, Simulation and Visualization Methods (MSV′10); 10–15 July 2010; Las Vegas: CSREA; 2010. p. 3–9.

[55] Hawick K.A. Characterising components and flocculation of sediment in a diffusion-limited cluster-cluster aggregation model. In: Proceedings of the International Conference on Modelling, Identification and Control (AsiaMIC 2013); 10–12 April 2013. Phuket, Thailand: IASTED; 2013. p. 1–8.

[56] Whitmer J.K., Luijten E. Sedimentation of aggregating colloids. J. Chem. Phys. 2011; 134: 034510:1–034519:10.

[57] Asakura S., Oosawa F. On interaction between two bodies immersed in a solution of macromolecules. J. Chem. Phys. 1954; 22: 1255–1256.

[58] Vrij A. Polymers at interfaces and the interactions in colloidal dispersions. Pure Appl. Chem. 1976; 48: 471–483.

[59] Yang Y.J., Kelkar A.V., Corti D.S., Franses E.I. Effect of interparticle interactions on agglomeration and sedimentation rates of colloidal silica microspheres. Langmuir. 2016; 32: 5111–5123.

[60] Buzzaccaro S., Secchi E., Brambilla G., Piazza R., Cipelletti L. Equilibrium concentration profiles and sedimentation kinetics of colloidal gels under gravitational stress. J. Phys.: Condens. Matter. 2012; 24: 284103:1–284103:8.

Chapter 11

Assembly of Nanoparticles into "Colloidal Molecules": Toward Complex and yet Defined Colloids with Exciting Perspectives

Claudia Simone Plüisch and Alexander Wittemann

Additional information is available at the end of the chapter

http://dx.doi.org/10.5772/65343

Abstract

In line with atoms being the elementary units of molecules and crystals, colloidal particles can be used as building blocks for organized materials. A major benefit in doing so is that joining colloids in a defined manner comes along with structuring. In view of opening avenues to more complex structural motifs, significant efforts must be geared to colloids with specific shapes and symmetries. A straightforward strategy is joining equal-sized spherical particles into stable clusters. Such clusters are called "colloidal molecules" because they may exhibit configurations resembling pretty much those of molecules. Their preparation can be based on the agglomeration of particles dispersed in an emulsion. The particles adsorb on the emulsion droplets and coagulate in a defined way during the evaporation of the droplet phase. Using this method originally applied to microscale particles, one can produce clusters with submicron-sized global dimensions. Variable parameters such as radii and concentration of cluster constituents provide the framework needed to obtain "colloidal molecules" that differ in size, shape, and physical properties. This opens up exciting perspectives for tailor-made colloids as building units for hierarchically organized materials. Moreover, new physical properties such as plasmonic "hotspots" may emerge from packing particles into assemblies of specific configurations.

Keywords: supracolloids, assemblies, nanoparticle clusters, heteroaggregates, colloidosomes, emulsion droplets, plasmonic colloids, depolarized dynamic light scattering, surface-enhanced Raman scattering

1. Introduction

The preparation of suitable building blocks is the first step for the creation of new hierarchically organized materials, which is comparable to the construction of a building from single bricks. The elementary units and their interactions not only determine decisive properties of the material, but also regulate its genesis. In this context, the shape of the building blocks is particularly important [1], as it might be used to manipulate particle organization into desired hierarchically organized superstructures and, ultimately, materials (**Figure 1**).

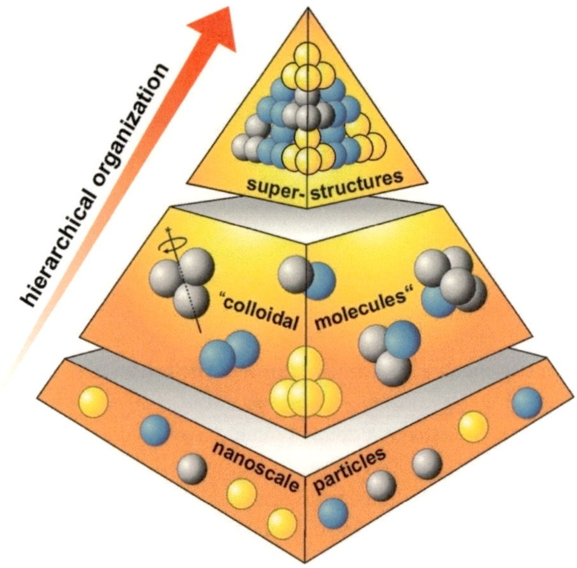

Figure 1. Perspectives of "colloidal molecules": Clusters of spherical elementary units have the potential to bridge the gap between single particles at the nanoscale and hierarchically organized materials at the macroscale. Moreover, they can serve as models for particles with complex shapes, or offer unique properties that emerge from packing nanoparticles into supraparticles.

However, minimization of surface tension favors the formation of spherical or quasi-spherical particle shapes. Under specific reaction conditions, which allow kinetically controlled growth, particle formation can be directed in an anisotropic dimension. Gold nanorods are probably the most prominent example of anisotropic inorganic colloids since the establishment of a wet-chemistry approach for their synthesis [2]. With regard to polymer colloids, dumbbell-shaped particles can be prepared by controlled phase separation during seeded polymerization (**Figure 2A**). These particles can be used again as seeds for the preparation of trimers or even tetramer particles [3]. Dumbbell-shaped particles were successfully used as elementary units to build dense colloidal crystals with a partial band gap and birefringence [4]. Nucleation and growth of polymer nodules on spherical seeds give access to multipod-like or raspberry-like colloids [5]. Because of the organized coordination of the polymer nodules with regard to the seed, such complex particles are often called "colloidal molecules," if their subunits are seen as their "colloidal atoms" [6, 7]. In this terminology, the number of polymer nodules per central sphere equals the number of "free valences" that if made "sticky" could act as docking sites to build up hierarchically organized superstructures [8]. Spherical "Janus particles", i.e., particles with hemispherical coatings, can be seen as the simplest representatives of "colloidal molecules."Janus colloids already have shown a rich variety of superstructures [9]. "Colloidal molecules" of higher complexity are often considered to be model systems for molecular matters but also to function as building units for new hierarchically organized materials [10].

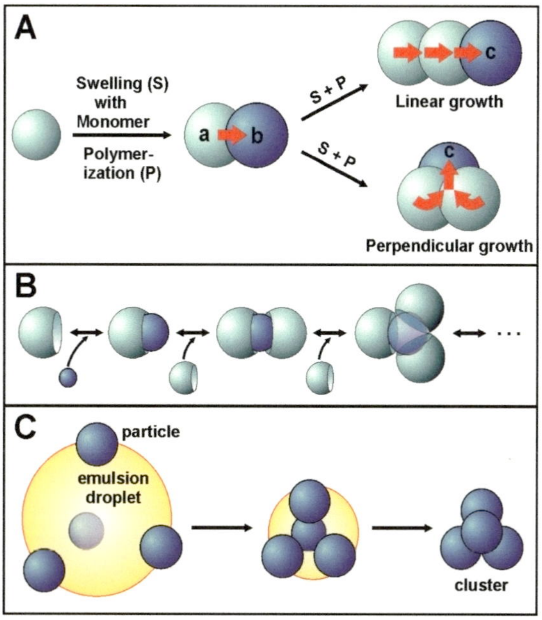

Figure 2. Routes toward "colloidal molecules": (A) Controlled phase separation during seeded emulsion polymerization: Dumbbell-shaped particles are obtained by swelling cross-linked polystyrene seeds with monomer and polymerizing it. Addition of a third bulb can follow either linear or perpendicular growth depending on cross-linking densities of bulbs a and b [3]. (B) A lock and key mechanism allows particles with complementary shapes to self-assemble into clusters in the presence of a depletant [13]. (C) Particle assembly assisted by emulsion droplets: Particles adsorb onto emulsion droplets driven by minimization of interfacial energy. Subsequent evaporation of the droplet phase forces particles to assemble into clusters, which are cemented together by van der Waals forces [15].

Chemical routes toward shape-tailored colloids are limited to specific systems and often demand careful handling of experimental parameters. Joining preformed colloidal subunits into "colloidal molecules" offers greater flexibility in the choice of elementary units. Hence, such an approach has the potential to provide access to an almost unlimited diversity of supracolloidal assemblies. Recently, first preparative strategies have been developed [1, 7]. The assembly into clusters can be mediated by physical interactions, often in combination with suitable templates. A successful assembling strategy must set a limit for the aggregation. Moreover, well-defined morphologies should result. This is challenging inasmuch "colloidal atoms," unlike true atoms, usually do not undergo directed interactions. For this purpose, microspheres having oppositely charged hemispheres were prepared. These particles spontaneously self-assemble into defined clusters, which, however, are just intermediates during the formation of larger aggregates because there is no limitation to the size of clusters formed through electrostatic interactions [11]. Another directional assembling route is adopting Fischer's famous lock-and-key concept [12] to the colloidal regime. Site-specific recognition and association among bowl-shaped (lock) and spherical (key) colloidal subunits can be triggered by short-range depletion attractions upon adding a non-adsorbing polymer [13]. This route was successfully tested for the formation of assemblies composed of a central sphere to which a limited number of bowl-shaped colloids are bound (**Figure 2B**). As an alternative to geometric recognition, directional assembly via depletion interactions can be also implemented by surface roughness that locally differs [14].

It is clear that these strategies are not likely to be applied for the fabrication of "colloidal molecules" on a larger scale. Particle assembly assisted by physical templates may present an important move in this direction. In such an approach, a limited number of particles are trapped in a confined space that is given by the template. Capillary forces that occur during evaporation of the dispersion medium force the particles to pack into particle clusters. The particle assemblies are kept together by van der Waals forces. One example of this is the assembly of particles within cylindrical holes as 2-D templates [16]. Similar to the methods described above, the use of 2-D templates does not provide access to "colloidal molecules" at larger scales. Moreover, the production of submicron-sized templates for the assembly of nanoparticles and the release of the assemblies from a solid mold present further obstacles.

A promising strategy is the use of liquid droplets as 3-D templates (**Figure 2C**). Driven by minimization of interfacial tension, particles adsorb onto emulsion droplets (Pickering effect) [17]. Subsequent evaporation of the dispersed droplet phase removes the liquid templates, while the assembly into particle clusters takes place [15]. Clusters made up of polystyrene (PS), poly(methyl methacrylate) (PMMA), or silica microspheres were prepared along these lines [18]. The supracolloids exhibit well-defined configurations, which are believed to result from an ordered arrangement of the particles at the droplet surface due to long-ranged dipole-dipole repulsion through the oil droplet and Coulomb interactions [19]. Constituents of colloidal clusters do not necessarily have to be spherical. Recently, clusters have been constructed from dumbbell-shaped particles as well [20]. This is a good example that there can be mutually complementary links between different approaches toward "colloidal molecules."

The number of particles in a given droplet determines the number of constituents of the cluster, which will result from this specific droplet. Hence, the distribution of the particles onto the droplets is directly related to the dispersity of the resulting clusters. A broad size distribution of the droplets would give rise to a broad range of different species, whereas the use of monodisperse emulsion droplets would limit the variance in the number of particles per droplet. Even monodisperse droplets do not enable the fabrication of a single set of clusters. The distribution of the particles on the droplets is random. Nonetheless, using monodisperse droplets significantly reduces the range of species and, therefore, increases the yields of distinct clusters. Some efforts have been made in this direction. If we disregard those limited to large particle assemblies [21, 22], one approach remains. In a Couette apparatus, a polydisperse precursor emulsion was transformed under shear into a narrowly dispersed emulsion bearing a limited number of particles on the droplets. After evaporation of the dispersed phase, high yields of particle doublets, triplets, and quadruplets were obtained [23].

Regardless of recent achievements, further efforts devoted to the fabrication of "colloidal molecules" at scales sufficient to build materials are still necessary to exploit their potential. Apart from the actual preparation, this is equally valid for sorting "colloidal molecules" into fractions of uniform configurations. If both prerequisites are satisfied, efficient strategies need to be established that allow "colloidal molecules" to assemble into well-defined superstructures.

All the above-mentioned facts indicate that properties of organized materials are strongly shaped by the dimensions of their elementary units. With regard to materials built from "colloidal molecules," the particles used for the fabrication of "colloidal molecules" represent the lowest level in hierarchy of materials (**Figure 1**). However, also the dimensions of the "colloidal molecules" at the middle level can be decisive, bearing already their assembly into superstructures in mind. Getting to the top level, namely to organized materials, may require "colloidal molecules" of dimensions comparable to the range of colloidal interactions allowing their assembly to proceed. This is one of the main reasons why this chapter addresses "colloidal molecules" made of nanoscale particles.

2. Preparation of "colloidal molecules"

The template-assisted strategies reported above are particularly suitable for the assembly of microscale particles [7, 10]. However, microscale particles and assemblies thereof are prone to sedimentation because of their dimensions. Their motion is different from Brownian motion because there is a preferred direction dictated by gravity, which prevents random diffusion. In order to assess this, the gravitational length l_g is an appropriate measure:

$$l_g = \frac{D_0}{v_s} = \frac{\kappa_B T}{g \Delta \rho V_p}, \tag{1}$$

where $k_B T$ is the thermal energy, g is the acceleration of gravity, $\Delta \rho$ is the difference between particle density and solvent density, and V_p is the particle volume. The gravitational length is defined as the ratio between the single-particle diffusion coefficient D_0 and the Stokes sedimentation velocity v_s. It can be regarded as the minimum distance an object has to settle before the sedimentation drift exceeds the root-mean-square displacement of Brownian motion [24]. By way of illustration, the gravitational length of a polystyrene microsphere with a diameter of 1.94 μm is already as short as its own diameter, whereas the gravitational length of a 100 nm-sized polystyrene particle exceeds its diameter by a factor of 141,000.

In addition to Brownian motion acting against settling, there is a second motivation to build colloidal assemblies from nanoscale particles. Clusters from a limited number of nanoparticles have global dimensions of the same order as the range of colloidal interactions (1–50 nm). This can become an essential requirement when developing strategies, in which the assembly of "colloidal molecules" into organized superstructures is mediated by colloidal interactions [25].

Similarly to the aforementioned assembly of microspheres [15], the assembly of nanoscale elementary particles can be assisted by evaporating emulsion droplets [26, 27]. However, the diameter of the emulsion droplets has to be drastically reduced in order to allow the assembly to proceed. In addition to smaller dimensions, the droplets should also exhibit a low dispersity in size to obtain a limited number of different cluster species. This can be achieved by making use of the basic concepts behind the formation of miniemulsions [28]. The crucial point here is that the droplet size distribution in a miniemulsion can be controlled by ultrasound (**Figure 3**). In the following sections, all essential aspects regarding the emulsion-assisted nanoparticle assembly will be presented, starting with general considerations on the choice of the building blocks.

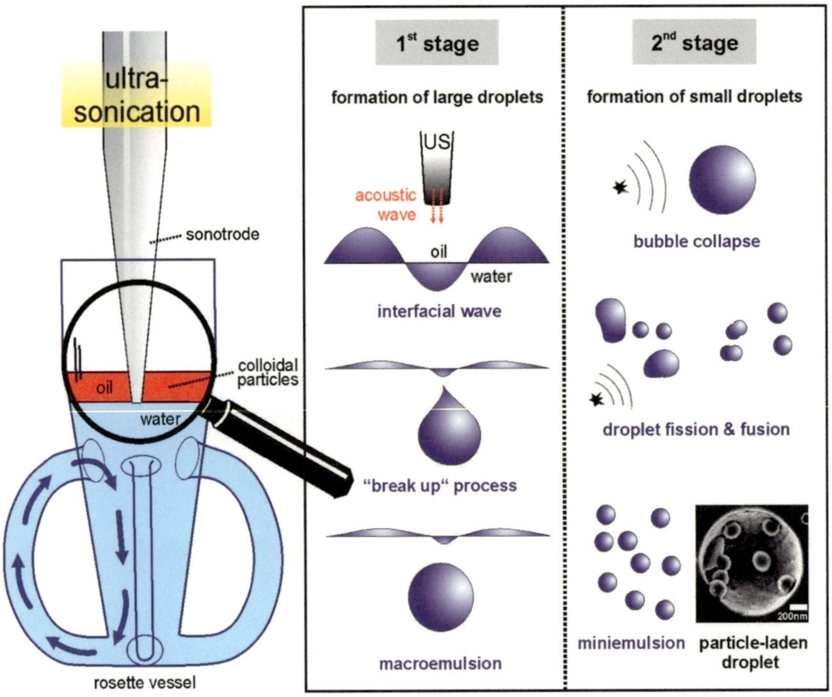

Figure 3. Schematic representation of emulsification by ultrasound: An aqueous solution containing an emulsifier is placed in a rosette vessel and overlaid with toluene. The particles to be packed into "colloidal molecules" were added either to the oil or to the water phase. The initial stage of emulsification starts with instabilities of interfacial waves that lead to eruption of the oil phase into the water phase, or vice versa, producing microscale droplets. The second stage involves fission of large droplets into even smaller ones by acoustic cavitation [36]. Collapse of cavitation bubbles creates shock waves that deform the emulsion droplets and finally cause their fission. A lower limit of the droplet size is set by droplet fusion driven by interfacial tension. The interplay of droplet fission and fusion leads to a steady state characterized by narrowly dispersed droplets. These droplets confine a certain number of particles at their surface, which marks the precondition for the preparation of "colloidal molecules" (cryo-FESEM micrograph at the lower right corner) [32]. See the text for further explanations.

2.1. Elementary units

The designation "colloidal molecules" should be reserved exclusively to colloidal structures that exhibit well-defined shapes [29]. Otherwise it would not be possible to exploit their potential as building blocks with specific symmetries for hierarchically organized materials [30].

With regard to "colloidal molecules" prepared by combining ex-situ formed particles, it is thus essential to use particles that are as uniform as possible. This equally applies to the shape and the size of the particles. Generally, centrosymmetric particles are used as building blocks to avoid directional interactions in particle assembly. Hence, the configurations of the resulting assemblies are solely regulated by packing criteria for spheres, such as the formation of energetically favorable states or optimization of packing efficiencies [31, 32].

Size restrictions are also relevant to the selection of suitable building blocks. The upper size limit is set by the colloidal regime for the reasons specified above. According to the classification made by IUPAC, a colloid should have at least in one direction a dimension below 1 μm. If transferred to "colloidal molecules," this criterion will apply to any direction, except for linear configurations. As long as the global dimensions are maintained below 1 μm, Brownian diffusion prevails over gravitational forces. It is precisely because uniform particles with diameters in the range of 50–250 nm are most suitable as elementary units of "colloidal molecules." Polystyrene particles (density: 1.054 gm^{-3}) of such dimensions exhibit gravitational lengths that exceed their own diameter by factors between several thousands and a few millions. In this consideration, the lower limit of particle size is set by the reduced affinity of small nanoparticles to adhere to emulsion droplets. The free energy associated with the detachment E of a spherical particle with radius r from the oil-water interface can be as follows [33]:

$$E = \prod r^2 \gamma_{ow} (1 - |\cos\theta|)^2, \tag{2}$$

where γ_{ow} denotes the interfacial tension between oil and water. Eq. (2) shows that the energy needed to remove a particle from the droplet surface rapidly increases with its size. In addition, the three-phase contact angle θ marks a second key parameter when dealing with attachment or detachment of particles to emulsion droplets. It is defined by the tangents to the particle surface and the oil-water interphase in each point of the contact line where the particle and the droplet surface meet [34].

Particles having a contact angle of 90° are most strongly bound to the droplet surface according to Eq. (2). They are equally wetted by oil and water because of a balance between hydrophobic and hydrophilic properties. Accordingly, ultrahydrophobic (contact angles near 0°) or ultrahydrophilic particles (contact angles near 180°) have low affinities to oil-water interphases. Hence, tuning their wetting behavior may be decisive to make particles adhere to a liquid template enabling their assembly into "colloidal molecules." Modification of the wetting behavior can be accomplished by inserting hydrophilic or hydrophobic groups into the particle surface. This can be achieved by (i) suitable comonomers during the synthesis of polymer colloids [27] or (ii) postmodification techniques such as silanization [19] or physisorption of amphiphiles (block copolymers, surfactants) [35]. Moreover, criteria other than the wetting behavior, such as swelling of polymer particles with the oil phase may also affect the affinity of the particles to stick to emulsion droplets [32].

Last but not least, stable suspensions of "colloidal molecules" should be obtained. In this regard, it is advantageous to use particles that bear surface charges. This yields electrostatically stabilized "colloidal molecules" without any need for post-modification [26]. Moreover, surface charges on the elementary units may be beneficial during the formation of "colloidal molecules" as well. Electrostatic repulsions among the particles support the agglomeration into energetically favorable packings and may thus contribute toward consistent configurations of "colloidal molecules" [32].

In what follows, the preparation of "colloidal molecules" built from surface-modified polystyrene and gold particles is presented. These particles meet all the above-mentioned demands. Moreover, with this selection of elementary units, it will be shown that the route toward "colloidal molecules" presented below is applicable to organic and inorganic colloids.

2.2. Ultrasound-assisted emulsification

The fabrication of "colloidal molecules" outlined in this chapter is based on the agglomeration of ex-situ prepared particles that are dispersed in an oil-in-water emulsion (**Figure 2C**). In this approach, evaporating oil droplets serve as physical templates to trap a limited number of particles and, consequently, allow their assembly into particle clusters [26]. This route toward defined particle clusters was established for the assembly of polymer and silica microspheres [18]. Droplets with diameters suitable for the assembly of microscale particles can be prepared by vigorous stirring using a rotor-stator homogenizer. However, it became apparent that the same procedure did not work out for the assembly of particles with diameters of about 100 nm [26]. No clusters were obtained. Most probably, the microscale droplets were split into smaller ones bearing just single particles at their surface during evaporation of the droplet phase. As a consequence, the diameters of the oil droplets had to be reduced by a comparable rate to the dimensions of the particles to be packed into clusters [27]. In addition, the dispersity of the emulsion droplets should be as as low as possible. Loading narrowly dispersed instead of polydisperse droplets by a random process will limit the range of particles on the droplets and, thus, the range of resulting assemblies [26].

Nanoparticle-laden emulsions were prepared as follows: At first, an aqueous solution of an oil-in-water emulsifier was overlaid with toluene. The elementary particles of the "colloidal molecules" were added either with the oil or the water phase. An ultrasonic horn was then dipped into the toluene until its tip touched the interphase between toluene and water, allowing propagation of ultrasound across the oil-water interphase (**Figure 3**). This, further assisted by the geometry of the vessel, ensured efficient mixing of the emulsion during sonication. Emulsification by ultrasound can be considered as a two-stage process, which starts with the formation of microscale droplets, which will then be split into smaller droplets with prolonged sonication [36]. The interphase between the two liquids is perturbed by ultrasound propagating along it because it acts as an acoustic boundary. As a consequence of instability

http://dx.doi.org/10.5772/65343

of the interphase, microscale droplets are formed in a sort of "breakup" mechanism. At this stage, the characteristic droplet diameters are widely governed by the interfacial ultrasound wavelength and its amplitude. The droplets with dimensions of about 100 μm are split into smaller ones by a mechanism that is called acoustic cavitation [36]. Above a certain threshold, sonication creates cavitation bubbles within a liquid. The bubbles grow bigger in size during rarefaction half-cycles than they shrink during compression half-cycles, resulting in an accumulative growth across the acoustic pressure cycles. On reaching a critical size, the bubbles instantaneously collapse, associated with the emission of shock waves. These, in turn, facilitate deformation of emulsion droplets and thus foster their fission into steadily smaller ones (**Figure 3**). Droplet fusion driven by interfacial energy sets a lower limit of the droplet size, which is characterized by a steady state among fission and fusion events. The interplay of fission and fusion yields narrowly dispersed emulsion droplets with diameters that typically fall in the range of 50–500 nm [28].

At the end of the ultrasonic treatment, it is important to suppress droplet growth mechanisms to preserve the average size and the narrow size distribution of the droplets. Making use of the Gibbs-Marangoni effect [37], collision growth was averted by addition of oil-in-water emulsifiers (Pluronic® F-68, Tween® 80, etc.) to the aqueous phase. Ostwald ripening, i.e., droplet growth originating from transfer of molecules among the droplets via diffusion through the continuous, favors the growth of larger droplets at the expense of smaller ones driven by the Laplace pressure [28]. Turbidity measurements revealed that Ostwald ripening was completely stopped by addition of dodecane [26, 27]. Because the hydrophobic molecules are poorly soluble in water they cannot be exchanged among the droplets. For this reason, the osmotic pressure created by the hydrophobe counteracts the Laplace pressure keeping droplet size constant.

For the preparation of "colloidal molecules," their elementary particles were either suspended in the oil or the water phase prior emulsification. Narrowly dispersed particle-laden emulsion droplets were obtained. The average sizes of the droplets were within 360 – 2000 nm, depending on emulsifier concentration, stabilizing agents, and sonication conditions [27]. Cryogenic field emission scanning electron microscopy (cryo-FESEM) allowed imaging nanoparticles while confined to the droplets [32]. Their distribution at the droplet surface was random. This finding is consistent with Monte Carlo (MC) computer simulations, which indicated that nanoparticles freely diffuse at the droplet surface [32]. Cross-linked polymer particles were found to be substantially swollen at the droplet interphase if the disperse phase presents a solvent for the polymer. This could be seen by a larger diameter together with a preferential immersion into the droplet phase (**Figure 3**).

2.3. Clusters of a single set of particles

Confining a limited number of particles to droplets forms the basis for their assembly into clusters. The latter can then be accomplished by gentle evaporation of the droplet phase. Since the droplets are carrying a statistically varying number of particles, this route yields assemblies that differ in their number of constituent particles. Any experimental parameter that affects size and number of the droplets (volume and viscosity of the disperse phase, concentration of emulsifier or other additives, sonication conditions) is at least as important to the range of species produced as the number of elementary particles present in the emulsion [27]. Hence, optimization of experimental conditions can thus substantially help to obtain higher total yields of "colloidal molecules" or to increase the yield of distinct species. Analysis aggregation numbers and yields of clusters built from the same number of particles can be carried out by differential centrifugal sedimentation (DCS) [27]. The experimental results were in quantitative accord to predictions from MC computer simulations for a comparable set of experimental parameters [32]. DCS was used not only to explore parameters relevant to cluster formation but also to improve cluster yields.

In the case of clusters built from cross-linked polystyrene particles, the total yield could be increased to up to 74 wt% with regard to constituent particles [27]. Polymer particles are usually prepared by heterophase polymerizations such as emulsion polymerization [38]. Hence, the particles are already suspended in water and can be added to the emulsion via the aqueous phase. Alternatively, freeze drying and redispersion in an organic solvent permit adding them with the oil phase. It was found that the amount of clusters is considerably larger in the latter scenario [27]. This finding is specific to assemblies built from cross-linked polymer particles and is most probably related to the swelling kinetics of polymer networks. In contrast to particles that cannot swell, the adhesion to the droplet surface is not solely determined by interfacial tension. Swelling with the organic solvent may result in a preferential immersion into the oil phase. If the particles are initially suspended in the aqueous phase, they start to swell when trapped to the interface. Conversely, they are already fully swollen, if initially suspended in the oil phase and may even partially deswell when getting trapped to the oil-water interface. Hence, the kinetics of swelling or deswelling may substantially affect agglomeration into clusters, as manifested in different cluster yields [27].

Figure 4 gives an overview of cluster configurations found in experiments together with predictions from MC computer simulations. It should be noted that the polystyrene particles as elementary units of the clusters had a low degree of polydispersity. Hence, polydispersity had no influence on the experimental configurations. Indeed, the structures found in experiments were identical to the ones in simulations assuming equal-sized constituent spheres [32]. Moreover, it did not make a difference to which phases of the emulsion the particles were initially added. The latter did affect yields, but did not result in other configurations. Minimal-energy clusters of spheres have been specified for various potentials. The most prominent ones are assemblies that globally optimize the Lennard-Jones potential [39]. Configurations of clusters with aggregation numbers $AN \leq 7$ are equivalent to Lennard-Jones clusters, expect for a small number of clusters with $AN = 5$ exhibiting a square pyramidal configuration. In addition, a second isomer was found for clusters with $AN = 6$ at large scales in specific cases, apart from the octahedral configuration [27]. This configuration can be described as a polytetrahedral packing (C_{2v} symmetry) of six spheres or, alternatively, as a pentagonal dipyramid with one particle missing. The configuration presents a minimal-energy packing of hard spheres [31], but was not found in computer simulations mimicking the droplet-based assembly route [32].

While packing efficiency is prevalent for small *AN*, the assembly route itself can have an influence on the configurations. The assembly proceeds at the surface of a spherical template that is concomitantly shrinking and has no particles inside (**Figure 2C**). Hence, a constraint originating from this strategy may hamper mere optimization of packing efficiency. In fact, predominant configurations were found for $8 \leq AN \leq 12$, both in experiments and simulations [32]. Packing of an equal number of hard spheres would permit more minimal-energy structures [31]. However, particle clusters templated by evaporating emulsion droplets did not exhibit a corresponding range of isomers, no matter whether built from nanoscale [27] or microscale particles [15]. This emphasizes that the template takes an active role when the "colloidal molecules" develop their final configurations [32].

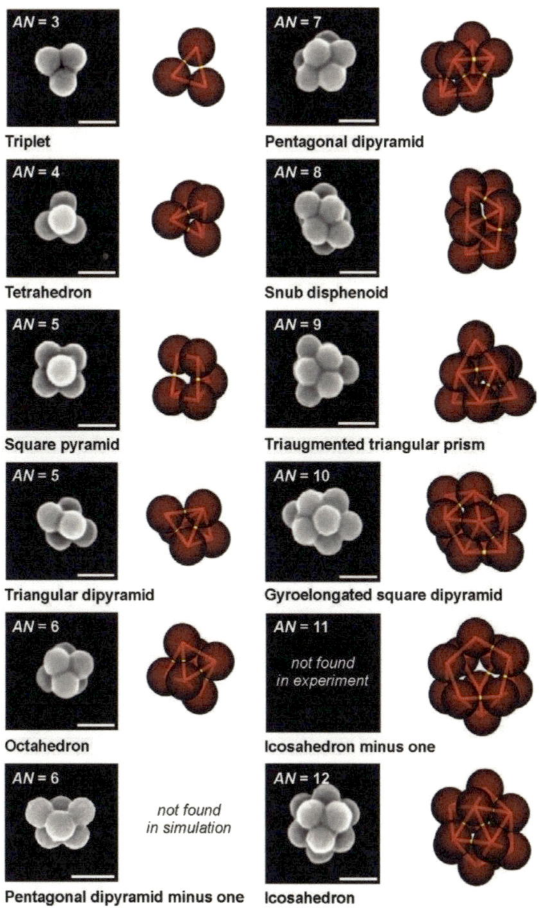

Figure 4. Joining polystyrene particles (average diameter: 154 nm) on droplets led to clusters with unique configurations, primarily characterized by the aggregation numbers *AN*. Cluster morphologies found in FESEM micrographs (left columns) agree with those found in MC computer simulations (right columns). The wireframe in the structures from simulations visualizes the geometric arrangement of the particle centers. The scale bars indicate 200 nm. Reproduced with permission from Ref. [32]. © 2011, AIP Publishing LLC.

The droplet-assisted strategy is not limited to polymer particles. It can be applied to the assembly of inorganic particles as well [35]. Nevertheless, there is a notable difference. Unlike polymer particles, inorganic particles do not swell. It would thus require hydrophobic particles to get them dispersed in the oil phase. In doing so, hydrophobic "colloidal molecules" that immediately self-assemble into large fractal structures would result. It is therefore advantageous to use inorganic particles that can be dispersed in the aqueous phase. Compatibility to water can be improved by surface coatings that make the particles more hydrophilic. It should be noted here, however, that the contact angles of the particles should not become too low to make them still adhere to the droplet surface (see Eq. (2)). Gold nanoparticles bearing a polymer coating were found to perfectly meet this requirement [35]. The conceptual expansion to "colloidal molecules" built from gold particles is motivated by intense plasmonic "hot spots" that may occur within such structures [40].

Gold nanoparticles were prepared by a kinetically controlled seed-mediated growth method in the presence of cetyltrimethylammonium bromide (CTAB) as a surfactant [41]. The multifaceted particles with average diameters of up to 70 nm exhibited low dispersities in size and shape, which is essential for the fabrication of "colloidal molecules." The CTAB molecules were exchanged by the commercial block copolymer (Pluronic® F-68) by repeated centrifugation and redispersion processes. The exchange of the surface layer was accompanied by a change of the zeta potential, which was used to check completeness of this process [35]. The polymer coating onto the particles was a key element for fabricating assemblies with intense plasmonic "hot spots."

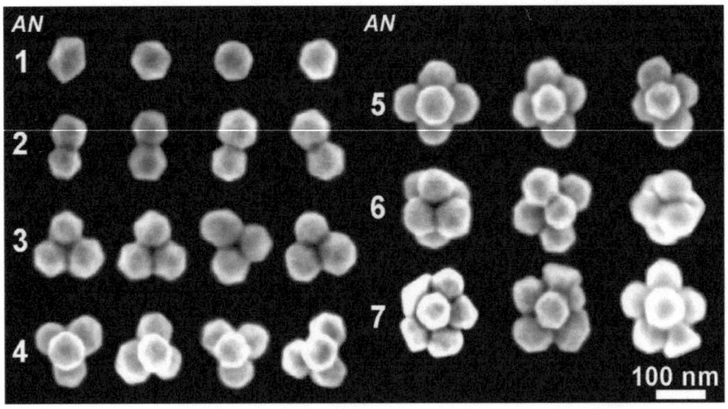

Figure 5. Configurations of clusters built from variable numbers (*AN*) of polymer-coated gold nanoparticles [35].

The assembly of polymer-coated gold nanoparticles was accomplished along similar lines to polymer particles. During emulsification by ultrasound, the disperse phase is broken down in a large number of small droplets. The volume of the dispersed toluene was thus kept as low as possible in view of the fact that such gold particles are available in limited quantities. This ensured that there were sufficient toluene droplets carrying two or more gold nanoparticles, allowing cluster formation to proceed. Evaporation of the droplet phase produced clusters of two to seven polymer-coated nanoparticles using the emulsion droplet-based technique described above [35]. As expected, shapes of "colloidal molecules" built from faceted particles were less regular than that built from amorphous spheres. Nonetheless, the plasmonic clusters had defined configurations (**Figure 5**). Applications of plasmonic "colloidal molecules" in surface-enhanced spectroscopies will be discussed in Section 3.2.

2.4. From hybrids to capsules

The range of accessible "colloidal molecules" can be further extended when different sets of particles are combined. The adsorption of nanoparticles onto ex-situ formed particle clusters (**Figure 6A**) imparted by electrostatic interactions is a straightforward route in this direction [42]. Long-term stable suspensions of composite particles were obtained at low and high surface coverage. Flocculation occurs near the isoelectric point of the composites, when surface charges of the support are balanced by the ones on the adsorbed nanoparticles. The versatility of the route was demonstrated for inorganic nanoparticles that differ in their chemical nature, size and surface charge [42]. Random sequential adsorption superimposed by electrostatic repulsions governs the positioning of the nanoparticles yielding spatially separated nanoparticles on the colloidal support (**Figure 6B**).

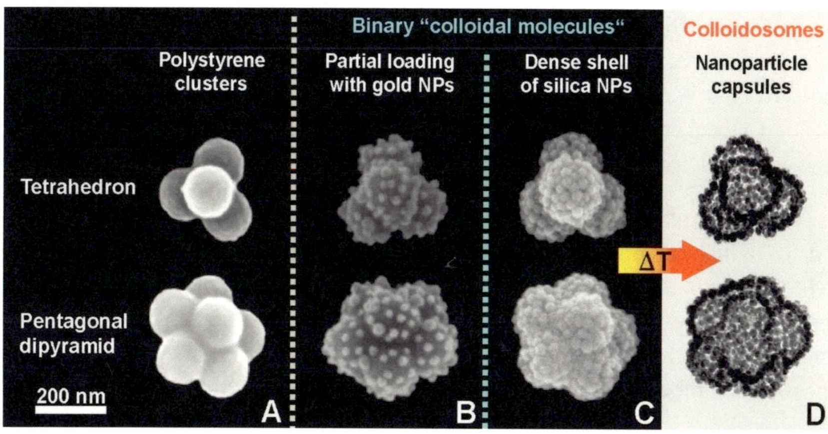

Figure 6. Formation of binary "colloidal molecules": Physisorption of negatively charged nanoparticles onto preformed clusters of positively charged polymer particles (A→B) [42]; co-assembly of polymer particles and oppositely charged nanoparticles at the surface of emulsion droplets yields core-shell heteroaggregates (C) that can act as precursors for the preparation of nanoparticle capsules with complex shapes (D) [43]. See the text for further explanations.

Dense nanoparticle shells on particle clusters were accessible by another strategy, which was based on the emulsion-assisted process [43]. In this approach, cross-linked polystyrene particles were co-assembled with oppositely charged silica nanoparticles, with the polymer particles added via the dispersed (toluene) phase, whereas the nanoparticles were suspended in the continuous (water) phase of the emulsion. This procedure allows oppositely charged particles to co-assemble at the oil-water interphase, while avoiding the formation of fractal aggregates.

Minimization of interfacial energy facilitates confinement of the larger polystyrene particles to the droplet surface. The affinity of silica nanoparticles to stick to the interface is a lot weaker due to their smaller dimensions (see Eq. (2)). This favored their deposition onto the surface of the oppositely charged polymer particles, evidently to the sides that were exposed to the continuous water phase. During adsorption of the nanoparticles, the polymer particles were substantially swollen with the dispersed toluene. Evaporation of the dispersed toluene induced deswelling of the polymer particles, and in parallel to this, their assembly into clusters. Deswelling of their support brought the nanoparticles closely together irrespective of electrostatic repulsions. MC computer simulations revealed that, given an appropriate number of nanoparticles per polymer particle, a dense shell of nanoparticles is formed tightly encasing a core of aggregated polymer spheres [43]. This is in agreement with experimental observations. The global shapes of the binary heteroaggregates were precisely determined by the number of polymer spheres forming the core (**Figure 6C**). As a result, the global shapes of the binary "colloidal molecules" were identical to aggregates built from a single set of particles.

The core-shell-type assemblies could be transferred into nanoporous capsules, simply by calcination at 500°C (air atmosphere). The polymer was completely removed during thermal treatment, as confirmed by elemental analysis [43]. Shaping the inner cavity by polymer particles yields compartmentalized nanoporous capsules (**Figure 6D**). Remarkably, all nanoparticle capsules were intact after the decomposition of the core despite they consist of a single layer of nanoparticles kept together by nanoscopic contacts. Hence, the capsules neither burst nor evaporate over the calcination. MC computer simulations revealed that interplay of short-range (van der Waals) attractions and long-range (electrostatic) repulsions can trap each nanoparticle in a local energy minimum [43]. Moreover, tiny bridges among the nanoparticles may form during calcination by the condensation of silanol groups at the silica surface.

Template-based assembling strategies can be effective tools for the preparation of complex supraparticles, which can be further used as precursors for tailor-made colloids. In this specific case, it gave access to a unique class of colloidosomes built from a single layer of nanoparticles. They are distinguished from other colloidosomes by shapes that are complex but still well-defined and have a high density of nanopores (**Figure 6D**). Such nanoparticle capsules might find applications as model systems for biological cells or as drug delivery vehicles.

2.5. Sorting "colloidal molecules"

In order to use "colloidal molecules" as buildings blocks for novel materials that require particles with specific symmetries or high complexity, it is crucial to have them sorted into uniform species. Sorting can be accomplished by density gradient ultracentrifugation (**Figure 7**) [26]. Rate zonal (density gradient) ultracentrifugation separates particle clusters according to their number of constituents, or more fundamentally by their sedimentation coefficients. This technique enables efficient sorting of "colloidal molecules" owing to high resolution because assemblies whose sedimentation coefficients differ by as small as 15% can be separated [44]. Besides superior resolving power, it allows sorting in suspension at a minimum of undesirable wall effects. In contrast to isopycnic centrifugation, which separates particles according to their densities, the role of density gradients in rate zonal centrifugation is secondary, but nevertheless necessary. It is used to eliminate currents (streaming) that have an adverse effect on banding particles according to their sedimentation rates in defined zones. Zonal centrifugation requires rotors, in which the density gradient is always parallel to the force onto which it is subjected [44]. This can be ensured by using high-speed swinging bucket rotors, which allowed efficient separation of particle clusters in scales that are sufficient for analytical purposes or practical applications that do not require large amounts [45]. Specific examples will be introduced and discussed in the following section.

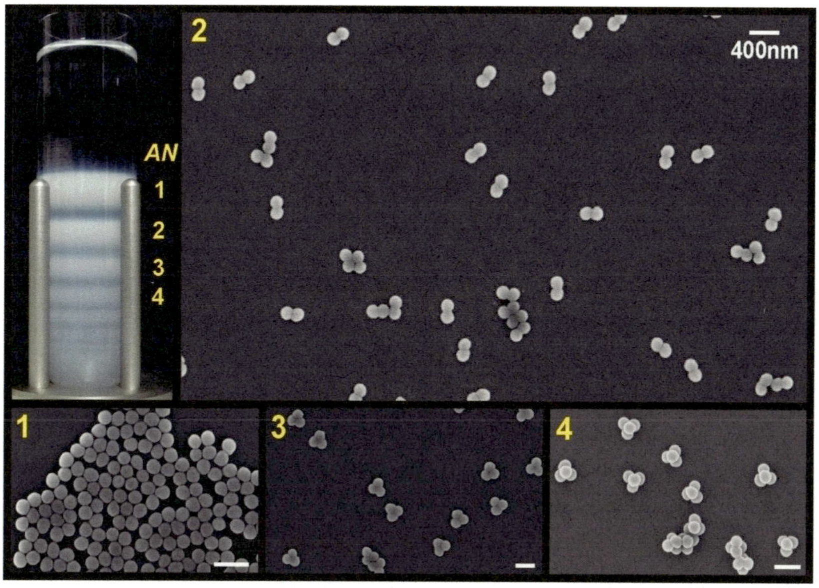

Figure 7. Sorting of submicron-sized clusters of polystyrene particles was accomplished by density gradient centrifugation using a high-speed swinging bucket rotor (photograph). Fractionation according to their sedimentation coefficients resulted in suspensions that contain an equal number of constituents: (1) single particles, (2) particle dimers, (3) particle trimers, and (4) particle tetramers. The scale bars in the FESEM micrographs indicate 400nm.

3. Specific applications

3.1. Model colloids

The diffusion of colloids with complex shapes is a key to a comprehensive understanding of practical phenomena such as biodistribution, sedimentation, coagulation, floatation, and rheology [46]. While the diffusion of particles with simple shapes such as spheres [47, 48], ellipsoids [49, 50], rods [51, 52], and platelets [53] was extensively studied by experiment [54], simulation [55], and theory [56]; studies on the diffusion of complex colloids are still quite scarce [57]. In this context, their defined geometries make "colloidal molecules" ideal model systems to explore the motion of colloidal objects with arbitrary shapes. This approach was introduced by Granick and co-workers using planar clusters of microspheres, which can be easily analyzed by optical microscopy. The motion of microscale objects is, however, affected by gravitational forces because of their rather small gravitational lengths (see Section 2). Moreover, analyzing particles near a solid wall, in the form of a microscope slide, can substantially change the diffusion coefficients [58]. These problems can be elegantly avoided if (i) the size of the clusters is maintained in the submicron regime with Brownian dynamics prevailing over gravitational forces, and if (ii) particle motion is studied within the bulk phase. The latter can be accomplished by light scattering techniques that allow studying the motion of shape-anisotropic particles over relevant length scales.

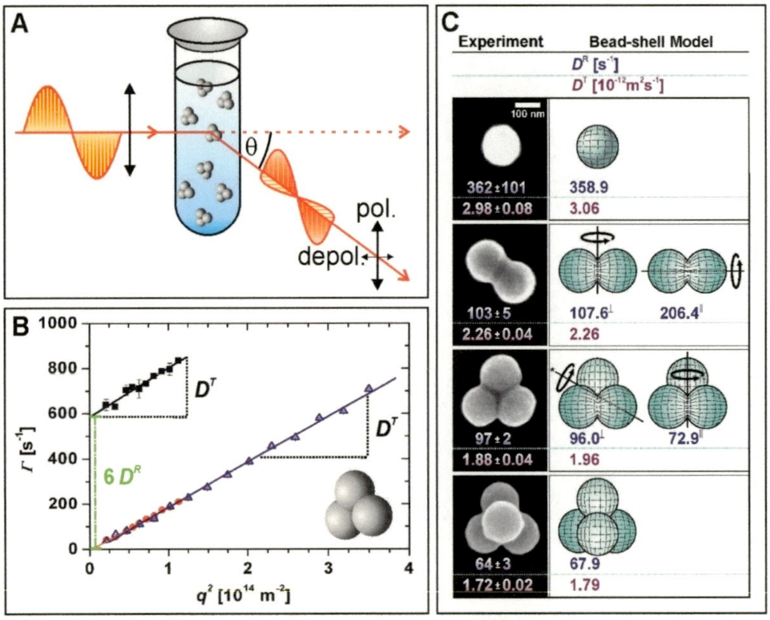

Figure 8. Diffusion of "colloidal molecules": Schematic representation of depolarized dynamic light scattering (DDLS); a horizontally oriented polarizer is used to measure the depolarized component of the scattered light, which provides insights into the translational and rotational diffusion of "colloidal molecules" (A). Relaxation rates Γ as the function of the square of the absolute values of the scattering vector q follow linear correlations that provide access to the translational (D_T) and rotational (D_R) diffusion coefficients (B). The fast mode is represented as black squares; the slow mode of DDLS (red spheres) agrees with results obtained from polarized DLS (blue triangles). Synopsis of experimental and theoretical diffusion coefficients (C): Underneath the experimental clusters that are oriented with their main body parallel to the plane of the figure, D_T and D_R as obtained by DDLS are given. These values agree well with theoretical predications for D_T and values for D_R perpendicular and parallel to the main symmetry axis using the bead-shell model [62], which considers the exact geometries of the clusters. Reproduced with permission from Ref. [45]. © 2009, American Chemical Society.

Dynamic light scattering (DLS) is the most common technique to probe the diffusion of colloidal particles [59, 60]. In the case of optically isotropic particles, decorrelation of scattered light is solely due to translational diffusion and thus cannot provide any insights into rotational motion. However, colloidal particles that are optically anisotropic in shape or composition may produce different orientations of incident and scattered electromagnetic fields. Hence, contributions of scattered light polarized in two directions perpendicular to each other can be detected using a polarizer setup [61]. It should be noted that the intensity of the scattered light that is polarized perpendicular to the intensity of the incident light is related to the mean-squared optical anisotropy of the particles and can be thus rather poor for colloids with low aspect ratios [51]. However, this mode providing insights into translational and rotational diffusion becomes accessible by depolarized dynamic light scattering (DDLS), which is based on a polarizer and analyzer principle (**Figure 8A**).

DDLS was used to probe rotational and translational dynamics of "colloidal molecules" that differ in their geometries defined by the aggregation number. The study was limited to fractions of clusters of not more than four 154 nm-sized polystyrene particles because such assemblies exhibited unique configurations if prepared by the droplet-assisted route (**Figure 4**). Rate-zonal density gradient centrifugation made the clusters available in sufficient quantities and sorting accuracy to measure well-defined intensity autocorrelation functions [45].

In the DDLS experiments, the depolarized autocorrelation functions (an analyzer perpendicular to a polarizer) were found to be the sum of two discrete exponential decays, whereas the polarized intensity autocorrelation functions (an analyzer and a polarizer aligned to each other) were consistent with single exponential decays [45]. Hence, two discrete modes associated with the motion of the clusters were observed in the depolarized analyzer-polarizer setup (**Figure 8A**). The slow mode of the depolarized experiment was equivalent to the single mode of the polarized experiment (**Figure 8B**). Moreover, the decay rates were in agreement with standard DLS experiments (in the absence of an analyzer). Hence, the slow mode of the depolarized is solely associated with the translation of the clusters and results from the limited extinction ratio of the analyzer. This yields a small portion of scattered light that is polarized equal to the incident light, which is in agreement with a standard DLS experiment, providing insights into translational dynamics. Accordingly, the fast mode represents the true depolarized signal. Its decay rate is associated with translational and rotational dynamics providing access to translational and rotational diffusion coefficients (**Figure 8B**). Simple scaling laws were applicable pointing to decoupling rotational and translational motion, which might be an immediate consequence of the small dimensions of the clusters [57]. In contrast to microscale particles, submicron-sized particles may travel distances equal to a multitude of their own diameter in bare translation movements before they undergo directional changes caused by rotation. Decoupling of rotational and translational motion produces linear scaling laws if the decay rates are plotted against the square of the absolute value of the scattering vector facilitating data analysis and making submicron-sized "colloidal molecules" ideal candidates to probe the diffusion of complex particles [45].

Interestingly, it turned out that even the spherical elementary units of the particle clusters produced a poor depolarized signal despite the fact that they should be almost optically isotropic in shape and composition. Nevertheless, the signal provided access to the rotational diffusion coefficient of the elementary units of the clusters, and thus gave a useful reference to the evaluation of the corresponding values derived for the clusters. In line with the correlation of the depolarized intensity to the optical anisotropy, the depolarized signal was strongest for dimer particles, whereas it was substantially lower for trimers and tetramers because of the concomitant decrease in the aspect ratio with the increasing aggregation number [45].

The translational diffusion was found to be widely governed by the global dimensions of the "colloidal molecules." This is primarily expressed by the observation that the translational diffusion coefficients are approximately governed by the reciprocal value of the average dimensions of the "colloidal molecules," in rough agreement with the Stokes-Einstein equation for spherical particles [59]. In contrast to translational motion, shape plays a prominent role in rotational dynamics. This is particularly visible in similar rotational diffusion coefficients of dimer and trimer particles, irrespective of dissimilar spatial dimensions (**Figure 8C**). DDLS is sensitive to the rotation of an object around the axis perpendicular to its main symmetry axis, whereas it cannot monitor the rotation around its main symmetry. This explains why the rotational diffusion coefficients of dimer and trimer particles are similar. The volume displaced by a rotation around the axis perpendicular to the main symmetry axis is equal for a dimer and trimer particle.

Further interpretation of the diffusion of "colloidal molecules" can be based on hydrodynamic modeling. Garcia de la Torre and co-workers established a bead-shell modeling strategy that is perfectly applicable to "colloidal molecules" [62, 63]. In this model, the shape of the "colloidal molecules" is mimicked by overlapping spheres. The surface exposed to the surrounding liquid is represented by a shell of touching but not overlapping small beads that act as friction elements. Hydrodynamic properties were calculated using such a treatment for various types of "colloidal molecules." The predictions for the rotational and translational diffusion coefficients were in perfect agreement with the experimental results (**Figure 8C**). This was achieved by assuming stick boundary conditions, whereas treatments based on slip boundary conditions did not reproduce the experimental results [45]. Substantially, lower orientation times of colloidal particles as compared to molecules make stick boundary conditions favorable over slip boundary conditions.

The study has demonstrated that "colloidal molecules" can act as models for particles with complex shapes. Understanding their diffusion can also contribute to gain deeper insights into the formation of matter because particle clusters are intermediates during the assembly of individual particles into macroscale structures.

3.2. Substrates for surface-enhanced spectroscopies

Beyond their use as model colloids, "colloidal molecules" may find other applications because of specific properties that emerge when nanoparticles are arranged in close configurations. Properties of densely packed nanoparticles are often quite distinct from properties of spatially isolated particles. It is thus of practical and fundamental importance to produce organized nanoparticle assemblies and make their collective properties, such as plasmon coupling, magnetic coupling, resonance energy transfer enhanced photoconductivity, available for applications [64]. An example of this is the surface-enhanced Raman scattering, which can be several orders in magnitude higher for nanoparticle assemblies than for single gold nanoparticles [35, 40, 65].

Raman spectroscopy is based on inelastic scattering of monochromatic light. It provides rich information on structure and composition of matter based on its vibrational modes. However, the Raman effect itself is rather poor. As for comparison, typical Raman cross-sections can be up to 14 orders in magnitude lower than relevant cross-sections in fluorescence spectroscopy [66]. For this reason, the invention of the laser marked an important milestone in the establishment of Raman spectroscopy in analytical laboratories. A comparable revolutionary coup might be based on the finding that if the analyte molecules are bound to plasmonic nanostructures, a large enhancement of the Raman stock signal will result [40]. Surface-enhanced Raman scattering (SERS) can be attributed to two different mechanisms. Firstly, strong magnifications of the fields of the incident light and the Raman scattered light result from interactions among the two optical fields and surface plasmons localized on the plasmonic substrate. Secondly, the electronic interaction among analyte molecules and the metallic substrate can itself modify the scattering process and produce efficiently larger Raman cross-sections [66]. The magnitude of signal enhancement that can be achieved is considerably influenced by the shape of the SERS substrate. This is supported by recent calculations of enhancement factor distributions along the surface of silver nanoparticles of distinct geometries [67]. A 30 nm-sized silver sphere has two rather broad plasmonic "hot spots" (i.e. location where the enhancement reaches a maximum) at the poles. At these sites a signal enhancement of five orders in magnitude is found. Over the entire sphere, the Raman signal is still four orders in magnitude higher than in the absence of a plasmonic substrate. A comparable ellipsoidal SERS substrate shows two distinct hot spots exactly at its tips, providing SERS enhancement of already seven orders in magnitude. In the case of two silver nanoparticles separated by a gap of 2 nm, an extraordinary signal enhancement of nine orders in magnitude is predicted [67]. This example clearly indicates that clusters of plasmonic nanoparticles are promising SERS substrates because they may pool several hot spots within one colloidal object. To produce intense hot spots, it is necessary to accurately adjust the interparticle distances by suitable spacers. Gap engineering is thus crucial to the production of efficient SERS substrates. On the one side, the gap dimensions should be not too large to produce hot spots and, on the other side, they must be large enough to provide access for analyte molecules. Based on the above-mentioned calculations, a gap size of 2 nm can be considered as a well-balanced compromise between the two contrary requirements.

High-resolution transmission electron microscopy (HRTEM) images (**Figure 9**) demonstrate that such interparticle distances are found for the clusters of polymer-coated gold nanoparticles described in Section 2.3. The use of a synthetic polymer coating for creating the required interparticle gaps has several advantages as compared to DNA, thiolated or covalent linkers. It allows efficient linking of colloidal particles with nanoscale interparticle distances [40, 65]. It also allows analyte molecules diffusing into the gap because of the low volume fraction of the polymer and, in specific cases, even concentrating there [68, 69]. Most importantly, polymers have low Raman cross-sections [70]. This avoids considerable falsifications of the vibrational signals by the gap material.

In the light of the above, "colloidal molecules" built from 50 nm-sized gold particles with 2 nm interparticle gaps created by a polymer coating combine high optical coupling with strong electromagnetic fields generated from particles of such diameters. The SERS enhancement was directly measured for clusters made up from 2 to 7 individual gold nanoparticles [35]. To this end, highly diluted suspensions of the assemblies were spin-coated on a microscope slide bearing a rectangular-coordinate grid pattern. This ensured sufficient spatial separation among the clusters and the retrieval of specific species allowing studies at the level of single plasmonic "colloidal molecules." Areas containing an isolated cluster were studied first by FESEM, then by dark-field optical microspectroscopy, and finally, by high-resolution SERS after adsorption of benzenethiol from the gas phase [35]. This procedure helped to determine configuration, optical response, and SERS enhancement of the very same plasmonic "colloidal molecule."

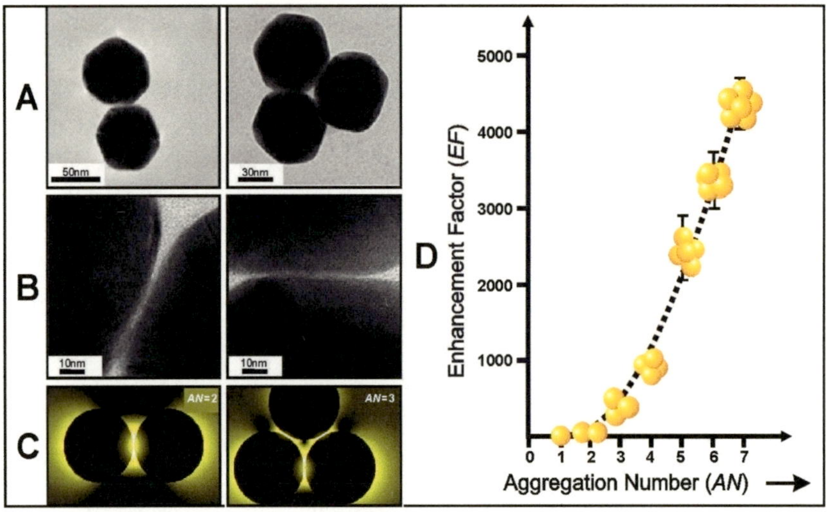

Figure 9. Plasmonic "colloidal molecules": HRTEM micrographs of gold nanoparticle dimers and trimers illustrating the 2 nm-sized interparticle gaps (A, B). Near-field intensities on a log thermal scale for dimer and trimer clusters along the plane passing through the particle centers showing the occurrence of intensive plasmonic "hot spots" right at the interparticle gaps (C). Optical enhancement factors that are achieved by combining single nanoparticles into clusters, normalized to the enhancement observed for a number of individual particles equal to the aggregation numbers (D).

Detailed information on the experimental optical response together with electromagnetic simulations can be found in Ref. [35]. At this place, the actual SERS enhancement that is achieved with specific cluster configurations shall be discussed. This section should be read bearing in mind that the Raman signals using single gold nanoparticles, that is to say, using the elementary units of the clusters as SERS substrates, are already five orders in magnitude higher than in the absence of any plasmonic colloids. If, however, the plasmonic nanoparticles are assembled into clusters of a rather small number of particles, the signal dramatically increases up to a level of three orders in magnitude (**Figure 9**). Even under off-resonant illumination conditions, a dimer particle can provide an intensity increase close to a 100-fold. Particles packed into a pentagonal bipyramid reach 5000-fold additional SERS enhancement. Interestingly, a simple scaling law can be used as a reference to predict the optical enhance-

http://dx.doi.org/10.5772/65343

ment factors *EF* as a function of the aggregation number *AN*, and, related to this, the number of interparticle gaps *GN*:

$$EF = AN + A * GN \tag{3}$$

Hence, the actual SERS enhancement is mainly governed by the number of interparticle gaps. This is fully understood when near-field distributions around the clusters are taken into account (**Figure 9**). They show intense plasmonic "hot spots" right at the gaps, which are almost exclusively responsible for the extraordinary enhancement, whereas other regions of the supracolloidal SERS substrate hardly contribute [35]. For this reason, the prefactor *A* in Eq. (3) is virtually identical to the contribution of a single plasmonic hot spot. Indeed it is found that this contribution is almost independent of the aggregation number of the substrate [35]. This clearly indicates that the hot spots form a regular pattern within the highly symmetric plasmonic "colloidal molecules," and act almost independently. It is evident that an even higher SERS enhancement may be achieved if clusters of higher aggregation numbers are used. However, larger assemblies of gold nanoparticles exhibit less regular configurations and are thus less suitable as model SERS substrates.

In light of their high SERS enhancements and their facile preparation, plasmonic "colloidal molecules" may find applications in the routine analytics of minute quantities of analyte molecules.

4. Conclusions

The results have shown that it is possible to build "colloidal molecules" even from nanoscale particles. Precise configurations are accessible given that the elementary particles have about the same size and shape. The preparation can be based on confining particles to evaporating emulsion droplets. Using ultrasound for emulsification provides control on droplet sizes and dispersity, which is an absolute prerequisite to the production of "colloidal molecules" from nanoscale building blocks. The emulsion-based route is quite versatile. It can be both applied to organic and inorganic particles. In both cases, assemblies with defined configurations can be obtained. Moreover, different sets of particles can be combined into binary supracolloids. It is also possible to co-assemble particles at the droplet surface that came into the emulsion via different phases.

Given that they are built from nanoparticles, "colloidal molecules" have dimensions of less than one micrometer. Consequently, they underlie Brownian motion, which prevails over sedimentation. Moreover, the diffusion of submicron-sized "colloidal molecules" can be studied by scattering techniques. This, combined with the knowledge of their precise geometry, makes clusters of nanoparticles to excellent model colloids aiming at a general understanding of the diffusion of particles with complex shapes, such as protein molecules. DDLS gave direct insights into the diffusion of "colloidal molecules" in dilute suspensions. Relationships for translational and rotational motion were identified for particle geometries, which cannot be described by means of the Stokes-Einstein equation for spherical particles. Knowing the geometry of the anisotropic colloids, the experimental diffusion coefficients could be compared to predictions from hydrodynamic models.

Beyond their use as model systems, "colloidal molecules" offer other exiting perspectives. Unique physical properties can emerge from combining functional subunits into assemblies with tailored shapes and symmetries. The occurrence of plasmonic hotspots in gold nanoparticle clusters is a clear example of a property that is not found in particles that are spatially separated from each other. Signal enhancements exceeding the ones achieved with single particles are found if plasmonic "colloidal molecules" are used as substrates in surface-enhanced spectroscopies. Moreover, varying the surface coating on the elementary units can be used to modify the plasmonic properties. In the course of further advancements with regard to plasmonic "colloidal molecules," more applications in sensing and nonlinear nanophotonics can be expected.

"Colloidal molecules" are particularly promising with regard to innovative applications that demand materials with specific symmetries or functionality. The anisotropic nature of "colloidal molecules" due to their non-spherical shapes could be exploited to their self-assembly into the next higher level of hierarchically organized materials. To this end, interactions among "colloidal molecules" need to be sufficiently weak but long-ranged and specific in order to allow the formation of close-packed equilibrium phases to proceed. Gelation, which prevents the formation of ordered assemblies, can be suppressed by external fields due to the anisotropic polarizability of non-spherical colloids. This technique was successfully applied to produce long-range ordered photonic crystals from dumbbell-shaped "colloidal molecules" [4]. Future efforts must be geared to giving "colloidal molecules" valences suitable for their directed assembly. A recent study about the spontaneous assembly of microspheres bearing two attractive patches at specific positions into well-defined Kagome-like structures is pointing in this direction [8].

Another challenge to be tackled is to equip "colloidal molecules" with tunable interparticle distances mimicking molecular flexibility. This would further contribute to bringing "colloidal molecules" closer to true molecules. Hierarchically organized materials made of such "colloidal molecules" could exhibit stimuli-regulated properties, such as self-healing, defect minimization, or switchable optical properties that are induced by changes in size and configuration.

Acknowledgements

Financial support from the Deutsche Forschungsgemeinschaft (DFG) within SFB 1214/A10 and the Struktur- und Inovationsfonds für die Forschung in Baden-Württemberg (SI-BW) is gratefully acknowledged.

Author details

Claudia Simone Plüisch and Alexander Wittemann*

*Address all correspondence to: alexander.wittemann@uni-konstanz.de

Colloid Chemistry, Department of Chemistry, University of Konstanz, Konstanz, Germany

References

[1] Sacanna S, Korpics M, Rodriguez K, Colón-Meléndez L, Kim S-H, Pine DJ, Yi GR. Shaping colloids for self-assembly. Nat Commun. 2013;4:1–6. DOI: 10.1038/ncomms2694

[2] Ye XC, Jin LH, Caglayan H, Chen J, Xing GZ, Zheng C, Doan-Nguyen V, Kang YJ, Engheta N, Kagan CR, Murray CB. Improved size-tunable synthesis of monodisperse gold nanorods through the use of aromatic additives. ACS Nano. 2012;6:2804–2817. DOI: 10.1021/nn300315j

[3] Kim JW, Larsen RJ, Weitz DA. Uniform nonspherical colloidal particles with tunable shapes. Adv Mater. 2007;19:2005–2009. DOI: 10.1002/adma.200602345

[4] Forster JD, Park JG, Mittal M, Noh H, Schreck CF, O'Hern CS, Cao H, Furst EM, Dufresne ER. Assembly of optical-scale dumbbells into dense photonic crystals. ACS Nano. 2011;5:6695–6700. DOI: 10.1021/nn202227f

[5] Taveau J-C, Nguyen D, Perro A, Ravaine S, Duguet E, Lambert O. New insights into the nucleation and growth of ps nodules on silica nanoparticles by 3d cryo-electron tomography. Soft Matter. 2008;4:311–315. DOI: 10.1039/B710137F

[6] van Blaaderen A. Materials science: Colloids get complex. Nature. 2006;439:545–546. DOI: 10.1038/439545a

[7] Duguet E, Desert A, Perro A, Ravaine S. Design and elaboration of colloidal molecules: An overview. Chem Soc Rev. 2011;40:941–960. DOI: 10.1039/c0cs00048e

[8] Chen Q, Bae SC, Granick S. Directed self-assembly of a colloidal kagome lattice. Nature. 2011;469:381–384. DOI: 10.1038/nature09713

[9] Yan J, Bae SC, Granick S. Colloidal superstructures programmed into magnetic janus particles. Adv Mater. 2015;27:874–879. DOI: 10.1002/adma.201403857

[10] Sacanna S, Pine DJ. Shape-anisotropic colloids: Building blocks for complex assemblies. Curr Opin Colloid Interf Sci. 2011;16:96–105. DOI: 10.1016/j.cocis.2011.01.003

[11] Hong L, Cacciuto A, Luijten E, Granick S. Clusters of charged janus spheres. Nano Lett. 2006;6:2510–2514. DOI: 10.1021/nl061857i

[12] Steed JW, Turner DR, Wallace KJ, editors. Core concepts in supramolecular chemistry and nanochemistry. Chichester: John Wiley & Sons Ltd.; 2007. 307 p. ISBN: 978-0-470-85867-7

[13] Sacanna S, Irvine WTM, Chaikin PM, Pine DJ. Lock and key colloids. Nature. 2010;464:575–578. DOI: 10.1038/nature08906

[14] Kraft DJ, Ni R, Smallenburg F, Hermes M, Yoon K, Weitz DA, van Blaaderen A, Groenewold J, Dijkstra M, Kegel WK. Surface roughness directed self-assembly of patchy particles into colloidal micelles. Proc Natl Acad Sci. 2012;109:10787–10792. DOI: 10.1073/pnas.1116820109

[15] Manoharan VN, Elsesser MT, Pine DJ. Dense packing and symmetry in small clusters of microspheres. Science. 2003;301:483–487. DOI: 10.1126/science.1086189

[16] Xia YN, Yin YD, Lu Y, McLellan J. Template-assisted self-assembly of spherical colloids into complex and controllable structures. Adv Funct Mater. 2003;13:907–918. DOI: 10.1002/adfm.200300002

[17] Pieranski P. Two-dimensional interfacial colloidal crystals. Phys Rev Lett. 1980;45:569–572. DOI: 10.1103/PhysRevLett.45.569

[18] Yi G-R, Manoharan VN, Michel E, Elsesser MT, Yang S-M, Pine DJ. Colloidal clusters of silica or polymer microspheres. Adv Mater. 2004;16:1204–1208. DOI: 10.1002/adma.200306638

[19] Horozov TS, Aveyard R, Clint JH, Binks BP. Order–disorder transition in monolayers of modified monodisperse silica particles at the octane-water interface. Langmuir. 2003;19:2822–2829. DOI: 10.1021/la020858x

[20] Peng B, Smallenburg F, Imhof A, Dijkstra M, van Blaaderen A. Colloidal clusters by using emulsions and dumbbell-shaped particles: Experiments and simulations. Angew Chem Int Edit. 2013;52:6709–6712. DOI: 10.1002/anie.201301520

[21] Kim S-H, Lee SY, Yi G-R, Pine DJ, Yang S-M. Microwave-assisted self-organization of colloidal particles in confining aqueous droplets. J Am Chem Soc. 2006;128:10897–10904. DOI: 10.1021/ja063528y

[22] Yi G-R, Thorsen T, Manoharan VN, Hwang M-J, Jeon S-J, Pine DJ, Quake SR, Yang S-M. Generation of uniform colloidal assemblies in soft microfluidic devices. Adv Mater. 2003;15:1300–1304. DOI: 10.1002/adma.200304890

[23] Zerrouki D, Rotenberg B, Abramson S, Baudry J, Goubault C, Leal-Calderon F, Pine DJ, Bibette M. Preparation of doublet, triangular, and tetrahedral colloidal clusters by controlled emulsification. Langmuir. 2006;22:57–62. DOI: 10.1021/la051765t

[24] Buzzaccaro S, Tripodi A, Rusconi R, Vigolo D, Piazza R. Kinetics of sedimentation in colloidal suspensions. J Phys Condens Mat. 2008;20:1–9. DOI: 10.1088/0953-8984/20/49/494219

[25] Plüisch CS, Wittemann A. Shape-tailored polymer colloids on the road to become structural motifs for hierarchically organized materials. Macromol Rapid Commun. 2013;34:1798–1814. DOI: 10.1002/marc.201300693

[26] Wagner CS, Lu Y, Wittemann A. Preparation of submicrometer-sized clusters from polymer spheres using ultrasonication. Langmuir. 2008;24:12126–12128. DOI: 10.1021/la802580k

[27] Wagner CS, Fischer B, May M, Wittemann A. Templated assembly of polymer particles into mesoscopic clusters with well-defined configurations. Colloid Polym Sci. 2010;288:487–498. DOI: 10.1007/s00396-009-2169-y

[28] Landfester K. Recent developments in miniemulsions - formation and stability mechanisms. Macromol Symp. 2000;150:171–178. DOI: 10.1002/1521-3900(200002)150:1<171::AID-MASY171>3.0.CO;2-D

[29] van Blaaderen A. Chemistry: Colloidal molecules and beyond. Science. 2003;301:470–471. DOI: 10.1126/science.1087140

[30] Frenkel D, Wales DJ. Colloidal self-assembly: Designed to yield. Nat Mater. 2011;10:410–411. DOI: 10.1038/nmat3037

[31] Meng G, Arkus N, Brenner MP, Manoharan VN. The free-energy landscape of clusters of attractive hard spheres. Science. 2010;327:560–563. DOI: 10.1126/science.1181263

[32] Schwarz I, Fortini A, Wagner CS, Wittemann A, Schmidt M. Monte Carlo computer simulations and electron microscopy of colloidal cluster formation via emulsion droplet evaporation. J Chem Phys. 2011;135:244501. DOI: 10.1063/1.3672106

[33] Binks BP. Particles as surfactants - similarities and differences. Curr Opin Colloid Interface. 2002;7:21–41. DOI: 10.1016/S1359-0294(02)00008-0

[34] Binks BP, Horozov TS, editors. Colloidal particles at liquid interfaces. Cambridge: Cambridge University Press; 2006. 501 p. ISBN: 978-0521-071314

[35] Pazos-Perez N, Wagner CS, Romo-Herrera JM, Liz-Marzán LM, García de Abajo FJ, Wittemann A, Fery A, Alvarez-Puebla RA. Organized plasmonic clusters with high coordination number and extraordinary enhancement in surface-enhanced raman scattering (sers). Angew Chem Int Edit. 2012;51:12688–12693. DOI: 10.1002/anie.201207019

[36] Radziuk D, Möhwald H. Ultrasonically treated liquid interfaces for progress in cleaning and separation processes. Phys Chem Chem Phys. 2016;18:21–46. DOI: 10.1039/c5cp05142h

[37] Tadros TF, editor. Emulsion science and technology: A general introduction. Weinheim: Wiley-VCH; 2009. 1–56 p. ISBN: 978-3-527-32525-2

[38] Gilbert RG. Emulsion polymerization: A mechanistic approach. London: Academic Press; 1995. ISBN: 978–0122830600

[39] Maranas CD, Floudas CA. A global optimization approach for Lennard-Jones microclusters. J Chem Phys. 1992;97:7667–7678. DOI: 10.1063/1.463486

[40] Romo-Herrera JM, Alvarez-Puebla RA, Liz-Marzán LM. Controlled assembly of plasmonic colloidal nanoparticle clusters. Nanoscale. 2011;3:1304–1315. DOI: 10.1039/c0nr00804d

[41] Rodriguez-Fernandez J, Perez-Juste J, García de Abajo FJ, Liz-Marzán LM. Seeded growth of submicron au colloids with quadrupole plasmon resonance modes. Langmuir. 2006;22:7007–7010. DOI: 10.1021/la060990n

[42] Wagner CS, Shehata S, Henzler K, Yuan J, Wittemann A. Towards nanoscale composite particles of dual complexity. J Colloid Interface Sci. 2011;355:115–123. DOI: 10.1016/j.jcis.2010.12.017

[43] Wagner CS, Fortini A, Hofmann E, Lunkenbein T, Schmidt M, Wittemann A. Particle nanosomes with tailored silhouettes. Soft Matter. 2012;8:1928–1933. DOI: 10.1039/C1SM07014B

[44] Fritsch A. Preparative density gradient centrifugations. Geneva: Beckman Instruments; 1975.

[45] Hoffmann M, Wagner CS, Harnau L, Wittemann A. 3D brownian diffusion of submicron-sized particle clusters. ACS Nano. 2009;3:3326–3334. DOI: 10.1021/nn900902b

[46] Myers D. Surfaces, interfaces, and colloids: Principles and applications. New York, USA: Wiley-VCH; 1999. 528 p. ISBN: 978-0-471-33060-8

[47] Degiorgio V, Piazza R. Rotational diffusion in concentrated colloidal dispersions of hard spheres. Phys Rev E. 1995;52:2707–2717. DOI: 10.1103/PhysRevE.52.2707

[48] Lettinga MP, Koenderink GH, Kuipers BWM, Bessels E, Philipse AP. Rotational dynamics of colloidal spheres probed with fluorescence recovery after photobleaching. J Chem Phys. 2004;120:4517–4529. DOI: 10.1063/1.1644799

[49] Quirantes A, Ben-Taleb A, Delgado AV. Determination of size/shape parameters of colloidal ellipsoids by photon correlation spectroscopy. Colloids Surf A. 1996;119:73–80. DOI: 10.1016/S0927-7757(96)03728-4

[50] Martchenko I, Dietsch H, Moitzi C, Schurtenberger P. Hydrodynamic properties of magnetic nanoparticles with tunable shape anisotropy: Prediction and experimental verification. J Phys Chem B. 2011;115:14838–14845. DOI: 10.1021/jp2078264

[51] de Souza Lima MM, Wong JT, Paillet M, Borsali R, Pecora R. Translational and rotational dynamics of rodlike cellulose whiskers. Langmuir. 2003;19:24–29. DOI: 10.1021/la020475z

[52] Kröger A, Deimede V, Belack J, Lieberwirth I, Fytas G, Wegner G. Equilibrium length and shape of rodlike polyelectrolyte micelles in dilute aqueous solutions. Macromolecules. 2007;40:105–115. DOI: 10.1021/ma061966j

[53] Jabbari-Farouji S, Eiser E, Wegdam GH, Bonn D. Ageing dynamics of translational and rotational diffusion in a colloidal glass. J Phys Condens Matter. 2004;16:L471–L477. DOI: 10.1088/0953-8984/16/41/L07

[54] Plum MA, Steffen W, Fytas G, Knoll W, Menges B. Probing dynamics at interfaces: Resonance enhanced dynamic light scattering. Opt Express. 2009;17:10364–10371. DOI: 10.1364/OE.17.010364

[55] Heyes DM, Nuevo MJ, Morales JJ, Branka AC. Translational and rotational diffusion of model nanocolloidal dispersions studied by molecular dynamics simulations. J Phys Condens Matter. 1998;10:10159–10178. DOI: 10.1088/0953-8984/10/45/005

[56] Aragon SR, Pecora R. Theory of dynamic light scattering from large anisotropic particles. J Chem Phys. 1977;66:2506–2516. DOI: 10.1063/1.434246

[57] Hoffmann M, Lu Y, Schrinner M, Ballauff M, Harnau L. Dumbbell-shaped polyelectrolyte brushes studied by depolarized dynamic light scattering. J Phys Chem B. 2008;112:14843–14850. DOI: 10.1021/jp806765y

[58] Anthony SM, Kim M, Granick S. Translation-rotation decoupling of colloidal clusters of various symmetries. J Chem Phys. 2008;129:244701–1–6. DOI: 10.1063/1.3043443

[59] Berne BJ, Pecora R. Dynamic light scattering: With applications to chemistry, biology, and physics. Mineola: Dover Publications; 2000. 384 p. ISBN: 978-0-486-41155-2

[60] Oetama RJ, Walz JY. Simultaneous investigation of sedimentation and diffusion of a single colloidal particle near an interface. J Chem Phys. 2006;124:164713–1–8. DOI: 10.1063/1.2189853

[61] Brogioli D, Salerno D, Cassina V, Sacanna S, Philipse AP, Croccolo F, Mantegazza F. Characterization of anisotropic nano-particles by using depolarized dynamic light scattering in the near field. Opt Express. 2009;17:1222–1233. DOI: 10.1364/OE.17.001222

[62] García de la Torre J, Del Rio Echenique G, Ortega A. Improved calculation of rotational diffusion and intrinsic viscosity of bead models for macromolecules and nanoparticles. J Phys Chem B. 2007;111:955–961. DOI: 10.1021/jp0647941

[63] Carrasco B, García de la Torre J. Hydrodynamic properties of rigid particles: Comparison of different modeling and computational procedures. Biophys J. 1999;76:3044–3057. DOI: 10.1016/S0006-3495(99)77457-6

[64] Liu K, Zhao NN, Kumacheva E. Self-assembly of inorganic nanorods. Chem Soc Rev. 2011;40:656–671. DOI: 10.1039/c0cs00133c

[65] Fan JA, Wu CH, Bao K, Bao JM, Bardhan R, Halas NJ, Manoharan VN, Nordlander P, Shvets G, Capasso F. Self-assembled plasmonic nanoparticle clusters. Science. 2010;328:1135–1138. DOI: 10.1126/science.1187949

[66] Kneipp K. Surface-enhanced raman scattering. Phys Today. 2007;60:40–46. DOI: 10.1063/1.2812122

[67] Le Ru EC, Etchegoin PG. Single-molecule surface-enhanced raman spectroscopy. Annu Rev Phys Chem. 2012;63:65–87. DOI: 10.1146/annurev-physchem-032511-143757

[68] Abdullin TI, Bondar OV, Shtyrlin YG, Kahraman M, Culha M. Pluronic block copolymer-mediated interactions of organic compounds with noble metal nanoparticles for sers analysis. Langmuir. 2010;26:5153–5159. DOI: 10.1021/la9036309

[69] Alvarez-Puebla R, Contreras-Ciceres R, Pastoriza-Santos I, Perez-Juste J, Liz-Marzán LM. Au@pnipam colloids as molecular traps for surface-enhanced, spectroscopic, ultrasensitive analysis. Angew Chem Int Ed. 2009;48:138–143. DOI: 10.1002/anie.200804059

[70] Alvarez-Puebla RA, Liz-Marzán LM. Traps and cages for universal sers detection. Chem Soc Rev. 2012;41:43–51. DOI: 10.1039/C1CS15155J

Chapter 12

Pretransitional Behavior and Dynamics in Liquid Crystal–Based Nanocolloids

Sylwester J. Rzoska, Szymon Starzonek and Aleksandra Drozd-Rzoska

Additional information is available at the end of the chapter

http://dx.doi.org/10.5772/64910

Abstract

The impact of nanoparticles on phase transitions in liquid crystal (LC)—nanoparticle nanocollids is still little known. This contribution results for dodecylcyanobiphenyl (12CB), pentylcyanobiphenyl (5CB), and hexyl isothiocyanatobiphenyl (6BT) as the LC host with the addition of $BaTiO_3$ barium titanate barium titanate nanoparticles. The latter has a strong impact on the value of dielectric constant, relaxation time, and the discontinuity of the isotropic–mesophase transitions. The first-ever high-pressure studies in such systems are also presented.

Keywords: liquid crystals, nanocolloids, dielectric constant

1. Introduction

Nanocolloids exhibit unique fundamental properties, important for a variety of innovative applications. They are obtained by adding to a liquid even a small amount of solid nanoparticles (NPs) [1, 2]. Liquid crystals (LC) have attracted particular attention as dispersing medium for nanoparticles due to the richness of phase transitions associated with the emergence of subsequent elements of symmetry, dominant importance of mesoscale structures, and the enormous sensitivity to exogenic impacts, such as pressure or electric field [3]. Consequently, one can expect that adding of nanoparticles can enable qualitative tailoring of properties of liquid crystals, yielding novel features without new chemical synthesis. It is expected that nanoparticles, with the length scale below 100 nm, can act as the specific additive moderator of molecular properties of the liquid crystalline host. The impact on the local field or geometrical hindrances associated with NPs can influence the local symmetry of the host, leading to the shift of phase transition temperatures or the decrease of the switching voltage [4]. The addition of NPs to a liquid host can also strongly change electric and heat conductivities [1, 2, 4]. All these are associated with a variety of topological defects (TD) introduced by NPs within mesomorphic fluid phases with different forms of ordering. Regarding application, the perspective type of LCs + NPs nanocolloids is related to highly birefringent nematics host with either positive or negative dielectric permittivity on increasing frequency from the kHz domain. This leads to so-called "Dual Frequency Functionality (DFN) materials" [4]. The last decade of study, still preliminary, clearly showed that there are clear links between material engineering-related issues for LCs + NPs hybrid composites with such challenging and seemingly distant areas as the physics of glass forming systems, quantum physics, superconductors physics, and even the cosmology. Consequently, nanocolloidal LCs + NPs soft hybrid composite systems can be considered as a unique *fundamental laboratory on the desk* [4].

All these led to the growth of theoretical and experimental interests in LC-based nanocolloids [1, 2, 4]. For the LC host materials, a particularly important issue constitute the impact of subsequent phase transitions on properties of subsequent mesophases This is associated with the range of pretransitional effects, the way of their characterization, and the discontinuity measure of predominantly weak discontinuous phase transitions. In fact, these phenomena constituted the background for the Landau–de Gennes model, one of the most basic and successful theoretical approaches for the physics of liquid crystals and more generally the soft matter [4, 5]. Regarding experimental insight, the broadband dielectric spectroscopy (BDS) plays the unique role in soft matter systems [4–7]. First, it can directly detect the pretransitional behavior associated with the emergence of mesomorphic fluctuations related to approaching subsequent mesophases. Second, BDS enables a study of the complex dynamics.

Surprisingly, these fundamentals for the physics of liquid crystal issues are still (very) weakly addressed for LCs + NPs nanocomposites [8]. In fact, the first clear evidence has only recently appeared [8]. This contribution presents the resume of this result with some new insights resume of this result with some new insights. Subsequently, new authors' results in other LC materials are communicated. The latter also includes results of the first-ever high-pressure studies for LCs + NPs nanocolloids.

2. Experimental

The tested samples of dodecylcyanobiphenyl (12CB), pentycyanobiphenyl (5CB), and hexylisothiocyanatobiphenyl (6BT) were synthesized and deeply purified to reach the minimal electric conductivity at Military University of Technology in Warsaw, Poland. They exhibit the following mesomorphisms: *Isotropic-Smectic-A-Solid* for 12CB, *Isotropic-Nematic-Solid* for 5CB, and *Isotropic-Smectic-E-Solid* for 6BT [3, 9]. The samples of 12CB and 5CB belongs to one of the most "classical" liquid crystalline homologous series: of n-cyanobiphenyls (nCB), regarding both fundamentals and practical implementations. All nCBs molecules have approximately a rod-like form, with a permanent dipole moment parallel to the long axis. Its approximate value

http://dx.doi.org/10.5772/64910

is equal to $\mu \approx 5D$ [9]. All these lead to a large anisotropy of dielectric permittivity for the perfectly ordered sample, namely, $\varepsilon_{\perp} = 4$ and $\varepsilon_{||} = 19$. $BaTiO_3$ nanoparticles (NPs) ($d = 50$ nm) were purchased from Research Nanomaterials (USA). All $BaTiO_3$ phases exhibit ferroelectricity except the cubic phase, which is associated with applied in this research nanoparticles. Mixtures of 12CB and $BaTiO_3$ NPs were sonificated with ultrasound frequency $f = 42$ kHz for a few hours in the isotropic phase (60°C) until obtaining the homogenous mixture for study. No sedimentation for at least 24 h was observed; hence, the tested nanocolloid did not contain additional stabilizing agents. The impedance analyzer (Solartron SI 1260) enabling high-resolution determination of dielectric constant with permanent five digits resolution was used. For the static domain between 1 kHz and 1 MHz, changes of real part of dielectric permittivity $\varepsilon'(f)$ were below 1%; so measurement of dielectric constant $\varepsilon = \varepsilon'(f)$ were carried out for $f =$ 50 kHz. The temperature was controlled by Julabo thermostat (with external circulation, 20 L volume). The temperature stability of samples was better than 0.02 K. Samples were placed in the measurement capacitor made from Invar, with $d = 0.2$ mm gap and diameter $2r = 20$ mm. Its design is given in references [3, 6]. The quartz ring was used as the spacer. This enable observation of the interior of the capacitor. The latter and the macroscopic gap of the capacitor made it possible to avoid bubbles, distorting results. For each concentration of nanoparticles, at least three series of measurements were carried out. Additionally, properties of samples were controlled after measurements (phase transition temperatures, sedimentation).

3. Results and discussion

The broadband dielectric spectroscopy (BDS) offers the possibility of the insight into processes: (i) coupled to leading intermolecular interactions and (ii) dynamics. The first issue can be

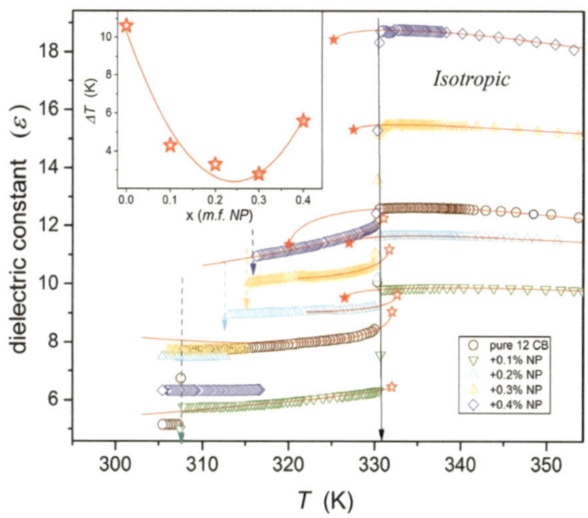

Figure 1. The behavior of dielectric constant ($f = 10$ KHz) in 12CB + $BaTiO_3$ nanocolloids in 12CB, covering the isotropic liquid, SmA, and solid mesophases. Solid red curves are described via Eq. (1) [8]. The inset shows the impact of nanoparticles concentration on the discontinuity of the I-SmA transition.

detected analyzing the evolution of dielectric constant. The second issue is related, for instance, to dielectric relaxation time. The first-ever study focusing on phase transitions *universal* properties in LCs + NPs nanocolloid are recalled in **Figure 1** [8]. They were conducted in dodecyl cyanobiphenyl (12CB) with the isotropic-smectic-A—solid mesomorphism. The rod-like structure, with the permanent dipole moment, located along the main molecular axis causes that studies of dielectric constant, i.e., the real (primary) component of dielectric permittivity in the low-frequency region, where $\varepsilon = \varepsilon'(f) \approx const$ allows to distinguish between parallel and antiparallel ordering dominated in the system. This fact and the appearance of premesomorphic fluctuation cause the notable pretransitional anomaly in the isotropic phase [8–11]:

$$\varepsilon(T) = \varepsilon^* + a^*\left(T - T^*\right) + A^*\left(T - T^*\right)^{\phi}, \quad T > T^C \tag{1}$$

where (ε^*, T^*) describe the loci of the hypothetical continuous phase transition, $T > T^C = T^* + \Delta T^*$, ΔT^* is the temperature metric of the discontinuity of the isotropic–mesophase transition, and T^C is for the isotropic–mesophase transition "clearing" temperature. The power exponent $\phi = 1 - \alpha$, where α is related to the specific heat "critical" anomaly. A^*, a^* are constant amplitudes.

Experiments for I-N, I-N*, I-SmA, and I-SmE phase transitions showed that for all cases the exponent $\phi \approx 0.5$ and then $\alpha \approx 0.5$. The exponent α is related to the pretransitional anomaly of the specific heat: $d\varepsilon/dT \propto c_p \propto (T - T^*)^{-\alpha}$ [10, 11]. Such behavior is shown in **Figure 1** via solid, red curves terminated with stars indicating the hypothetical continuous phase transition. As shown in Ref. [8], the addition of BaTiO3 nanoparticles did not change the value of the exponent $\alpha \approx 0.5$. However, the value of dielectric constant can be decreased or increased even by 50% in comparison to pure 12CB. The presence of NPs has a notable impact on the smectic-A mesophase, or rather SmA modified by NPs mesophase, but the very weak influence on the clearing temperature. These features are well visible in **Figure 1**, where additionally the inset presents the very strong influence of NPs amount on the value of the discontinuity ΔT^*, being one of the most important characteristics of the isotropic–mesophase transition.

Subsequently, new results regarding LCs + NPs nanocolloids are preliminary discussed. 12CB belongs to the n-cyanobiphenyls (nCB), one of the most classical liquid crystalline series regarding both fundamentals and applications. Its key and the most known representative is pentylcyanobiphenyl (5CB) with isotropic–nematic–crystal mesomorphism. On compressing, the clearing temperature usually increases and such an evolution for 5CB is shown in **Figure 2**, which includes also the extension into the negative pressures domain. Consequently, one can approach the I-N transition as the function of temperature or of pressure. The behavior of dielectric constant for both paths in 5CB and 5CB + $BaTiO_3$ nanocolloids is presented in **Figures 3** and **4**.

http://dx.doi.org/10.5772/64910

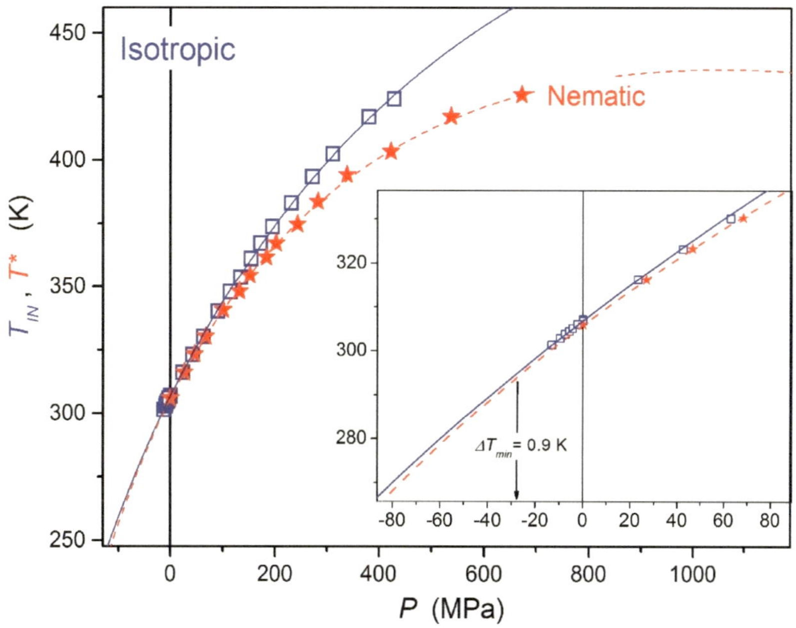

Figure 2. The pressure dependence of the isotropic–nematic (I-N) "clearing" temperature in 5CB [12]. The stars and the dotted curve are for the hypothetical continuous phase transition.

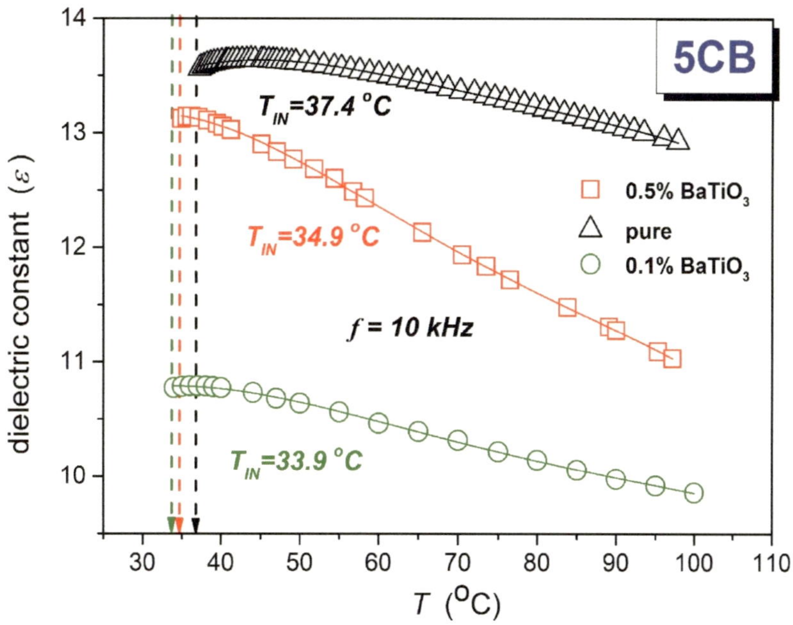

Figure 3. The pretransitional behavior of dielectric constant in 5CB + $BaTiO_3$ nanocolloids. The solid curves are related to Eq. (1).

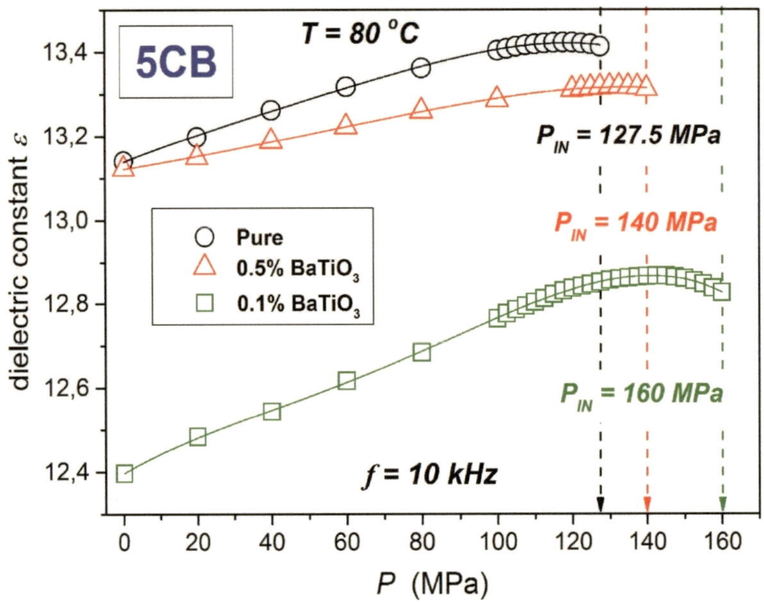

Figure 4. The pretransitional behavior of dielectric constant in 5CB + $BaTiO_3$ nanocolloids for the isothermal path of approaching the clearing point. The solid curves are related to Eq. (2).

The pretransitional behavior dominates the behavior of dielectric constant in the whole tested temperature range, up to $T_{10} + 70K$. It is associated with an increasing impact of prenematic fluctuations characterized by $\vec{n}$ and $-\vec{n}$ equivalence for the director indicating the orientational ordering [3]. This causes the ordering of permanent dipole moments in the antiparallel way, and consequently, the dielectric permittivity within prenematic fluctuations is governed by its perpendicular component ($\varepsilon_{\perp} \approx 4$ for 5CB). On approaching the clearing temperature, the volume occupied by prenematic fluctuations increases leading to the crossover $d\varepsilon/dT < 0 \rightarrow d\varepsilon/dT > 0$. One can estimate that for the crossover at $d\varepsilon/dT = 0$: $N_{fluct} \cdot \varepsilon_{fluct} = N_{bckg} \cdot \varepsilon_{bckg} \rightarrow N_{fluct} \cdot \varepsilon_{\perp} + N_{bckg} \cdot \left[(1/3)\, \varepsilon_{||} + (2/3)\varepsilon_{\perp}\right]$

where N is for the number of molecules, $\varepsilon_{\perp}$ and $\varepsilon_{||}$ are for the perpendicular and parallel components of dielectric constant, indexes *fluct* and *bckg* denote fluctuations, and the fluid-like background with chaotically arranged rod-like molecules. For 5CB one can approximate, $\varepsilon_{\perp} \approx 4$ and $\varepsilon_{||} \approx 19$, and the substitution of these values lead to the estimation $N_{fluct}/N_{bckg} \approx 0.3$, i.e., 23% of the molecules are contained within prenematic fluctuations. It is visible that in 5CB doped with $BaTiO_3$, the number of molecules within fluctuations is reached closer to the I-N transition that in the "native" material what can be linked to steric distortions introduced by nanoparticles within prenematic fluctuations.

It is visible in **Figure 3** that the addition of nanoparticles has a weak impact on the clearing temperature but notably changes the discontinuity ΔT^*. **Figure 4** presents the pretransitional effect in 5CB + $BaTiO_3$ nanocolloids for the isothermal compression. In this case, the addition of NPs has a strong impact on the loci of the clearing point, the discontinuity of the phase transition ΔP^*, the loci of $d\varepsilon/dT$ crossover, and the value of the isotropic–nematic transition P_{IN}. In this case, the pretransitional anomaly is described by the pressure counterpart of Eq. (1) [12, 13]:

$$\varepsilon(P) = \varepsilon^* + a_P^*\left(\left|P^* - P\right|\right) + A_P^*\left(\left|P^* - P\right|\right)^{\phi}, P < P^C \text{ and } T = const \tag{2}$$

where $(\varepsilon^*, T^*, P^*)$ describe the loci of the hypothetical continuous phase transition, $P < P^C = P^* - \Delta P^*$, ΔP^* is the temperature metric of the discontinuity of the isotropic–mesophase transition, and P^C is for the isotropic–mesophase transition "clearing" temperature. The power exponent $\phi = 1 - \alpha$, where α is related to the specific heat "critical" anomaly. A_P^*, a_P^* are constant amplitudes.

When comparing the isothermal and the isobaric behavior, it is worth recalling that the shift of temperature influences mainly the activation energy, whereas the compression changes the free volume. The impact of nanoparticles on more complex mesophases is shown in **Figures 5**–**7**. which focuses on the Smectic E phase in hexyl-isothiocyanatobiphenyl (6BT). The SmE phase belongs to the group of the most complex mesophases which in some classification is located even beyond the family of liquid crystals. The SmE phase is characterized by both the complete orientational and translational ordering, but the specific of the latter causes the viscoelasticity of SmE to be close to the "liquid-type" border [3, 9].

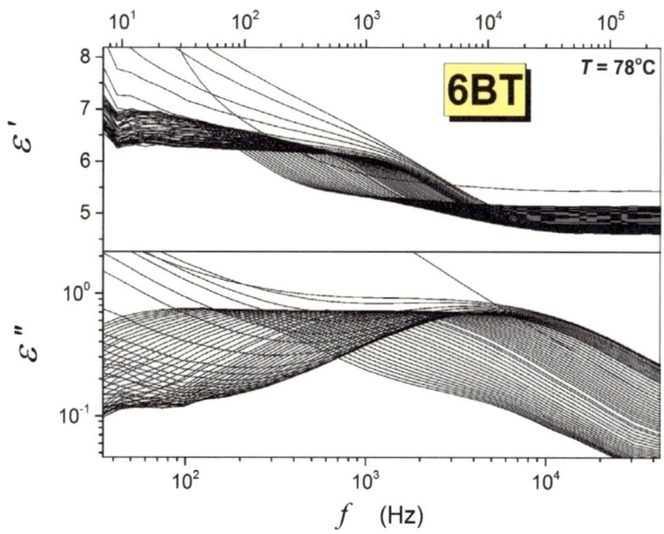

Figure 5. Dielectric spectra showing the real and imaginary parts of dielectric permittivity in the SmE phase of 6BT.

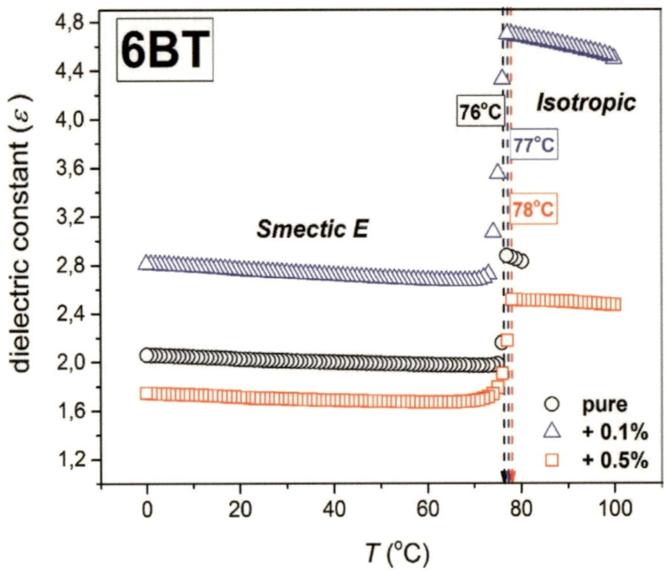

Figure 6. Changes of dielectric constant in 6BT + BaTiO3 nanocolloids, focused on the Smectic E phase.

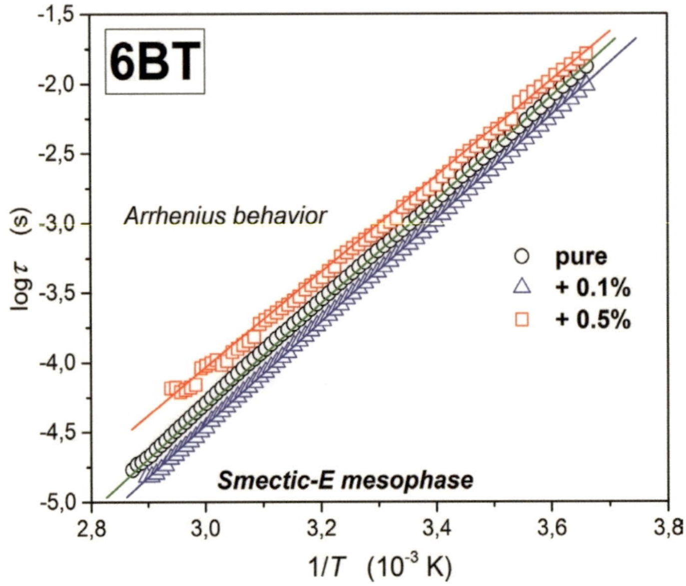

Figure 7. The "Arrhenius evolution" of the relaxation time for the main relaxation process in the SmE phase for 6BT + BaTiO3 nanocolloids.

Figure 5 presents dielectric spectra in 6BT, both for the imaginary (ε'') and real components (ε'') of dielectric permittivity. For the latter, the "flat part" for $f > 3$ kHz, indicates the static domain $\varepsilon = \varepsilon' \approx const.$

For the imaginary part of dielectric permittivity, the peak of the primary relaxation loss curve makes it possible to estimate the key relaxation time in the given system $\tau = 1/\left(2\pi f_{peak}\right)$. **Figure 6** shows the evolution of dielectric constant in the SmE mesophase, for the pure 6BT and its nanocolloids with BaTiO3, and the similar behavior for the relaxation time is given in **Figure 7**. Contrary to the case of 12CB based nanocolloids (**Figure 1**), the small concentration strongly increases dielectric constant, whereas it decreases when increasing the concentration.

Figure 7 shows the impact of $BaTiO_3$ nanoparticles on the key relaxation time in the 6BT matrix. In the broad range of temperatures, exceeding 80 K, is purely Arrhenius, i.e., described by the Arrhenius relation $\tau(T) = \tau_0 exp\left(\Delta E_a/RT\right)$. This issue is worth stressing since in complex viscoelastic systems most often the super-Arrhenius evolution occurs, with the temperature dependent on the apparent activation energy $\Delta E_a(T)$. It is notable from **Figure 7** that the small addition of BaTiO3 speeds up the key relaxation process where the increase of the dopant slows the relaxation.

Concluding, although the studies of LCs + NPs nanocolloids have already a notable history, the impact of nanoparticles on basic properties of phase transitions, relevant for modeling within the physics of liquid crystals is almost unknown. The only report focusing on this topic has appeared only very recently. This contribution presents the resume of these results and presents preliminary insights into nanocolloids based on 5CB, with the isotropic–nematic transition and for 6BT with the long-range Smectic E phase, although the addition of nanoparticles seems to have no impact on the value of the leading "critical" exponent. It influences very strongly on the value of the isotropic–mesophase discontinuity. There is also a qualitative impact on the value of dielectric constant or dielectric relaxation time. However, for the given type of nanoparticles such an impact depends both on the type of liquid crystalline materials and the mesophase.

Author details

Sylwester J. Rzoska*, Szymon Starzonek and Aleksandra Drozd-Rzoska

*Address all correspondence to: sylwester.rzoska@gmail.com

Laboratory of Ceramic Glassy Materials, Institute of High Pressure Physics, Polish Academy of Sciences, Warsaw, Poland

References

[1] S. K. Das, S. U. Choi, W. Yu, T. Pradeep, Nanofluids: Science and Technology (Wiley&Sons, NY, 2008).

[2] B. Takabi, M. Alizadeh, Nanofluid and Hybrid Nanofluid (Lambert Acad. Publ., NY, 2014).

[3] G. Vertogen, W. H. de Jeu, Thermotropic Liquid Crystals—Fundamentals, Springer Series in Chemical Physics (Springer, Berlin, 2008).

[4] Q. Li (ed.), Liquid Crystals beyond Displays: Chemistry, Physics and Applications (Wiley, NY, 2012).

[5] R. A. L. Jones, Introduction to Soft Matter (Oxford Univ. Press, Oxford, 2005).

[6] F. Kremer, Broadband Dielectric Spectroscopy (Springer, Berlin, 2010).

[7] S. Starzonek, S. J. Rzoska, A. Drozd-Rzoska, S. Pawlus, E. Biała, J.-C. Martinez-Garcia & L. Kistersky, Fractional Debye–Stokes–Einstein behaviour in an ultraviscous nanocolloid: glycerol and silver nanoparticles, Soft Matter 11, 5554 (2015)

[8] S. J. Rzoska, S, Starzonek, A. Drozd-Rzoska, K. Czupryński, K. Chmiel, G. Gaura, A. Michulec, B. Szczypek, W. Walas, The impact of $BaTiO_3$ nanonoparticles on pretransitional effects in liquid crystalline dodecylcyanobiphenyl, Phys. Rev. E 93, 534 (2016).

[9] D. Demus, J. Goodby, G. W. Gray, H.-W. Spiess, V. Vill, The Handbook of Liquid Crystals: Fundamentals (Wiley-VCH, NY, 2008).

[10] A. Drozd-Rzoska, S.J. Rzoska, J. Ziolo, Critical behaviour of dielectric permittivity in the isotropic phase of nematogens, Phys. Rev. E 54, 6452 (1996).

[11] S. J. Rzoska, M. Paluch, S. Pawlus, A. Drozd-Rzoska, J. Jadzyn, K. Czupryński, R. Dąbrowski, Complex dielectric relaxation in supercooling and superpressing liquid-crystalline ciral isopentycyanobiphenyl, Phys. Rev. E 68, 031705 (2003).

[12] S. J. Rzoska, P. K. Mukherjee, M. Rutkowska, Does the characteristic value of the discontinuity of the isotropic–mesophase transition in n-cyanobiphenyls exist?, J. Phys. Condensed Matter 24(10), 395101 (2012).

[13] S. J. Rzoska, A. Drozd-Rzoska, P. K. Mukherjee, D. O. Lopez, J. C. Martinez-Garcia, Distortions-sensitive analysis of pretransional behavior in n-octyloxycyanobiphenyl (8OCB), J. Phys. Cond. Matter 25, 245105 (2013).